Mastering Data Analysis with R

Gain clear insights into your data and solve real-world data science problems with R – from data munging to modeling and visualization

Gergely Daróczi

[PACKT] open source*
PUBLISHING   community experience distilled

BIRMINGHAM - MUMBAI

Mastering Data Analysis with R

First published: September 2015

Production reference: 1280915

Published by Packt Publishing Ltd.
Livery Place
35 Livery Street
Birmingham B3 2PB, UK.

ISBN 978-1-78398-202-8

www.packtpub.com

Credits

Author
Gergely Daróczi

Reviewers
Krishna Gawade
Alexey Grigorev
Mykola Kolisnyk
Mzabalazo Z. Ngwenya
Mohammad Rafi

Commissioning Editor
Akram Hussain

Acquisition Editor
Meeta Rajani

Content Development Editor
Nikhil Potdukhe

Technical Editor
Mohita Vyas

Copy Editors
Stephen Copestake
Angad Singh

Project Coordinator
Sanchita Mandal

Proofreader
Safis Editing

Indexer
Tejal Soni

Graphics
Jason Monteiro

Production Coordinator
Manu Joseph

Cover Work
Manu Joseph

About the Author

Gergely Daróczi is a former assistant professor of statistics and an enthusiastic R user and package developer. He is the founder and CTO of an R-based reporting web application at `http://rapporter.net` and a PhD candidate in sociology. He is currently working as the lead R developer/research data scientist at `https://www.card.com/` in Los Angeles.

Besides maintaining around half a dozen R packages, mainly dealing with reporting, Gergely has coauthored the books *Introduction to R for Quantitative Finance* and *Mastering R for Quantitative Finance* (both by Packt Publishing) by providing and reviewing the R source code. He has contributed to a number of scientific journal articles, mainly in social sciences but in medical sciences as well.

I am very grateful to my family, including my wife, son, and daughter, for their continuous support and understanding, and for missing me while I was working on this book—a lot more than originally planned.

I am also very thankful to Renata Nemeth and Gergely Toth for taking over the modeling chapters. Their professional and valuable help is highly appreciated. David Gyurko also contributed some interesting topics and preliminary suggestions to this book. And last but not least, I received some very useful feedback from the official reviewers and from Zoltan Varju, Michael Puhle, and Lajos Balint on a few chapters that are highly related to their field of expertise—thank you all!

About the Reviewers

Krishna Gawade is a data analyst and senior software developer with Saint-Gobain's S.A. IT development center. Krishna discovered his passion for computer science and data analysis while at Mumbai University where he holds a bachelor's degree in computer science. He has been awarded multiple times from Saint-Gobain for his contribution on various data driven projects.

He has been a technical reviewer on *R Data Analysis Cookbook* (ISBN: 9781783989065). His current interests are data analysis, statistics, machine learning, and artificial intelligence. He can be reached at gawadesk@gmail.com, or you can follow him on Twitter at @gawadesk.

Alexey Grigorev is an experienced software developer and data scientist with five years of professional experience. In his day-to-day job, he actively uses R and Python for data cleaning, data analysis, and modeling.

Mykola Kolisnyk has been involved in test automation since 2004 through various activities, including creating test automation solutions from the scratch, leading test automation teams, and performing consultancy regarding test automation processes. In his career, he has had experience of different test automation tools, such as Mercury WinRunner, MicroFocus SilkTest, SmartBear TestComplete, Selenium-RC, WebDriver, Appium, SoapUI, BDD frameworks, and many other engines and solutions. Mykola has experience with multiple programming technologies based on Java, C#, Ruby, and more. He has worked for different domain areas, such as healthcare, mobile, telecommunications, social networking, business process modeling, performance and talent management, multimedia, e-commerce, and investment banking.

He has worked as a permanent employee at ISD, GlobalLogic, Luxoft, and Trainline.com. He also has experience in freelancing activities and was invited as an independent consultant to introduce test automation approaches and practices to external companies.

Currently, he works as a mobile QA developer at the Trainline.com. Mykola is one of the authors (together with Gennadiy Alpaev) of the online *SilkTest Manual* (http://silktutorial.ru/) and participated in the creation of the TestComplete tutorial at http://tctutorial.ru/, which is one of the biggest related documentation available at RU.net.

Besides this, he participated as a reviewer on *TestComplete Cookbook* (ISBN: 9781849693585) and *Spring Batch Essentials, Packt Publishing* (ISBN: 9781783553372).

Mzabalazo Z. Ngwenya holds a postgraduate degree in mathematical statistics from the University of Cape Town. He has worked extensively in the field of statistical consulting and currently works as a biometrician at a research and development entity in South Africa. His areas of interest are primarily centered around statistical computing, and he has over 10 years of experience with the use of R for data analysis and statistical research. Previously, he was involved in reviewing *Learning RStudio for R Statistical Computing, Mark P.J. van der Loo* and *Edwin de Jonge*; *R Statistical Application Development by Example Beginner's Guide, Prabhanjan Narayanachar Tattar*; *R Graph Essentials, David Alexandra Lillis*; *R Object-oriented Programming, Kelly Black*; and *Mastering Scientific Computing with R, Paul Gerrard* and *Radia Johnson*. All of these were published by Packt Publishing.

Mohammad Rafi is a software engineer who loves data analytics, programming, and tinkering with anything he can get his hands on. He has worked on technologies such as R, Python, Hadoop, and JavaScript. He is an engineer by day and a hardcore gamer by night.

He was one of the reviewers on *R for Data Science*. Mohammad has more than 6 years of highly diversified professional experience, which includes app development, data processing, search expert, and web data analytics. He started with a web marketing company. Since then, he has worked with companies such as Hindustan Times, Google, and InMobi.

www.PacktPub.com

Support files, eBooks, discount offers, and more

For support files and downloads related to your book, please visit www.PacktPub.com.

Did you know that Packt offers eBook versions of every book published, with PDF and ePub files available? You can upgrade to the eBook version at www.PacktPub.com and as a print book customer, you are entitled to a discount on the eBook copy. Get in touch with us at service@packtpub.com for more details.

At www.PacktPub.com, you can also read a collection of free technical articles, sign up for a range of free newsletters and receive exclusive discounts and offers on Packt books and eBooks.

▯] PACKT ℠

https://www2.packtpub.com/books/subscription/packtlib

Do you need instant solutions to your IT questions? PacktLib is Packt's online digital book library. Here, you can search, access, and read Packt's entire library of books.

Why subscribe?
- Fully searchable across every book published by Packt
- Copy and paste, print, and bookmark content
- On demand and accessible via a web browser

Free access for Packt account holders

If you have an account with Packt at www.PacktPub.com, you can use this to access PacktLib today and view 9 entirely free books. Simply use your login credentials for immediate access.

Table of Contents

Preface

R has become the *lingua franca* of statistical analysis, and it's already actively and heavily used in many industries besides the academic sector, where it originated more than 20 years ago. Nowadays, more and more businesses are adopting R in production, and it has become one of the most commonly used tools by data analysts and scientists, providing easy access to thousands of user-contributed packages.

Mastering Data Analysis with R will help you get familiar with this open source ecosystem and some statistical background as well, although with a minor focus on mathematical questions. We will primarily focus on how to get things done practically with R.

As data scientists spend most of their time fetching, cleaning, and restructuring data, most of the first hands-on examples given here concentrate on loading data from files, databases, and online sources. Then, the book changes its focus to restructuring and cleansing data—still not performing actual data analysis yet. The later chapters describe special data types, and then classical statistical models are also covered, with some machine learning algorithms.

What this book covers

Chapter 1, Hello, Data!, starts with the first very important task in every data-related task: loading data from text files and databases. This chapter covers some problems of loading larger amounts of data into R using improved CSV parsers, pre-filtering data, and comparing support for various database backends.

Chapter 2, Getting Data from the Web, extends your knowledge on importing data with packages designed to communicate with Web services and APIs, shows how to scrape and extract data from home pages, and gives a general overview of dealing with XML and JSON data formats.

Chapter 3, Filtering and Summarizing Data, continues with the basics of data processing by introducing multiple methods and ways of filtering and aggregating data, with a performance and syntax comparison of the deservedly popular `data.table` and `dplyr` packages.

Chapter 4, Restructuring Data, covers more complex data transformations, such as applying functions on subsets of a dataset, merging data, and transforming to and from long and wide table formats, to perfectly fit your source data with your desired data workflow.

Chapter 5, Building Models (authored by Renata Nemeth and Gergely Toth), is the first chapter that deals with real statistical models, and it introduces the concepts of regression and models in general. This short chapter explains how to test the assumptions of a model and interpret the results via building a linear multivariate regression model on a real-life dataset.

Chapter 6, Beyond the Linear Trend Line (authored by Renata Nemeth and Gergely Toth), builds on the previous chapter, but covers the problems of non-linear associations of predictor variables and provides further examples on generalized linear models, such as logistic and Poisson regression.

Chapter 7, Unstructured Data, introduces new data types. These might not include any information in a structured way. Here, you learn how to use statistical methods to process such unstructured data through some hands-on examples on text mining algorithms, and visualize the results.

Chapter 8, Polishing Data, covers another common issue with raw data sources. Most of the time, data scientists handle dirty-data problems, such as trying to cleanse data from errors, outliers, and other anomalies. On the other hand, it's also very important to impute or minimize the effects of missing values.

Chapter 9, From Big to Smaller Data, assumes that your data is already loaded, clean, and transformed into the right format. Now you can start analyzing the usually high number of variables, to which end we cover some statistical methods on dimension reduction and other data transformations on continuous variables, such as principal component analysis, factor analysis, and multidimensional scaling.

Chapter 10, Classification and Clustering, discusses several ways of grouping observations in a sample using supervised and unsupervised statistical and machine learning methods, such as hierarchical and k-means clustering, latent class models, discriminant analysis, logistic regression and the k-nearest neighbors algorithm, and classification and regression trees.

Chapter 11, A Social Network Analysis of the R Ecosystem, concentrates on a special data structure and introduces the basic concept and visualization techniques of network analysis, with a special focus on the `igraph` package.

Chapter 12, Analyzing a Time Series, shows you how to handle time-date objects and analyze related values by smoothing, seasonal decomposition, and ARIMA, including some forecasting and outlier detection as well.

Chapter 13, Data around Us, covers another important dimension of data, with a primary focus on visualizing spatial data with thematic, interactive, contour, and Voronoi maps.

Chapter 14, Analyzing the R Community, provides a more complete case study that combines many different methods from the previous chapters to highlight what you have learned in this book and what kind of questions and problems you might face in future projects.

Appendix, References, gives references to the used R packages and some further suggested readings for each aforementioned chapter.

What you need for this book

All the code examples provided in this book should be run in the R console, which needs to be installed on your computer. You can download the software for free and find the installation instructions for all major operating systems at `http://r-project.org`.

Although we will not cover advanced topics, such as how to use R in Integrated Development Environments (IDE), there are awesome plugins and extensions for Emacs, Eclipse, vi, and Notepad++, besides other editors. Also, we highly recommend that you try RStudio, which is a free and open source IDE dedicated to R, at `https://www.rstudio.com/products/RStudio`.

Besides a working R installation, we will also use some user-contributed R packages. These can easily be installed from the Comprehensive R Archive Network (CRAN) in most cases. The sources of the required packages and the versions used to produce the output in this book are listed in *Appendix, References*.

To install a package from CRAN, you will need an Internet connection. To download the binary files or sources, use the `install.packages` command in the R console, like this:

```
> install.packages('pander')
```

Some packages mentioned in this book are not (yet) available on CRAN, but may be installed from Bitbucket or GitHub. These packages can be installed via the `install_bitbucket` and the `install_github` functions from the `devtools` package. Windows users should first install `rtools` from `https://cran.r-project.org/bin/windows/Rtools`.

After installation, the package should be loaded to the current R session before you can start using it. All the required packages are listed in the appendix, but the code examples also include the related R command for each package at the first occurrence in each chapter:

```
> library(pander)
```

We highly recommend downloading the code example files of this book (refer to the *Downloading the example code* section) so that you can easily copy and paste the commands in the R console without the R prompt shown in the printed version of the examples and output in the book.

If you have no experience with R, you should start with some free introductory articles and manuals from the R home page, and a short list of suggested materials is also available in the appendix of this book.

Who this book is for

If you are a data scientist or an R developer who wants to explore and optimize their use of R's advanced features and tools, then this is the book for you. Basic knowledge of R is required, along with an understanding of database logic. If you are a data scientist, engineer, or analyst who wants to explore and optimize your use of R's advanced features, this is the book for you. Although a basic knowledge of R is required, the book can get you up and running quickly by providing references to introductory materials.

Conventions

You will find a number of styles of text that distinguish between different kinds of information. Here are some examples of these styles, and an explanation of their meaning.

Function names, arguments, variables and other code reference in text are shown as follows: "The header argument of the read.big.matrix function defaults to FALSE."

Any command-line input or output that is shown in the R console is written as follows:

```
> set.seed(42)
> data.frame(
+    A = runif(2),
+    B = sample(letters, 2))
         A B
1 0.9148060 h
2 0.9370754 u
```

The > character represents the prompt, which means that the R console is waiting for commands to be evaluated. Multiline expressions start with the same symbol on the first line, but all other lines have a + sign at the beginning to show that the last R expression is not complete yet (for example, a closing parenthesis or a quote is missing). The output is returned without any extra leading character, with the same monospaced font style.

New terms and **important words** are shown in bold.

> Warnings or important notes appear in a box like this.

> Tips and tricks appear like this.

Reader feedback

Feedback from our readers is always welcome. Let us know what you think about this book—what you liked or disliked. Reader feedback is important for us as it helps us develop titles that you will really get the most out of.

To send us general feedback, simply e-mail feedback@packtpub.com, and mention the book's title in the subject of your message.

If there is a topic that you have expertise in and you are interested in either writing or contributing to a book, see our author guide at www.packtpub.com/authors.

Customer support

Now that you are the proud owner of a Packt book, we have a number of things to help you to get the most from your purchase.

Downloading the example code

You can download the example code files from your account at http://www.packtpub.com for all the Packt Publishing books you have purchased. If you purchased this book elsewhere, you can visit http://www.packtpub.com/support and register to have the files e-mailed directly to you.

Downloading the color images of this book

We also provide you with a PDF file that has color images of the screenshots/ diagrams used in this book. The color images will help you better understand the changes in the output. You can download this file from `http://www.packtpub.com/ sites/default/files/downloads/1234OT_ColorImages.pdf`.

Errata

Although we have taken every care to ensure the accuracy of our content, mistakes do happen. If you find a mistake in one of our books—maybe a mistake in the text or the code—we would be grateful if you could report this to us. By doing so, you can save other readers from frustration and help us improve subsequent versions of this book. If you find any errata, please report them by visiting `http://www.packtpub. com/submit-errata`, selecting your book, clicking on the **Errata Submission Form** link, and entering the details of your errata. Once your errata are verified, your submission will be accepted and the errata will be uploaded to our website or added to any list of existing errata under the Errata section of that title.

To view the previously submitted errata, go to `https://www.packtpub.com/books/ content/support` and enter the name of the book in the search field. The required information will appear under the **Errata** section.

Piracy

Piracy of copyrighted material on the Internet is an ongoing problem across all media. At Packt, we take the protection of our copyright and licenses very seriously. If you come across any illegal copies of our works in any form on the Internet, please provide us with the location address or website name immediately so that we can pursue a remedy.

Please contact us at `copyright@packtpub.com` with a link to the suspected pirated material.

We appreciate your help in protecting our authors and our ability to bring you valuable content.

Questions

If you have a problem with any aspect of this book, you can contact us at `questions@packtpub.com`, and we will do our best to address the problem.

1
Hello, Data!

Most projects in R start with loading at least some data into the running R session. As R supports a variety of file formats and database backend, there are several ways to do so. In this chapter, we will not deal with basic data structures, which are already familiar to you, but will concentrate on the performance issue of loading larger datasets and dealing with special file formats.

> For a quick overview on the standard tools and to refresh your knowledge on importing general data, please see *Chapter 7* of the official *An Introduction to R* manual of CRAN at `http://cran.r-project.org/doc/manuals/R-intro.html#Reading-data-from-files` or Rob Kabacoff's Quick-R site, which offers keywords and cheat-sheets for most general tasks in R at `http://www.statmethods.net/input/importingdata.html`. For further materials, please see the *References* section in the *Appendix*.

Although R has its own (serialized) binary `RData` and `rds` file formats, which are extremely convenient to use for all R users as these also store R object meta-information in an efficient way, most of the time we have to deal with other input formats—provided by our employer or client.

One of the most popular data file formats is flat files, which are simple text files in which the values are separated by white-space, the pipe character, commas, or more often by semi-colon in Europe. This chapter will discuss several options R has to offer to load these kinds of documents, and we will benchmark which of these is the most efficient approach to import larger files.

Sometimes we are only interested in a subset of a dataset; thus, there is no need to load all the data from the sources. In such cases, database backend can provide the best performance, where the data is stored in a structured way preloaded on our system, so we can query any subset of that with simple and efficient commands. The second section of this chapter will focus on the three most popular databases (MySQL, PostgreSQL, and Oracle Database), and how to interact with those in R.

Besides some other helper tools and a quick overview on other database backend, we will also discuss how to load Excel spreadsheets into R — without the need to previously convert those to text files in Excel or Open/LibreOffice.

Of course this chapter is not just about data file formats, database connections, and such boring internals. But please bear in mind that data analytics always starts with loading data. This is unavoidable, so that our computer and statistical environment know the structure of the data before doing some real analytics.

Loading text files of a reasonable size

The title of this chapter might also be *Hello, Big Data!*, as now we concentrate on loading relatively large amount of data in an R session. But what is Big Data, and what amount of data is problematic to handle in R? What is reasonable size?

R was designed to process data that fits in the physical memory of a single computer. So handling datasets that are smaller than the actual accessible RAM should be fine. But please note that the memory required to process data might become larger while doing some computations, such as principal component analysis, which should be also taken into account. I will refer to this amount of data as reasonable sized datasets.

Loading data from text files is pretty simple with R, and loading any reasonable sized dataset can be achieved by calling the good old `read.table` function. The only issue here might be the performance: how long does it take to read, for example, a quarter of a million rows of data? Let's see:

```
> library('hflights')
> write.csv(hflights, 'hflights.csv', row.names = FALSE)
```

> As a reminder, please note that all R commands and the returned output are formatted as earlier in this book. The commands starts with > on the first line, and the remainder of multi-line expressions starts with +, just as in the R console. To copy and paste these commands on your machine, please download the code examples from the Packt homepage. For more details, please see the *What you need for this book* section in the *Preface*.

Yes, we have just written an 18.5 MB text file to your disk from the `hflights` package, which includes some data on all flights departing from Houston in 2011:

```
> str(hflights)
'data.frame':   227496 obs. of  21 variables:
 $ Year             : int  2011 2011 2011 2011 2011 2011 2011 ...
 $ Month            : int  1 1 1 1 1 1 1 1 1 ...
 $ DayofMonth       : int  1 2 3 4 5 6 7 8 9 10 ...
 $ DayOfWeek        : int  6 7 1 2 3 4 5 6 7 1 ...
 $ DepTime          : int  1400 1401 1352 1403 1405 1359 1359 ...
 $ ArrTime          : int  1500 1501 1502 1513 1507 1503 1509 ...
 $ UniqueCarrier    : chr  "AA" "AA" "AA" "AA" ...
 $ FlightNum        : int  428 428 428 428 428 428 428 428 428 ...
 $ TailNum          : chr  "N576AA" "N557AA" "N541AA" "N403AA" ...
 $ ActualElapsedTime: int  60 60 70 70 62 64 70 59 71 70 ...
 $ AirTime          : int  40 45 48 39 44 45 43 40 41 45 ...
 $ ArrDelay         : int  -10 -9 -8 3 -3 -7 -1 -16 44 43 ...
 $ DepDelay         : int  0 1 -8 3 5 -1 -1 -5 43 43 ...
 $ Origin           : chr  "IAH" "IAH" "IAH" "IAH" ...
 $ Dest             : chr  "DFW" "DFW" "DFW" "DFW" ...
 $ Distance         : int  224 224 224 224 224 224 224 224 224 ...
 $ TaxiIn           : int  7 6 5 9 9 6 12 7 8 6 ...
 $ TaxiOut          : int  13 9 17 22 9 13 15 12 22 19 ...
 $ Cancelled        : int  0 0 0 0 0 0 0 0 0 0 ...
 $ CancellationCode : chr  "" "" "" "" ...
 $ Diverted         : int  0 0 0 0 0 0 0 0 0 0 ...
```

The `hflights` package provides an easy way to load a subset of the huge Airline Dataset of the Research and Innovation Technology Administration at the Bureau of Transportation Statistics. The original database includes the scheduled and actual departure/arrival times of all US flights along with some other interesting information since 1987, and is often used to demonstrate machine learning and Big Data technologies. For more details on the dataset, please see the column description and other meta-data at http://www.transtats.bts.gov/DatabaseInfo.asp?DB_ID=120&Link=0.

We will use this 21-column data to benchmark data import times. For example, let's see how long it takes to import the CSV file with `read.csv`:

```
> system.time(read.csv('hflights.csv'))
   user  system elapsed
  1.730   0.007   1.738
```

It took a bit more than one and a half seconds to load the data from an SSD here. It's quite okay, but we can achieve far better results by identifying then specifying the classes of the columns instead of calling the default `type.convert` (see the docs in `read.table` for more details or search on StackOverflow, where the performance of `read.csv` seems to be a rather frequent and popular question):

```
> colClasses <- sapply(hflights, class)
> system.time(read.csv('hflights.csv', colClasses = colClasses))
   user  system elapsed
  1.093   0.000   1.092
```

It's much better! But should we trust this one observation? On our way to mastering data analysis in R, we should implement some more reliable tests — by simply replicating the task *n* times and providing a summary on the results of the simulation. This approach provides us with performance data with multiple observations, which can be used to identify statistically significant differences in the results. The `microbenchmark` package provides a nice framework for such tasks:

```
> library(microbenchmark)
> f <- function() read.csv('hflights.csv')
> g <- function() read.csv('hflights.csv', colClasses = colClasses,
+                          nrows = 227496, comment.char = '')
> res <- microbenchmark(f(), g())
> res
Unit: milliseconds
 expr       min        lq    median        uq       max neval
  f() 1552.3383 1617.8611 1646.524 1708.393 2185.565   100
  g()  928.2675  957.3842  989.467 1044.571 1284.351   100
```

So we defined two functions: f stands for the default settings of `read.csv` while, in the g function, we passed the aforementioned column classes along with two other parameters for increased performance. The `comment.char` argument tells R not to look for comments in the imported data file, while the `nrows` parameter defined the exact number of rows to read from the file, which saves some time and space on memory allocation. Setting `stringsAsFactors` to `FALSE` might also speed up importing a bit.

> Identifying the number of lines in the text file could be done with some third-party tools, such as `wc` on Unix, or a slightly slower alternative would be the `countLines` function from the `R.utils` package.

But back to the results. Let's also visualize the median and related descriptive statistics of the test cases, which was run 100 times by default:

```
> boxplot(res, xlab = '',
+    main = expression(paste('Benchmarking ', italic('read.table'))))
```

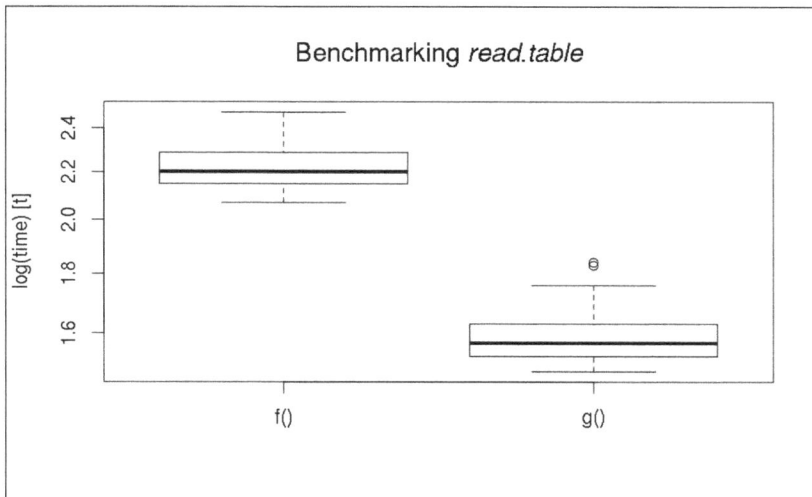

The difference seems to be significant (please feel free to do some statistical tests to verify that), so we made a 50+ percent performance boost simply by fine-tuning the parameters of `read.table`.

Data files larger than the physical memory

Loading a larger amount of data into R from CSV files that would not fit in the memory could be done with custom packages created for such cases. For example, both the `sqldf` package and the `ff` package have their own solutions to load data from chunk to chunk in a custom data format. The first uses SQLite or another SQL-like database backend, while the latter creates a custom data frame with the `ffdf` class that can be stored on disk. The `bigmemory` package provides a similar approach. Usage examples (to be benchmarked) later:

```
> library(sqldf)
> system.time(read.csv.sql('hflights.csv'))
```

```
   user  system elapsed
  2.293   0.090   2.384
> library(ff)
> system.time(read.csv.ffdf(file = 'hflights.csv'))
   user  system elapsed
  1.854   0.073   1.918
> library(bigmemory)
> system.time(read.big.matrix('hflights.csv', header = TRUE))
   user  system elapsed
  1.547   0.010   1.559
```

Please note that the header defaults to FALSE with read.big.matrix from the bigmemory package, so be sure to read the manual of the referenced functions before doing your own benchmarks. Some of these functions also support performance tuning like read.table. For further examples and use cases, please see the *Large memory and out-of-memory data* section of the *High-Performance and Parallel Computing with R* CRAN Task View at http://cran.r-project.org/web/views/HighPerformanceComputing.html.

Benchmarking text file parsers

Another notable alternative for handling and loading reasonable sized data from flat files to R is the data.table package. Although it has a unique syntax differing from the traditional S-based R markup, the package comes with great documentation, vignettes, and case studies on the indeed impressive speedup it can offer for various database actions. Such uses cases and examples will be discussed in the *Chapter 3, Filtering and Summarizing Data* and *Chapter 4, Restructuring Data*.

The package ships a custom R function to read text files with improved performance:

```
> library(data.table)
> system.time(dt <- fread('hflights.csv'))
   user  system elapsed
  0.153   0.003   0.158
```

Loading the data was extremely quick compared to the preceding examples, although it resulted in an R object with a custom data.table class, which can be easily transformed to the traditional data.frame if needed:

```
> df <- as.data.frame(dt)
```

Or by using the `setDF` function, which provides a very fast and in-place method of object conversion without actually copying the data in the memory. Similarly, please note:

```
> is.data.frame(dt)
[1] TRUE
```

This means that a `data.table` object can fall back to act as a `data.frame` for traditional usage. Leaving the imported data as is or transforming it to `data.frame` depends on the latter usage. Aggregating, merging, and restructuring data with the first is faster compared to the standard data frame format in R. On the other hand, the user has to learn the custom syntax of `data.table` — for example, `DT[i, j, by]` stands for "from DT subset by `i`, then do `j` grouped by by". We will discuss it later in the *Chapter 3, Filtering and Summarizing Data.*

Now, let's compare all the aforementioned data import methods: how fast are they? The final winner seems to be `fread` from `data.table` anyway. First, we define some methods to be benchmarked by declaring the test functions:

```
> .read.csv.orig    <- function() read.csv('hflights.csv')
> .read.csv.opt     <- function() read.csv('hflights.csv',
+     colClasses = colClasses, nrows = 227496, comment.char = '',
+     stringsAsFactors = FALSE)
> .read.csv.sql     <- function() read.csv.sql('hflights.csv')
> .read.csv.ffdf    <- function() read.csv.ffdf(file = 'hflights.csv')
> .read.big.matrix  <- function() read.big.matrix('hflights.csv',
+     header = TRUE)
> .fread            <- function() fread('hflights.csv')
```

Now, let's run all these functions 10 times each instead of several hundreds of iterations like previously — simply to save some time:

```
> res <- microbenchmark(.read.csv.orig(), .read.csv.opt(),
+   .read.csv.sql(), .read.csv.ffdf(), .read.big.matrix(), .fread(),
+   times = 10)
```

And print the results of the benchmark with a predefined number of digits:

```
> print(res, digits = 6)
Unit: milliseconds
             expr      min      lq   median       uq      max neval
   .read.csv.orig() 2109.643 2149.32 2186.433 2241.054 2421.392    10
```

`.read.csv.opt()`	1525.997	1565.23	1618.294	1660.432	1703.049	10
`.read.csv.sql()`	2234.375	2265.25	2283.736	2365.420	2599.062	10
`.read.csv.ffdf()`	1878.964	1901.63	1947.959	2015.794	2078.970	10
`.read.big.matrix()`	1579.845	1603.33	1647.621	1690.067	1937.661	10
`.fread()`	153.289	154.84	164.994	197.034	207.279	10

Please note that now we were dealing with datasets fitting in actual physical memory, and some of the benchmarked packages are designed and optimized for far larger databases. So it seems that optimizing the `read.table` function gives a great performance boost over the default settings, although if we are after really fast importing of reasonable sized data, using the `data.table` package is the optimal solution.

Loading a subset of text files

Sometimes we only need some parts of the dataset for an analysis, stored in a database backend or in flat files. In such situations, loading only the relevant subset of the data frame will result in much more speed improvement compared to any performance tweaks and custom packages discussed earlier.

Let's imagine we are only interested in flights to Nashville, where the annual *useR!* conference took place in 2012. This means we need only those rows of the CSV file where the `Dest` equals `BNA` (this International Air Transport Association airport code stands for Nashville International Airport).

Instead of loading the whole dataset in 160 to 2,000 milliseconds (see the previous section) and then dropping the unrelated rows (see in *Chapter 3, Filtering and Summarizing Data*), let's see the possible ways of filtering the data while loading it.

The already mentioned `sqldf` package can help with this task by specifying a SQL statement to be run on the temporary SQLite database created for the importing task:

```
> df <- read.csv.sql('hflights.csv',
+   sql = "select * from file where Dest = '\"BNA\"'")
```

This `sql` argument defaults to `"select * from file"`, which means loading all fields of each row without any filters. Now we extended that with a `filter` statement. Please note that in our updated SQL statements, we also added the double quotes to the search term, as `sqldf` does not automatically recognize the quotes as special; it regards them as part of the fields. One may overcome this issue also by providing a custom filter argument, such as the following example on Unix-like systems:

```
> df <- read.csv.sql('hflights.csv',
+    sql = "select * from file where Dest = 'BNA'",
+    filter = 'tr -d ^\\" ')
```

The resulting data frame holds only 3,481 observations out of the 227,496 cases in the original dataset, and filtering inside the temporary SQLite database of course speeds up data importing a bit:

```
> system.time(read.csv.sql('hflights.csv'))
   user   system elapsed
  2.117    0.070   2.191
> system.time(read.csv.sql('hflights.csv',
+    sql = "select * from file where Dest = '\"BNA\"'"))
   user   system elapsed
  1.700    0.043   1.745
```

The slight improvement is due to the fact that both R commands first loaded the CSV file to a temporary SQLite database; this process of course takes some time and cannot be eliminated from this process. To speed up this part of the evaluation, you can specify dbname as NULL for a performance boost. This way, the SQLite database would be created in memory instead of a tempfile, which might not be an optimal solution for larger datasets.

Filtering flat files before loading to R

Is there a faster or smarter way to load only a portion of such a text file? One might apply some regular expression-based filtering on the flat files before passing them to R. For example, grep or ack might be a great tool to do so in a Unix environment, but it's not available by default on Windows machines, and parsing CSV files by regular expressions might result in some unexpected side-effects as well. Believe me, you never want to write a CSV, JSON, or XML parser from scratch!

Anyway, a data scientist nowadays should be a real jack-of-all-trades when it comes to processing data, so here comes a quick and dirty example to show how one could read the filtered data in less than 100 milliseconds:

```
> system.time(system('cat hflights.csv | grep BNA', intern = TRUE))
   user   system elapsed
  0.040    0.050   0.082
```

Well, that's a really great running time compared to any of our previous results! But what if we want to filter for flights with an arrival delay of more than 13.5 minutes?

Another way, and probably a more maintainable approach, would be to first load the data into a database backend, and query that when any subset of the data is needed. This way we could for example, simply populate a SQLite database in a file only once, and then later we could fetch any subsets in a fragment of `read.csv.sql`'s default run time.

So let's create a persistent SQLite database:

```
> sqldf("attach 'hflights_db' as new")
```

This command has just created a file named to `hflights_db` in the current working directory. Next, let's create a table named `hflights` and populate the content of the CSV file to the database created earlier:

```
> read.csv.sql('hflights.csv',
+    sql = 'create table hflights as select * from file',
+    dbname = 'hflights_db')
```

No benchmarking was made so far, as these steps will be run only once, while the queries for sub-parts of the dataset will probably run multiple times later:

```
> system.time(df <- sqldf(
+    sql = "select * from hflights where Dest = '\"BNA\"'",
+    dbname = "hflights_db"))
   user  system elapsed
  0.070   0.027   0.097
```

And we have just loaded the required subset of the database in less than 100 milliseconds! But we can do a lot better if we plan to often query the persistent database: why not dedicate a real database instance for our dataset instead of a simple file-based and server-less SQLite backend?

Loading data from databases

The great advantage of using a dedicated database backend instead of loading data from the disk on demand is that databases provide:

* Faster access to the whole or selected parts of large tables
* Powerful and quick ways to aggregate and filter data before loading it to R
* Infrastructure to store data in a relational, more structured scheme compared to the traditional matrix model of spreadsheets and R objects
* Procedures to join and merge related data

- Concurrent and network access from multiple clients at the same time
- Security policies and limits to access the data
- A scalable and configurable backend to store data

The DBI package provides a database interface, a communication channel between R and various **relational database management systems** (**RDBMS**), such as MySQL, PostgreSQL, MonetDB, Oracle, and for example Open Document Databases, and so on. There is no real need to install the package on its own because, acting as an interface, it will be installed anyway as a dependency, if needed.

Connecting to a database and fetching data is pretty similar with all these backends, as all are based on the relational model and using SQL to manage and query data. Please be advised that there are some important differences between the aforementioned database engines and that several more open-source and commercial alternatives also exist. But we will not dig into the details on how to choose a database backend or how to build a data warehouse and **extract, transform, and load** (ETL) workflows, but we will only concentrate on making connections and managing data from R.

> SQL, originally developed at IBM, with its more than 40 years of history, is one of the most important programming languages nowadays — with various dialects and implementations. Being one of the most popular declarative languages all over the world, there are many online tutorials and free courses to learn how to query and manage data with SQL, which is definitely one of the most important tools in every data scientist's Swiss army knife.
>
> So, besides R, it's really worth knowing your way around RDBMS, which are extremely common in any industry you may be working at as a data analyst or in a similar position.

Setting up the test environment

Database backends usually run on servers remote from the users doing data analysis, but for testing purposes, it might be a good idea to install local instances on the machine running R. As the installation process can be extremely different on various operating systems, we will not enter into any details of the installation steps, but we will rather refer to where the software can be downloaded from and some further links to great resources and documentation for installation.

Please note that installing and actually trying to load data from these databases is totally optional and you do not have to follow each step—the rest of the book will not depend on any database knowledge or prior experience with databases. On the other hand, if you do not want to mess your workspace with temporary installation of multiple database applications for testing purposes, using virtual machines might be an optimal workaround. Oracle's `VirtualBox` provides a free and easy way of running multiple virtual machines with their dedicated operating system and userspace.

> For detailed instructions on how to download then import a `VirtualBox` image, see the *Oracle* section.

This way you can quickly deploy a fully functional, but disposable, database environment to test-drive the following examples of this chapter. In the following image, you can see `VirtualBox` with four installed virtual machines, of which three are running in the background to provide some database backends for testing purposes:

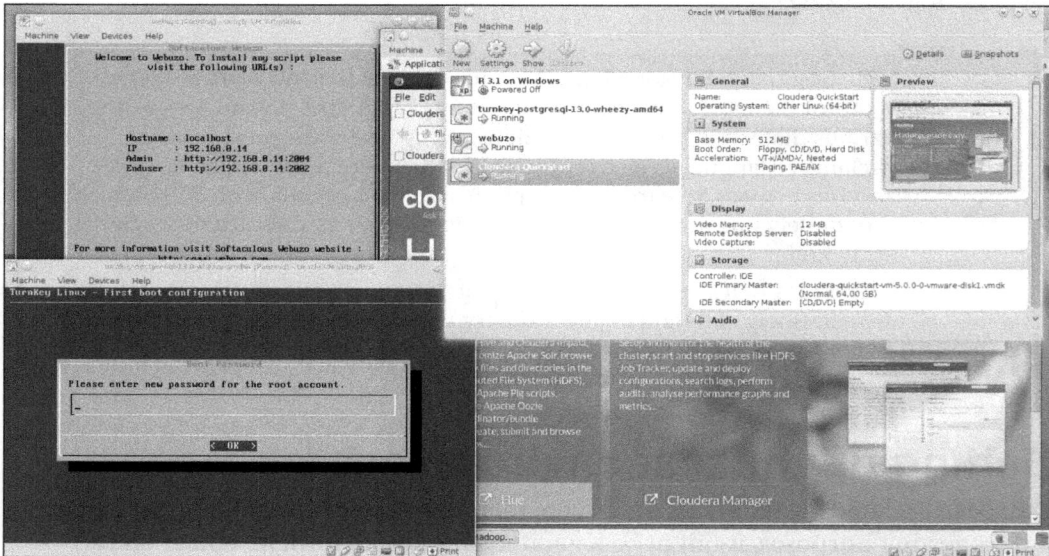

> VirtualBox can be installed by your operating system's package manager on Linux or by downloading the installation binary/sources from https://www.virtualbox.org/wiki/Downloads. For detailed and operating-system specific installation information, please refer to the *Chapter 2, Installation details* of the manual: http://www.virtualbox.org/manual/.

Nowadays, setting up and running a virtual machine is really intuitive and easy; basically you only need a virtual machine image to be loaded and launched. Some virtual machines, so called appliances, include the operating system, with a number of further software usually already configured to work, for simple, easy and quick distribution.

> Once again, if you do not enjoy installing and testing new software or spending time on learning about the infrastructure empowering your data needs, the following steps are not necessary and you can freely skip these optional tasks primarily described for full-stack developers/data scientists.

Such pre-configured virtual machines to be run on any computer can be downloaded from various providers on the Internet in multiple file formats, such as OVF or OVA. General purpose VirtualBox virtual appliances can be downloaded for example from http://virtualboximages.com/vdi/index or http://virtualboxes.org/images/.

> Virtual appliances should be imported in VirtualBox, while non-OVF/OVA disk images should be attached to newly created virtual machines; thus, some extra manual configuration might also be needed.

Oracle also has a repository with a bunch of useful virtual images for data scientist apprentices and other developers at http://www.oracle.com/technetwork/community/developer-vm/index.html, with for example the Oracle Big Data Lite VM developer virtual appliance featuring the following most important components:

- Oracle Database
- Apache Hadoop and various tools in Cloudera distribution
- The Oracle R Distribution
- Build on Oracle Enterprise Linux

Disclaimer: Oracle wouldn't be my first choice personally, but they did a great job with their platform-independent virtualization environment, just like with providing free developer VMs based on their commercial products. In short, it's definitely worth using the provided Oracle tools.

> If you cannot reach your installed virtual machines on the network, please update your network settings to use *Host-only adapter* if no Internet connection is needed, or *Bridged networking* for a more robust setup. The latter setting will reserve an extra IP on your local network for the virtual machine; this way, it becomes accessible easily. Please find more details and examples with screenshots in the *Oracle database* section.

Another good source of virtual appliances created for open-source database engines is the Turnkey GNU/Linux repository at `http://www.turnkeylinux.org/database`. These images are based on Debian Linux, are totally free to use, and currently support the MySQL, PostgreSQL, MongoDB, and CouchDB databases.

A great advantage of the Turnkey Linux media is that it includes only open-source, free software and non-proprietary stuff. Besides, the disk images are a lot smaller and include only the required components for one dedicated database engine. This also results in far faster installation with less overhead in terms of the required disk and memory space.

Further similar virtual appliances are available at `http://www.webuzo.com/sysapps/databases` with a wider range of database backends, such as Cassandra, HBase, Neo4j, Hypertable, or Redis, although some of the Webuzo appliances might require a paid subscription for deployment.

And as the new cool being Docker, I even more suggest you to get familiar with its concept on deploying software containers incredibly fast. Such container can be described as a standalone filesystem including the operating system, libraries, tools, data and so is based on abstraction layers of Docker images. In practice this means that you can fire up a database including some demo data with a one-liner command on your localhost, and developing such custom images is similarly easy. Please see some simple examples and further references at my R and Pandoc-related Docker images described at `https://github.com/cardcorp/card-rocker`.

MySQL and MariaDB

MySQL is the most popular open-source database engine all over the world based on the number of mentions, job offers, Google searches, and so on, summarized by the DB-Engines Ranking: http://db-engines.com/en/ranking. Mostly used in Web development, the high popularity is probably due to the fact that MySQL is free, platform-independent, and relatively easy to set up and configure—just like its drop-in replacement fork called **MariaDB**.

> MariaDB is a community-developed, fully open-source fork of MySQL, started and led by the founder of MySQL, Michael Widenius. It was later merged with SkySQL; thus further ex-MySQL executives and investors joined the fork. MariaDB was created after Sun Microsystems bought MySQL, currently owned by Oracle, and the development of the database engine changed.

We will refer to both engines as MySQL in the book to keep it simple, as MariaDB can be considered as a drop-in replacement for MySQL, so please feel free to reproduce the following examples with either MySQL or MariaDB.

Although the installation of a MySQL server is pretty straightforward on most operating systems (https://dev.mysql.com/downloads/mysql/), one might rather prefer to have the database installed in a virtual machine. Turnkey Linux provides small but fully configured, virtual appliances for free: http://www.turnkeylinux. org/mysql.

R provides multiple ways to query data from a MySQL database. One option is to use the RMySQL package, which might be a bit tricky for some users to install. If you are on Linux, please be sure to install the development packages of MySQL along with the MySQL client, so that the package can compile on your system. And, as there are no binary packages available on CRAN for Windows installation due to the high variability of MySQL versions, Windows users should also compile the package from source:

```
> install.packages('RMySQL', type = 'source')
```

Windows users might find the following blog post useful about the detailed installation steps: http://www.ahschulz.de/2013/07/23/installing-rmysql-under-windows/.

> For the sake of simplicity, we will refer to the MySQL server as localhost listening on the default 3306 port; user will stand as user and password as password in all database connections. We will work with the hflights table in the hflights_db database, just like in the SQLite examples a few pages earlier. If you are working in a remote or virtual server, please modify the host, username, and so on arguments of the following code examples accordingly.

After successfully installing and starting the MySQL server, we have to set up a test database, which we could later populate in R. To this end, let us start the MySQL command-line tool to create the database and a test user.

Please note that the following example was run on Linux, and a Windows user might have to also provide the path and probably the exe file extension to start the MySQL command-line tool:

```
daroczig : zsh - Konsole
File   Edit   View   Bookmarks   Settings   Help
[daroczig@nevermind][~]% mysql -u root
Welcome to the MariaDB monitor.  Commands end with ; or \g.
Your MariaDB connection id is 9
Server version: 5.5.36-MariaDB-log MariaDB Server

Copyright (c) 2000, 2014, Oracle, Monty Program Ab and others.

Type 'help;' or '\h' for help. Type '\c' to clear the current input statement.

MariaDB [(none)]> create database hflights_db;
Query OK, 1 row affected (0.00 sec)

MariaDB [(none)]> grant all privileges on hflights_db.* to 'user'@'localhost' identified by 'password';
Query OK, 0 rows affected (0.00 sec)

MariaDB [(none)]> exit
Bye
[daroczig@nevermind][~]% mysql -u user -ppassword
Welcome to the MariaDB monitor.  Commands end with ; or \g.
Your MariaDB connection id is 11
Server version: 5.5.36-MariaDB-log MariaDB Server

Copyright (c) 2000, 2014, Oracle, Monty Program Ab and others.

Type 'help;' or '\h' for help. Type '\c' to clear the current input statement.

MariaDB [(none)]> show databases;
+--------------------+
| Database           |
+--------------------+
| information_schema |
| hflights_db        |
| test               |
+--------------------+
3 rows in set (0.01 sec)

MariaDB [(none)]> exit
Bye
daroczig : zsh
```

This quick session can be seen in the previous screenshot, where we first connected to the MySQL server in the command-line as the `root` (admin) user. Then we created a database named `hflights_db`, and granted all privileges and permissions of that database to a new user called `user` with the password set to `password`. Then we simply verified whether we could connect to the database with the newly created user, and we exited the command-line MySQL client.

To load data from a MySQL database into R, first we have to connect and also often authenticate with the server. This can be done with the automatically loaded `DBI` package when attaching `RMySQL`:

```
> library(RMySQL)
Loading required package: DBI
> con <- dbConnect(dbDriver('MySQL'),
+     user = 'user', password = 'password', dbname = 'hflights_db')
```

Now we can refer to our MySQL connection as `con`, where we want to deploy the `hflights` dataset for later access:

```
> dbWriteTable(con, name = 'hflights', value = hflights)
[1] TRUE
> dbListTables(con)
[1] "hflights"
```

The `dbWriteTable` function wrote the `hflights` data frame with the same name to the previously defined connection. The latter command shows all the tables in the currently used databases, equivalent to the SHOW TABLES SQL command. Now that we have our original CVS file imported to MySQL, let's see how long it takes to read the whole dataset:

```
> system.time(dbReadTable(con, 'hflights'))
   user  system elapsed
  0.993   0.000   1.058
```

Or we can do so with a direct SQL command passed to `dbGetQuery` from the same `DBI` package:

```
> system.time(dbGetQuery(con, 'select * from hflights'))
   user  system elapsed
  0.910   0.000   1.158
```

And, just to keep further examples simpler, let's get back to the `sqldf` package, which stands for "SQL select on data frames". As a matter of fact, `sqldf` is a convenient wrapper around DBI's `dbSendQuery` function with some useful defaults, and returns `data.frame`. This wrapper can query various database engines, such as SQLite, MySQL, H2, or PostgreSQL, and defaults to the one specified in the global `sqldf.driver` option; or, if that's `NULL`, it will then check if any R packages have been loaded for the aforementioned backends.

As we have already loaded `RMySQL`, now `sqldf` will default to using MySQL instead of SQLite. But we still have to specify which connection to use; otherwise the function will try to open a new one—without any idea about our complex username and password combination, not to mention the mysterious database name. The connection can be passed in each `sqldf` expression or defined once in a global option:

```
> options('sqldf.connection' = con)
> system.time(sqldf('select * from hflights'))
   user  system elapsed
  0.807   0.000   1.014
```

The difference in the preceding three versions of the same task does not seem to be significant. That 1-second timing seems to be a pretty okay result compared to our previously tested methods—although loading the whole dataset with `data.table` still beats this result. What about if we only need a subset of the dataset? Let's fetch only those flights ending in Nashville, just like in our previous SQLite example:

```
> system.time(sqldf('SELECT * FROM hflights WHERE Dest = "BNA"'))
   user  system elapsed
  0.000   0.000   0.281
```

This does not seem to be very convincing compared to our previous SQLite test, as the latter could reproduce the same result in less than 100 milliseconds. But please also note that that both the user and system elapsed times are zero, which was not the case with SQLite.

The returned elapsed time by `system.time` means the number of milliseconds passed since the start of the evaluation. The user and system times are a bit trickier to understand; they are reported by the operating system. More or less, `user` means the CPU time spent by the called process (like R or the MySQL server), while `system` reports the CPU time required by the kernel and other operating system processes (such as opening a file for reading). See `?proc.time` for further details.

This means that no CPU time was used at all to return the required subset of data, which took almost 100 milliseconds with SQLite. How is it possible? What if we index the database on `Dest`?

```
> dbSendQuery(con, 'CREATE INDEX Dest_idx ON hflights (Dest(3));')
```

This SQL query stands for creating an index named `Dest_idx` in our table based on the `Dest` column's first three letters.

> SQL index can seriously boost the performance of a SELECT statement with WHERE clauses, as MySQL this way does not have to read through the entire database to match each row, but it can determine the position of the relevant search results. This performance boost becomes more and more spectacular with larger databases, although it's also worth mentioning that indexing only makes sense if subsets of data are queried most of the time. If most or all data is needed, sequential reads would be faster.

Live example:

```
> system.time(sqldf('SELECT * FROM hflights WHERE Dest = "BNA"'))
   user  system elapsed
  0.024   0.000   0.034
```

It seems to be a lot better! Well, of course, we could have also indexed the SQLite database, not just the MySQL instance. To test it again, we have to revert the default `sqldf` driver to SQLite, which was overridden by loading the `RMySQL` package:

```
> options(sqldf.driver = 'SQLite')
> sqldf("CREATE INDEX Dest_idx ON hflights(Dest);",
+    dbname = "hflights_db"))
NULL
> system.time(sqldf("select * from hflights where
+    Dest = '\"BNA\"'", dbname = "hflights_db"))
   user  system elapsed
  0.034   0.004   0.036
```

So it seems that both database engines are capable of returning the required subset of data in a fraction of a second, which is a lot better even compared to what we achieved with the impressive `data.table` before.

Although SQLite proved to be faster than MySQL in some earlier examples, there are many reasons to choose the latter in most situations. First, SQLite is a file-based database, which simply means that the database should be on a filesystem attached to the computer running R. This usually means having the SQLite database and the running R session on the same computer. Similarly, MySQL can handle larger amount of data; it has user management and rule-based control on what they can do, and concurrent access to the same dataset. The smart data scientist knows how to choose his weapon—depending on the task, another database backend might be the optimal solution. Let's see what other options we have in R!

PostgreSQL

While MySQL is said to be the most popular open-source relational database management system, PostgreSQL is famous for being "the world's most advanced open source database". This means that PostgreSQL is often considered to have more features compared to the simpler but faster MySQL, including analytic functions, which has led to PostgreSQL often being described as the open-source version of Oracle.

This sounds rather funny now, as Oracle owns MySQL today. So a bunch of things have changed in the past 20-30 years of RDBMS history, and PostgreSQL is not so slow any more. On the other hand, MySQL has also gained some nice new features—for example MySQL also became ACID-compliant with the InnoDB engine, allowing rollback to previous states of the database. There are some other differences between the two popular database servers that might support choosing either of them. Now let's see what happens if our data provider has a liking for PostgreSQL instead of MySQL!

Installing PostgreSQL is similar to MySQL. One may install the software with the operating system's package manager, download a graphical installer from http://www.enterprisedb.com/products-services-training/pgdownload, or run a virtual appliance with, for example, the free Turnkey Linux, which provides a small but fully configured disk image for free at http://www.turnkeylinux.org/postgresql.

After successfully installing and starting the server, let's set up the test database — just like we did after the MySQL installation:

```
                          daroczig : bash - Konsole
 File   Edit   View   Bookmarks   Settings   Help
[postgres@nevermind ~]$ createuser --pwprompt user
Enter password for new role:
Enter it again:
[postgres@nevermind ~]$ createdb hflights_db
[postgres@nevermind ~]$ psql -U postgres
psql (9.3.4)
Type "help" for help.

postgres=# \du
                            List of roles
 Role name |                  Attributes                      | Member of
-----------+--------------------------------------------------+-----------
 postgres  | Superuser, Create role, Create DB, Replication   | {}
 user      |                                                  | {}

postgres=# \list
                              List of databases
    Name    |  Owner   | Encoding |  Collate    |   Ctype     |   Access privileges
------------+----------+----------+-------------+-------------+----------------------
 hflights_db | postgres | UTF8    | en_US.UTF-8 | en_US.UTF-8 |
 postgres   | postgres | UTF8     | en_US.UTF-8 | en_US.UTF-8 |
 template0  | postgres | UTF8     | en_US.UTF-8 | en_US.UTF-8 | =c/postgres         +
            |          |          |             |             | postgres=CTc/postgres
 template1  | postgres | UTF8     | en_US.UTF-8 | en_US.UTF-8 | =c/postgres         +
            |          |          |             |             | postgres=CTc/postgres
(4 rows)

postgres=# grant all privileges on hflights_db to user
postgres-# \q
      daroczig : bash
```

The syntax is a bit different in some cases, and we have used some command-line tools for the user and database creation. These helper programs are shipped with PostgreSQL by default, and MySQL also have some similar functionality with `mysqladmin`.

After setting up the initial test environment, or if we already have a working database instance to connect, we can repeat the previously described data management tasks with the help of the RPostgreSQL package:

```
> library(RPostgreSQL)
Loading required package: DBI
```

> If your R session starts to throw strange error messages in the following examples, it's highly possible that the loaded R packages are conflicting. You could simply start a clean R session, or detach the previously attached packages — for example, detach('package:RMySQL', unload = TRUE).

Connecting to the database (listening on the default port number 5432) is again familiar:

```
> con <- dbConnect(dbDriver('PostgreSQL'), user = 'user',
+    password = 'password', dbname = 'hflights_db')
```

Let's verify that we are connected to the right database instance, which should be currently empty without the hflights table:

```
> dbListTables(con)
character(0)
> dbExistsTable(con, 'hflights')
[1] FALSE
```

Then let's write our demo table in PostgreSQL and see if the old rumor about it being slower than MySQL is still true:

```
> dbWriteTable(con, 'hflights', hflights)
[1] TRUE
> system.time(dbReadTable(con, 'hflights'))
   user  system elapsed
  0.590   0.013   0.921
```

Seems to be impressive! What about loading partial data?

```
> system.time(dbGetQuery(con,
+ statement = "SELECT * FROM hflights WHERE \"Dest\" = 'BNA';"))
   user  system elapsed
  0.026   0.000   0.082
```

Just under 100 milliseconds without indexing! Please note the extra escaped quotes around Dest, as the default PostgreSQL behavior folds unquoted column names to lower case, which would result in a column dest does not exist error. Creating an index and running the preceding query with much improved speed can be easily reproduced based on the MySQL example.

Oracle database

Oracle Database Express Edition can be downloaded and installed from http://www.oracle.com/technetwork/database/database-technologies/express-edition/downloads/index.html. Although this is not a full-featured Oracle database, and it suffers from serious limitations, the Express Edition is a free and not too resource-hungry way to build a test environment at home.

> Oracle database is said to be the most popular database management system in the world, although it is available only with a proprietary license, unlike the previous two discussed RDBMSs, which means that Oracle offers the product with term licensing. On the other hand, the paid license also comes with priority support from the developer company, which is often a strict requirement in enterprise environments. Oracle Database has supported a variety of nice features since its first release in 1980, such as sharding, master-master replication, and full ACID properties.

Another way of getting a working Oracle database for testing purposes is to download an Oracle Pre-Built Developer VM from `http://www.oracle.com/technetwork/community/developer-vm/index.html`, or a much smaller image custom created for *Hands-on Database Application Development* at *Oracle Technology Network Developer Day*: `http://www.oracle.com/technetwork/database/enterprise-edition/databaseappdev-vm-161299.html`. We will follow the instructions from the latter source.

After accepting the License Agreement and registering for free at Oracle, we can download the `OTN_Developer_Day_VM.ova` virtual appliance. Let's import it to VirtualBox via **Import appliance** in the **File** menu, then choose the `ova` file, and click **Next**:

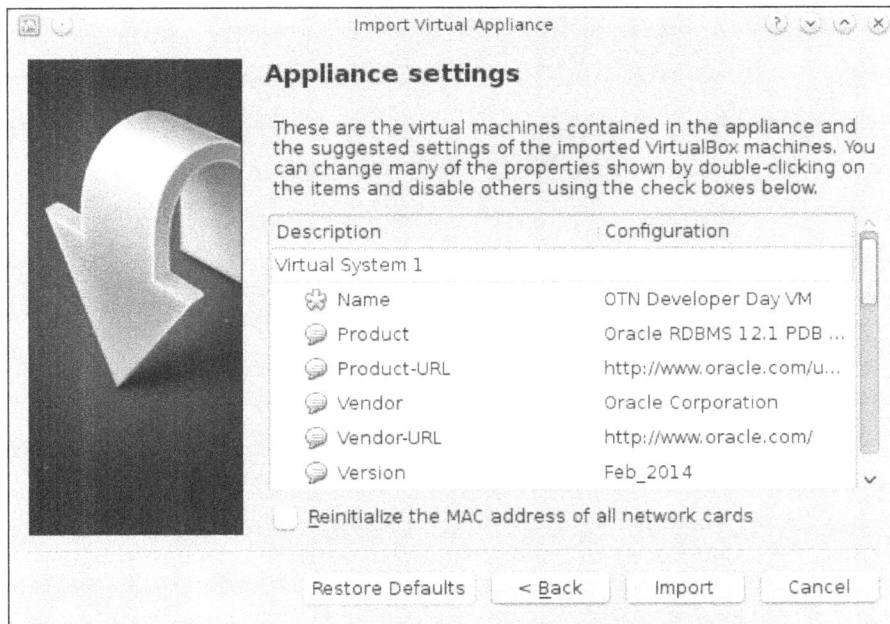

After clicking **Import**, you will have to agree again to the Software License Agreement. Importing the virtual disk image (15 GB) might take a few minutes:

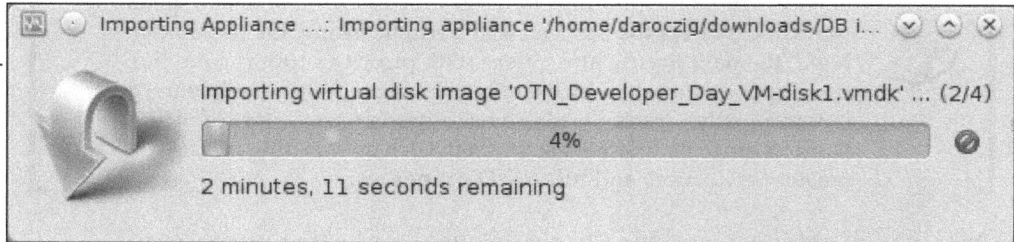

After importing has finished, we should first update the networking configuration so that we can access the internal database of the virtual machine from outside. So let's switch from **NAT** to **Bridged Adapter** in the settings:

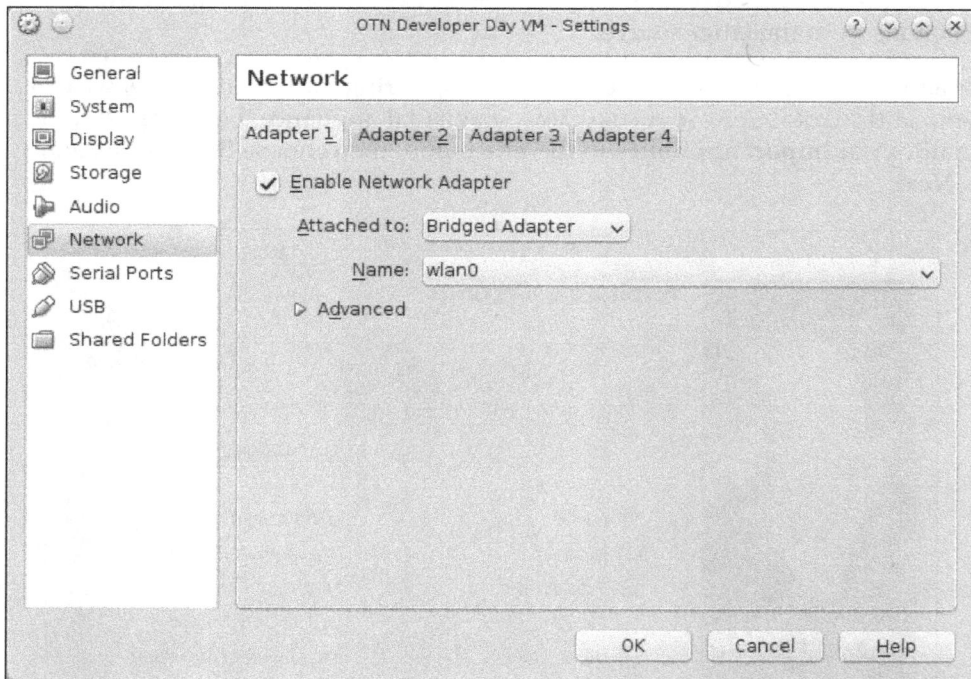

Then we can simply start the newly created virtual machine in VirtualBox. After Oracle Linux has booted, we can log in with the default `oracle` password.

Although we have set a bridged networking interface for our virtual machine, which means that the VM is directly connected to our real sub-network with a real IP address, the machine is not yet accessible over the network. To connect with the default DHCP settings, simply navigate to the top red bar and look for the networking icon, then select **System eth0**. After a few seconds the VM is accessible from your host machine, as the guest system should be connected to your network. You can verify that by running the `ifconfig` or `ip addr show eth0` command in the already running console:

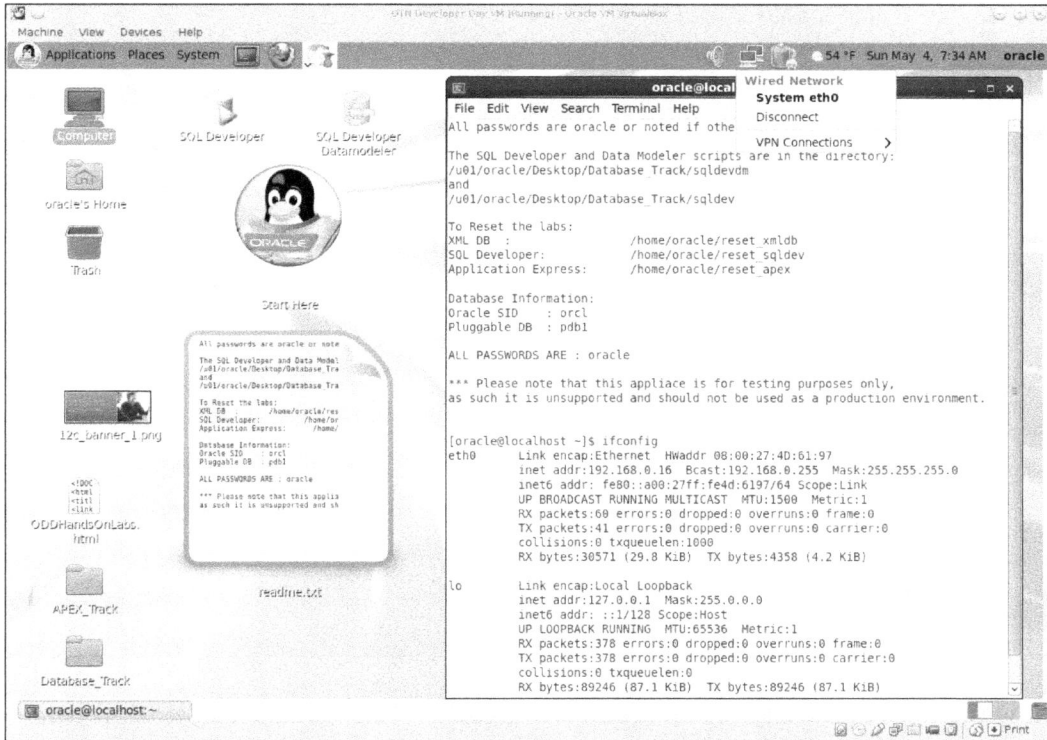

Unfortunately, this already running Oracle database is not yet accessible outside the guest machine. The developer VM comes with a rather strict firewall by default, which should be disabled first. To see the rules in effect, run the standard `iptables -L -n` command and, to flush all rules, execute `iptables -F`:

```
[root@localhost oracle]# iptables -L -n
Chain INPUT (policy ACCEPT)
target     prot opt source               destination
ACCEPT     all  --  0.0.0.0/0            0.0.0.0/0            state RELATED,ESTABLISHED
ACCEPT     icmp --  0.0.0.0/0            0.0.0.0/0
ACCEPT     all  --  0.0.0.0/0            0.0.0.0/0
ACCEPT     tcp  --  0.0.0.0/0            0.0.0.0/0            state NEW tcp dpt:22
REJECT     all  --  0.0.0.0/0            0.0.0.0/0            reject-with icmp-host-prohibited

Chain FORWARD (policy ACCEPT)
target     prot opt source               destination
REJECT     all  --  0.0.0.0/0            0.0.0.0/0            reject-with icmp-host-prohibited

Chain OUTPUT (policy ACCEPT)
target     prot opt source               destination
[root@localhost oracle]# iptables -F
[root@localhost oracle]# iptables -L -n
Chain INPUT (policy ACCEPT)
target     prot opt source               destination

Chain FORWARD (policy ACCEPT)
target     prot opt source               destination

Chain OUTPUT (policy ACCEPT)
target     prot opt source               destination
```

Now that we have a running and remotely accessible Oracle database, let's prepare the R client side. Installing the `ROracle` package might get tricky on some operating systems, as there are no prebuilt binary packages and you have to manually install the Oracle Instant Client Lite and SDK libraries before compiling the package from source. If the compiler complained about the path of your previously installed Oracle libraries, please pass the `--with-oci-lib` and `--with-oci-inc` arguments with your custom paths with the `--configure-args` parameter. More details can be found in the package installation document: `http://cran.r-project.org/web/packages/ROracle/INSTALL`.

For example, on Arch Linux you can install the Oracle libs from AUR, then run the following command in `bash` after downloading the R package from CRAN:

```
# R CMD INSTALL --configure-args='--with-oci-lib=/usr/include/    \
>   --with-oci-inc=/usr/share/licenses/oracle-instantclient-basic' \
>   ROracle_1.1-11.tar.gz
```

After installing and loading the package, opening a connection is extremely similar to the pervious examples with `DBI::dbConnect`. We only pass an extra parameter here. First, let us specify the hostname or direct IP address of the Oracle database included in the `dbname` argument. Then we can connect to the already existing PDB1 database of the developer machine instead of the previously used `hflights_db` — just to save some time and space in the book on slightly off-topic database management tasks:

```
> library(ROracle)
Loading required package: DBI
> con <- dbConnect(dbDriver('Oracle'), user = 'pmuser',
+   password = 'oracle', dbname = '//192.168.0.16:1521/PDB1')
```

And we have a working connection to Oracle RDBMS:

```
> summary(con)
User name:              pmuser
Connect string:         //192.168.0.16:1521/PDB1
Server version:         12.1.0.1.0
Server type:            Oracle RDBMS
Results processed:      0
OCI prefetch:           FALSE
Bulk read:              1000
Statement cache size:   0
Open results:           0
```

Let's see what we have in the bundled database on the development VM:

```
> dbListTables(con)
[1] "TICKER_G" "TICKER_O" "TICKER_A" "TICKER"
```

So it seems that we have a table called TICKER with three views on tick data of three symbols. Saving the hflights table in the same database will not do any harm, and we can also instantly test the speed of the Oracle database when reading the whole table:

```
> dbWriteTable(con, 'hflights', hflights)
[1] TRUE
> system.time(dbReadTable(con, 'hflights'))
   user  system elapsed
  0.980   0.057   1.256
```

And the extremely familiar subset with 3,481 cases:

```
> system.time(dbGetQuery(con,
+ "SELECT * FROM \"hflights\" WHERE \"Dest\" = 'BNA'"))
   user  system elapsed
  0.046   0.003   0.131
```

Please note the quotes around the table name. In the previous examples with MySQL and PostgreSQL, the SQL statements run fine without those. However, the quotes are needed in the Oracle database, as we have saved the table with an all-lowercase name, and the default rule in Oracle DB is to store object names in upper case. The only other option is to use double quotes to create them, which is what we did; thus we have to refer to the table with quotes around the lowercase name.

> We started with unquoted table and column names in MySQL, then had to add escaped quotes around the variable name in the PostgreSQL query run from R, and now in Oracle database we have to put both names between quotes—which demonstrates the slight differences in the various SQL flavors (such as MySQL, PostgreSQL, PL/SQL of Oracle or Microsoft's Transact-SQL) on top of ANSI SQL.
>
> And more importantly: do not stick to one database engine with all your projects, but rather choose the optimal DB for the task if company policy doesn't stop you doing so.

These results were not so impressive compared to what we have seen by PostgreSQL, so let's also see the results of an indexed query:

```
> dbSendQuery(con, 'CREATE INDEX Dest_idx ON "hflights" ("Dest")')
Statement:          CREATE INDEX Dest_idx ON "hflights" ("Dest")
Rows affected:      0
Row count:          0
```

```
Select statement:       FALSE
Statement completed:    TRUE
OCI prefetch:           FALSE
Bulk read:              1000
> system.time(dbGetQuery(con, "SELECT * FROM \"hflights\"
+ WHERE \"Dest\" = 'BNA'"))
   user   system  elapsed
  0.023   0.000    0.069
```

I leave the full-scale comparative testing and benchmarking to you, so that you can run custom queries in the tests fitting your exact needs. It is highly possible that the different database engines perform differently in special use cases.

To make this process a bit more seamless and easier to implement, let's check out another R way of connecting to databases, although probably with a slight performance trade-off. For a quick scalability and performance comparison on connecting to Oracle databases with different approaches in R, please see https://blogs.oracle.com/R/entry/r_to_oracle_database_connectivity.

ODBC database access

As mentioned earlier, installing the native client software, libraries, and header files for the different databases so that the custom R packages can be built from source can be tedious and rather tricky in some cases. Fortunately, we can also try to do the opposite of this process. An alternative solution can be installing a middleware **Application Programming Interface (API)** in the databases, so that R, or as a matter of fact any other tool, could communicate with them in a standardized and more convenient way. However, please be advised that this more convenient way impairs performance due to the translation layer between the application and the DBMS.

The RODBC package implements access to such a layer. The **Open Database Connectivity (ODBC)** driver is available for most database management systems, even for CSV and Excel files, so RODBC provides a standardized way to access data in almost any databases if the ODBC driver is installed. This platform-independent interface is available for SQLite, MySQL, MariaDB, PostgreSQL, Oracle database, Microsoft SQL Server, Microsoft Access, and IBM DB2 on Windows and on Linux.

For a quick example, let's connect to MySQL running on `localhost` (or on a virtual machine). First, we have to set up a **Database Source Name** (**DSN**) with the connection details, such as:

- Database driver
- Host name or address and port number, optionally a Unix socket
- Database name
- Optionally the username and password to be used for the connection

This can be done in the command line by editing the `odbc.ini` and `odbcinst.ini` files on Linux after installing the `unixODBC` program. The latter should include the following configuration for the MySQL driver in your `/etc` folder:

```
[MySQL]
Description     = ODBC Driver for MySQL
Driver          = /usr/lib/libmyodbc.so
Setup           = /usr/lib/libodbcmyS.so
FileUsage       = 1
```

The `odbc.ini` file includes the aforementioned DSN configuration for the exact database and server:

```
[hflights]
Description     = MySQL hflights test
Driver          = MySQL
Server          = localhost
Database        = hflights_db
Port            = 3306
Socket          = /var/run/mysqld/mysqld.sock
```

Or use a graphical user interface on Mac OS or Windows, as shown in the following screenshot:

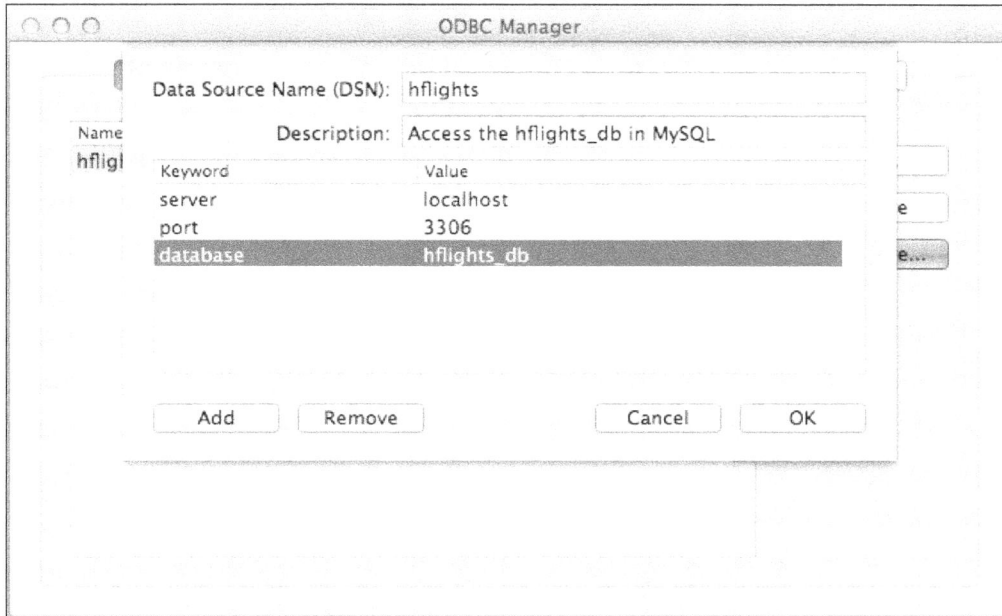

After configuring a DSN, we can connect with a one-line command:

```
> library(RODBC)
> con <- odbcConnect("hflights", uid = "user", pwd = "password")
```

Let's fetch the data we saved in the database before:

```
> system.time(hflights <- sqlQuery(con, "select * from hflights"))
   user   system  elapsed
  3.180    0.000    3.398
```

Well, it took a few seconds to finish. That's the trade-off for using a more convenient and high-level interface to interact with the database. Removing and uploading data to the database can be done with similar high-level functions (such as sqlFetch) besides the odbc* functions, providing low-level access to the database. Quick examples:

```
> sqlDrop(con, 'hflights')
> sqlSave(con, hflights, 'hflights')
```

You can use the exact same commands to query any of the other supported database engines; just be sure to set up the DSN for each backend, and to close your connections if not needed any more:

```
> close(con)
```

The RJDBC package can provide a similar interface to database management systems with a **Java Database Connectivity (JDBC)** driver.

Using a graphical user interface to connect to databases

Speaking of high-level interfaces, R also has a graphical user interface to connect to MySQL in the dbConnect package:

```
> library(dbConnect)
Loading required package: RMySQL
Loading required package: DBI
Loading required package: gWidgets
> DatabaseConnect()
Loading required package: gWidgetsRGtk2
Loading required package: RGtk2
```

No parameters, no custom configuration in the console, just a simple dialog window:

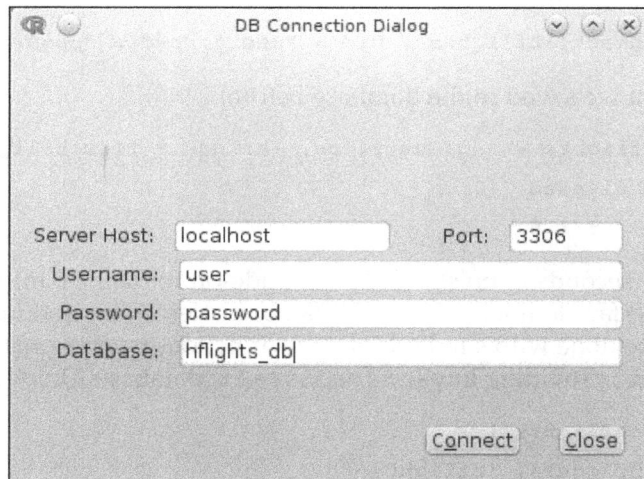

After providing the required connection information, we can easily view the raw data and the column/variable types, and run custom SQL queries. A basic query builder can also help novice users to fetch subsamples from the database:

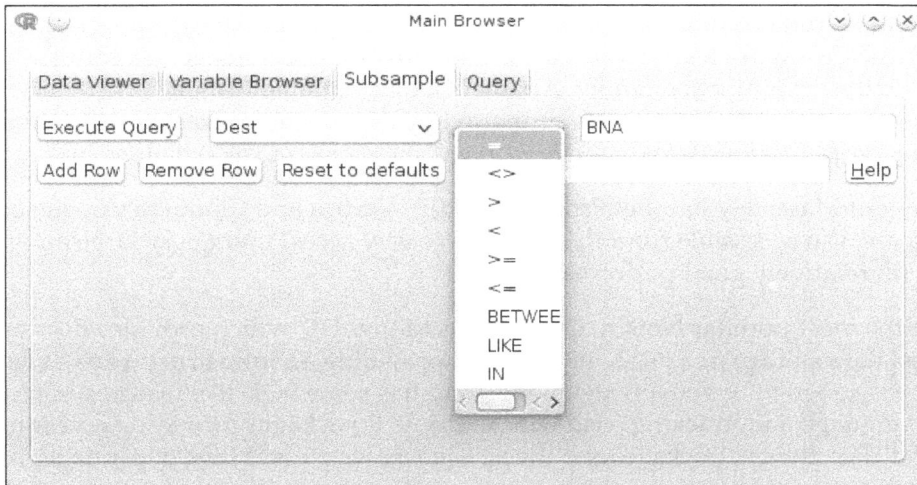

The package ships with a handy function called sqlToR, which can turn the SQL results into R objects with a click in the GUI. Unfortunately, dbConnect relies heavily on RMySQL, which means it's a MySQL-only package, and there is no plan to extend the functionality of this interface.

Other database backends

Besides the previously mentioned popular databases, there are several other implementations that we cannot discuss here in detail.

For example, column-oriented database management systems, such as MonetDB, are often used to store large datasets with millions of rows and thousands of columns to provide the backend for high-performance data mining. It also has great R support with the MonetDB.R package, which was among the most exciting talks at the useR! 2013 conference.

The ever-growing popularity of the NoSQL ecosystem also provides similar approaches, although usually without supporting SQL and providing a schema-free data storage. Apache Cassandra is a good example of such a similar, column-oriented, and primarily distributed database management system with high availably and performance, run on commodity hardware. The `RCassandra` package provides access to the basic Cassandra features and the Cassandra Query Language in a convenient way with the `RC.*` function family. Another Google Bigtable-inspired and similar database engine is HBase, which is supported by the `rhbase` package, part of the `RHadoop` project: https://github.com/RevolutionAnalytics/RHadoop/wiki.

Speaking of Massively Parallel Processing, HP's Vertica and Cloudera's open-source Impala are also accessible from R, so you can easily access and query large amount of data with relatively good performance.

One of the most popular NoSQL databases is MongoDB, which provides document-oriented data storage in a JSON-like format, providing an infrastructure to dynamic schemas. MongoDB is actively developed and has some SQL-like features, such as a query language and indexing, also with multiple R packages providing access to this backend. The `RMongo` package uses the *mongo-java-driver* and thus depends on Java, but provides a rather high-level interface to the database. Another implementation, the `rmongodb` package, is developed and maintained by the MongoDB Team. The latter has more frequent updates and more detailed documentation, but the R integration seems to be a lot more seamless with the first package as `rmongodb` provides access to the raw MongoDB functions and BSON objects, instead of concentrating on a translation layer for general R users. A more recent and really promising package supporting MongoDB is `mongolite` developed by Jeroen Ooms.

CouchDB, my personal favorite for most schema-less projects, provides very convenient document storage with JSON objects and HTTP API, which means that integrating in applications, such as any R script, is really easy with, for example, the `RCurl` package, although you may find the `R4CouchDB` more quick to act in interacting with the database.

Google BigQuery also provides a similar, REST-based HTTP API to query even terabytes of data hosted in the Google infrastructure with an SQL-like language. Although the `bigrquery` package is not available on CRAN yet, you may easily install it from GitHub with the `devtools` package from the same author, Hadley Wickham:

```
> library(devtools)
> install_github('bigrquery', 'hadley')
```

To test-drive the features of this package and Google BigQuery, you can sign up for a free account to fetch and process the demo dataset provided by Google, respecting the 10,000 requests per day limitation for free usage. Please note that the current implementation is a read-only interface to the database.

For rather similar database engines and comparisons, see for example `http://db-engines.com/en/systems`. Most of the popular databases already have R support but, if not, I am pretty sure that someone is already working on it. It's worth checking the CRAN packages at `http://cran.r-project.org/web/packages/available_packages_by_name.html` or searching on GitHub or on `http://R-bloggers.com` to see how other R users manage to interact with your database of choice.

Importing data from other statistical systems

In a recent academic project, where my task was to implement some financial models in R, I got the demo dataset to be analyzed as Stata `dta` files. Working as a contractor at the university, without access to any Stata installations, it might have been problematic to read the binary file format of another statistical software, but as the `dta` file format is documented and the specification is publicly available at `http://www.stata.com/help.cgi?dta`, some members of the Core R Team have already implemented an R parser in the form of the `read.dta` function in the `foreign` package.

To this end, loading (and often writing) Stata — or for example SPSS, SAS, Weka, Minitab, Octave, or dBase files — just cannot be easier in R. Please see the complete list of supported file formats and examples in the package documentation or in the *R Data Import/Export* manual: `http://cran.r-project.org/doc/manuals/r-release/R-data.html#Importing-from-other-statistical-systems`.

Loading Excel spreadsheets

One of the most popular file formats to store and transfer relatively small amounts of data in academic institutions and businesses (besides CSV files) is still Excel `xls` (or `xlsx`, more recently). The first is a proprietary binary file format from Microsoft, which is exhaustively documented (the `xls` specification is available in a document of more than 1,100 pages and 50 megabytes!), but importing multiple sheets, macros, and formulas is not straightforward even nowadays. This section will only cover the most used platform-independent packages to interact with Excel.

One option is to use the previously discussed `RODBC` package with the Excel driver to query an Excel spreadsheet. Other ways of accessing Excel data depend on third-party tools, such as using Perl to automatically convert the Excel file to CSV then importing it into R as the `read.xls` function from the `gdata` package. But installing Perl on Windows sometimes seems to be tedious; thus, `RODBC` might be a more convenient method on that platform.

Some platform-independent, Java-based solutions also provide a way to not just read, but also write Excel files, especially to the xlsx, the Office Open XML file, format. Two separate implementations exist on CRAN to read and write Excel 2007 and the 97/2000/XP/2003 file formats: the xlConnect and the xlsx packages. Both are actively maintained, and use the Apache POI Java API project. This latter means that it runs on any platform that supports Java, and there is no need to have Microsoft Excel or Office on the computer; both packages can read and write Excel files on their own.

On the other hand, if you would rather not depend on Perl or Java, the recently published openxlsx package provides a platform-independent (C++-powered) way of reading and writing xlsx files. Hadley Wickham released a similar package, but with a slightly modified scope: the readxl package can read (but not write) both the xls and xlsx file formats.

Remember: pick the most appropriate tool for your needs! For example to read Excel files without many external dependencies, I'd choose readxl; but, for writing Excel 2003 spreadsheets with cell formatting and more advanced features, probably we cannot save the Java dependency and should use the xlConnect or xlsx packages over the xlsx-only openxlsx package.

Summary

This chapter focused on some rather boring, but important tasks that we usually do every day. Importing data is among the first steps of every data science projects, thus mastering data analysis should start with how to load data into the R session in an efficient way.

But efficiency is an ambiguous term in this sense: loading data should be quick in a technical point of view so as not to waste our time, although coding for long hours to speed up the importing process does not make much sense either.

The chapter gave a general overview on the most popular available options to read text files, to interact with databases, and to query subsets of data in R. Now you should be able to deal with all the most often used different data sources, and probably you can also choose which data source would be the ideal candidate in your projects and then do the benchmarks on your own, as we did previously.

The next chapter will extend this knowledge further by providing use cases for fetching data from the Web and different APIs. This simply means that you will be able to use public data in your projects, even if you do not yet have those in binary dataset files or on database backends.

2

Getting Data from the Web

It happens pretty often that we want to use data in a project that is not yet available in our databases or on our disks, but can be found on the Internet. In such situations, one option might be to get the IT department or a data engineer at our company to extend our data warehouse to scrape, process, and load the data into our database as shown in the following diagram:

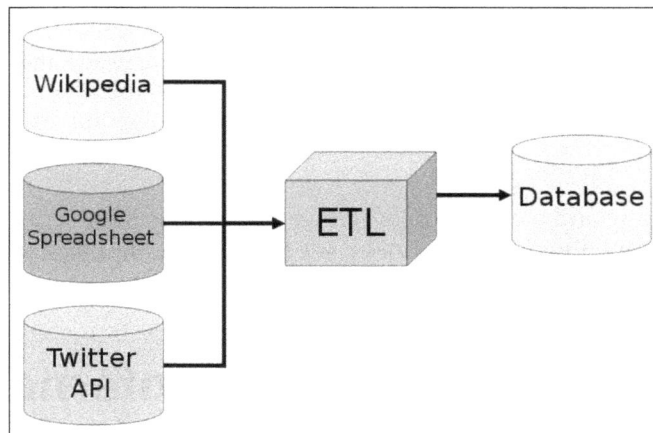

On the other hand, if we have no *ETL* system (*to Extract, Transform, and Load data*) or simply just cannot wait a few weeks for the IT department to implement our request, we are on our own. This is pretty standard for the data scientist, as most of the time we are developing prototypes that can be later transformed into products by software developers. To this end, a variety of skills are required in the daily round, including the following topics that we will cover in this chapter:

- Downloading data programmatically from the Web
- Processing XML and JSON formats

- Scraping and parsing data from raw HTML sources
- Interacting with APIs

Although being a *data scientist* was referred to as the sexiest job of the 21st century (Source: `https://hbr.org/2012/10/data-scientist-the-sexiest-job-of-the-21st-century/`), most data science tasks have nothing to do with data analysis. Worse, sometimes the job seems to be boring, or the daily routine requires just basic IT skills and no machine learning at all. Hence, I prefer to call this role a *data hacker* instead of *data scientist*, which also means that we often have to get our hands dirty.

For instance, scraping and scrubbing data is the least sexy part of the analysis process for sure, but it's one of the most important steps; it is also said, that around 80 percent of data analysis is spent cleaning data. There is no sense in running the most advanced machine learning algorithm on junk data, so be sure to take your time to get useful and tidy data from your sources.

> This chapter will also depend on extensive usage of Internet browser debugging tools with some R packages. These include Chrome `DevTools` or `FireBug` in Firefox. Although the steps to use these tools will be straightforward and also shown on screenshots, it's definitely worth mastering these tools for future usage; therefore, I suggest checking out a few tutorials on these tools if you are into fetching data from online sources. Some starting points are listed in the *References* section of the *Appendix* at the end of the book.

For a quick overview and a collection of relevant R packages for scraping data from the Web and to interact with Web services, see the *Web Technologies and Services CRAN Task View* at `http://cran.r-project.org/web/views/WebTechnologies.html`.

Loading datasets from the Internet

The most obvious task is to download datasets from the Web and load those into our R session in two manual steps:

1. Save the datasets to disk.
2. Read those with standard functions, such as `read.table` or for example `foreign::read.spss`, to import `sav` files.

But we can often save some time by skipping the first step and loading the flat text data files directly from the URL. The following example fetches a comma-separated file from the **Americas Open Geocode (AOG)** database at `http://opengeocode.org`, which contains the government, national statistics, geological information, and post office websites for the countries of the world:

```
> str(read.csv('http://opengeocode.org/download/CCurls.txt'))
'data.frame':  249 obs. of  5 variables:
 $ ISO.3166.1.A2               : Factor w/ 248 levels "AD" ...
 $ Government.URL              : Factor w/ 232 levels ""  ...
 $ National.Statistics.Census..URL: Factor w/ 213 levels ""  ...
 $ Geological.Information.URL  : Factor w/ 116 levels ""  ...
 $ Post.Office.URL            : Factor w/ 156 levels ""  ...
```

In this example, we passed a hyperlink to the `file` argument of `read.table`, which actually downloaded the text file before processing. The `url` function, used by `read.table` in the background, supports HTTP and FTP protocols, and can also handle proxies, but it has its own limitations. For example `url` does not support **Hypertext Transfer Protocol Secure (HTTPS)** except for a few exceptions on Windows, which is often a must to access Web services that handle sensitive data.

> HTTPS is not a separate protocol alongside HTTP, but instead HTTP over an encrypted SSL/TLS connection. While HTTP is considered to be insecure due to the unencrypted packets travelling between the client and server, HTTPS does not let third-parties discover sensitive information with the help of signed and trusted certificates.

In such situations, it's wise, and used to be the only reasonable option, to install and use the `RCurl` package, which is an R client interface to curl: `http://curl.haxx.se`. Curl supports a wide variety of protocols and URI schemes and handles cookies, authentication, redirects, timeouts, and even more.

For example, let's check the U.S. Government's open data catalog at `http://catalog.data.gov/dataset`. Although the general site can be accessed without SSL, most of the generated download URLs follow the HTTPS URI scheme. In the following example, we will fetch the **Comma Separated Values (CSV)** file of the Consumer Complaint Database from the Consumer Financial Protection Bureau, which can be accessed at `http://catalog.data.gov/dataset/consumer-complaint-database`.

> This CSV file contains metadata on around a quarter of a million of complaints about financial products and services since 2011. Please note that the file is around 35-40 megabytes, so downloading it might take some time, and you would probably not want to reproduce the following example on mobile or limited Internet. If the `getURL` function fails with a certificate error (this might happen on Windows), please provide the path of the certificate manually by `options(RCurlOptions = list(cainfo = system.file("CurlSSL", "cacert.pem", package = "RCurl")))` or try the more recently published `curl` package by Jeroen Ooms or `httr` (RCurl front-end) by Hadley Wickham—see later.

Let's see the distribution of these complaints by product type after fetching and loading the CSV file directly from R:

```
> library(RCurl)
Loading required package: bitops
> url <- 'https://data.consumerfinance.gov/api/views/x94z-ydhh/rows.
csv?accessType=DOWNLOAD'
> df   <- read.csv(text = getURL(url))
> str(df)
'data.frame':  236251 obs. of   14 variables:
 $ Complaint.ID         : int   851391 851793 ...
 $ Product              : Factor w/ 8 levels ...
 $ Sub.product          : Factor w/ 28 levels ...
 $ Issue                : Factor w/ 71 levels "Account opening ...
 $ Sub.issue            : Factor w/ 48 levels "Account status" ...
 $ State                : Factor w/ 63 levels "","AA","AE",,..
 $ ZIP.code             : int   14220 64119 ...
 $ Submitted.via        : Factor w/ 6 levels "Email","Fax" ...
 $ Date.received        : Factor w/ 897 levels  ...
 $ Date.sent.to.company : Factor w/ 847 levels "","01/01/2013" ...
 $ Company              : Factor w/ 1914 levels ...
 $ Company.response     : Factor w/ 8 levels "Closed" ...
 $ Timely.response.     : Factor w/ 2 levels "No","Yes" ...
 $ Consumer.disputed.   : Factor w/ 3 levels "","No","Yes" ...
> sort(table(df$Product))
```

Money transfers	Consumer loan	Student loan
965	6564	7400
Debt collection	Credit reporting	Bank account or service
24907	26119	30744
Credit card	Mortgage	
34848	104704	

Although it's nice to know that most complaints were received about mortgages, the point here was to use curl to download the CSV file with a HTTPS URI and then pass the content to the `read.csv` function (or any other parser we discussed in the previous chapter) as text.

> Besides GET requests, you can easily interact with RESTful API endpoints via POST, DELETE, or PUT requests as well by using the postForm function from the RCurl package or the httpDELETE, httpPUT, or httpHEAD functions — see details about the httr package later.

Curl can also help to download data from a secured site that requires authorization. The easiest way to do so is to login to the homepage in a browser, save the cookie to a text file, and then pass the path of that to cookiefile in getCurlHandle. You can also specify useragent among other options. Please see http://www.omegahat.org/RCurl/RCurlJSS.pdf for more details and an overall (and very useful) overview on the most important RCurl features.

Although curl is extremely powerful, the syntax and the numerous options with the technical details might be way too complex for those without a decent IT background. The httr package is a simplified wrapper around RCurl with some sane defaults and much simpler configuration options for common operations and everyday actions.

For example, cookies are handled automatically by sharing the same connection across all requests to the same website; error handling is much improved, which means easier debugging if something goes wrong; the package comes with various helper functions to, for instance, set headers, use proxies, and easily issue GET, POST, PUT, DELETE, and other methods. Even more, it also handles authentication in a much more user-friendly way — along with OAuth support.

> OAuth is the open standard for authorization with the help of intermediary service providers. This simply means that the user does not have to share actual credentials, but can rather delegate rights to access some of the stored information at the service providers. For example, one can authorize Google to share the real name, e-mail address, and so on with a third-party without disclosing any other sensitive information or any need for passwords. Most generally, OAuth is used for password-less login to various Web services and APIs. For more information, please see the *Chapter 14, Analyzing the R Community*, where we will use OAuth with Twitter to authorize the R session for fetching data.

But what if the data is not available to be downloaded as CSV files?

Other popular online data formats

Structured data is often available in XML or JSON formats on the Web. The high popularity of these two formats is due to the fact that both are human-readable, easy to handle from a programmatic point of view, and can manage any type of hierarchical data structure, not just a simple tabular design, as CSV files are.

> JSON is originally derived from *JavaScript Object Notation*, which recently became one of the top, most-used standards for human-readable data exchange format. JSON is considered to be a low-overhead alternative to XML with attribute-value pairs, although it also supports a wide variety of object types such as number, string, boolean, ordered lists, and associative arrays. JSON is highly used in Web applications, services, and APIs.

Of course, R also supports loading (and saving) data in JSON. Let's demonstrate that by fetching some data from the previous example via the Socrata API (more on that later in the *R packages to interact with data source APIs* section of this chapter), provided by the Consumer Financial Protection Bureau. The full documentation of the API is available at http://www.consumerfinance.gov/complaintdatabase/technical-documentation.

The endpoint of the API is a URL where we can query the background database without authentication is http://data.consumerfinance.gov/api/views. To get an overall picture on the structure of the data, the following is the returned JSON list opened in a browser:

As JSON is extremely easy to read, it's often very helpful to skim through the structure manually before parsing. Now let's load that tree list into R with the `rjson` package:

```
> library(rjson)
> u <- 'http://data.consumerfinance.gov/api/views'
> fromJSON(file = u)
[[1]]
[[1]]$id
[1] "25ei-6bcr"


[[1]]$name
[1] "Credit Card Complaints"


[[1]]$averageRating
[1] 0

...
```

Well, it does not seem to be the same data we have seen before in the comma-separated values file! After a closer look at the documentation, it's clear that the endpoint of the API returns metadata on the available views instead of the raw tabular data that we saw in the CSV file. So let's see the view with the ID of `25ei-6bcr` now for the first five rows by opening the related URL in a browser:

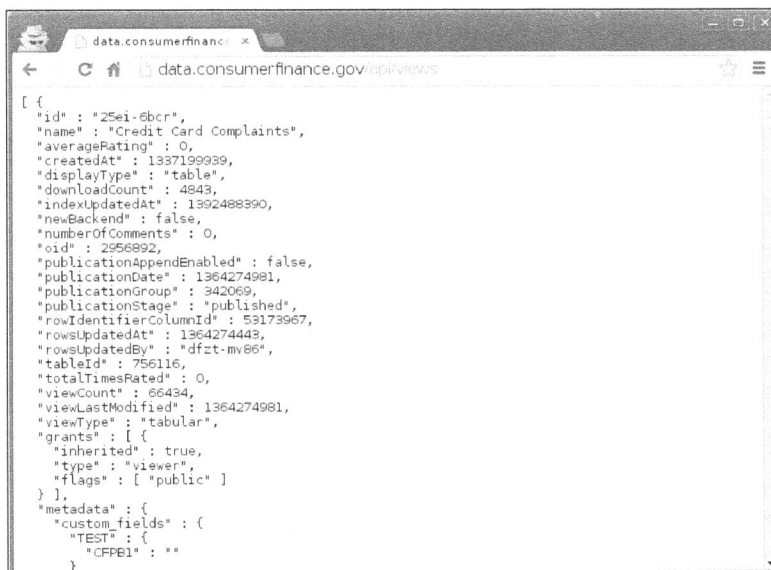

The structure of the resulting JSON list has changed for sure. Now let's read that hierarchical list into R:

```
> res <- fromJSON(file = paste0(u,'/25ei-6bcr/rows.json?max_rows=5'))
> names(res)
[1] "meta" "data"
```

We managed to fetch the data along with some further meta-information on the view, columns, and so on, which is not something that we are interested in at the moment. As `fromJSON` returned a `list` object, we can simply drop the metadata and work with the `data` rows from now on:

```
> res <- res$data
> class(res)
[1] "list"
```

This is still a `list`, which we usually want to transform into a `data.frame` instead. So we have `list` with five elements, each holding 19 nested children. Please note that one of those, the 13th sub element, is list again with 5-5 vectors. This means that transforming the tree list into tabular format is not straightforward, even less so when we realize that one of those vectors holds multiple values in an unprocessed JSON format. So, for the sake of simplicity and proof of a concept demo, let's simply ditch the location-related values now and transform all other values to `data.frame`:

```
> df <- as.data.frame(t(sapply(res, function(x) unlist(x[-13]))))
> str(df)
'data.frame':   5 obs. of  18 variables:
 $ V1 : Factor w/ 5 levels "16756","16760",..: 3 5 ...
 $ V2 : Factor w/ 5 levels "F10882C0-23FC-4064-979C-07290645E64B" ...
 $ V3 : Factor w/ 5 levels "16756","16760",..: 3 5 ...
 $ V4 : Factor w/ 1 level "1364270708": 1 1 ...
 $ V5 : Factor w/ 1 level "403250": 1 1 ...
 $ V6 : Factor w/ 5 levels "1364274327","1364274358",..: 5 4 ...
 $ V7 : Factor w/ 1 level "546411": 1 1 ...
 $ V8 : Factor w/ 1 level "{\n}": 1 1 ...
 $ V9 : Factor w/ 5 levels "2083","2216",..: 1 2 ...
 $ V10: Factor w/ 1 level "Credit card": 1 1 ...
 $ V11: Factor w/ 2 levels "Referral","Web": 1 1 ...
 $ V12: Factor w/ 1 level "2011-12-01T00:00:00": 1 1 ...
 $ V13: Factor w/ 5 levels "Application processing delay",..: 5 1 ...
```

```
$ V14: Factor w/ 3 levels "2011-12-01T00:00:00",..: 1 1 ...
$ V15: Factor w/ 5 levels "Amex","Bank of America",..: 2 5 ...
$ V16: Factor w/ 1 level "Closed without relief": 1 1 ...
$ V17: Factor w/ 1 level "Yes": 1 1 ...
$ V18: Factor w/ 2 levels "No","Yes": 1 1 ...
```

So we applied a simple function that drops location information from each element of the list (by removing the 13th element of each *x*), automatically simplified to `matrix` (by using `sapply` instead of `lapply` to iterate though each element of the list), transposed it (via `t`), and then coerced the resulting object to `data.frame`.

Well, we can also use some helper functions instead of manually tweaking all the list elements, as earlier. The `plyr` package (please find more details in *Chapter 3*, *Filtering and Summarizing Data* and *Chapter 4*, *Restructuring Data*) includes some extremely useful functions to split and combine data:

```
> library(plyr)
> df <- ldply(res, function(x) unlist(x[-13]))
```

It looks a lot more familiar now, although we miss the variable names, and all values were converted to character vectors or factors—even the dates that were stored as UNIX timestamps. We can easily fix these problems with the help of the provided metadata (`res$meta`): for example, let's set the variable names by extracting (via the `[` operator) the name field of all columns except for the dropped (13th) location data:

```
> names(df) <- sapply(res$meta$view$columns, `[`, 'name')[-13]
```

One might also identify the object classes with the help of the provided metadata. For example, the `renderTypeName` field would be a good start to check, and using `as.numeric` for number and `as.POSIXct` for all `calendar_date` fields would resolve most of the preceding issues.

Well, did you ever hear that around 80 percent of data analysis is spent on data preparation?

Parsing and restructuring JSON and XML to `data.frame` can take a long time, especially when you are dealing with hierarchical lists primarily. The `jsonlite` package tries to overcome this issue by transforming R objects into a conventional JSON data structure and vice-versa instead of raw conversion. This means from a practical point of view that `jsonlite::fromJSON` will result in `data.frame` instead of raw list if possible, and it makes the interchange data format even more seamless. Unfortunately, we cannot always transform lists to a tabular format; in such cases, the list transformations can be speeded up by for example the `rlist` package. Please find more details on list manipulations in *Chapter 14*, *Analyzing the R Community*.

> **Extensible Markup Language** (**XML**) was originally developed by the World Wide Web Consortium in 1996 to store documents in a both human-readable and machine-readable format. This popular syntax is used in for example the Microsoft Office Open XML and Open/LibreOffice OpenDocument file formats, in RSS feeds, and in various configuration files. As the format is also highly used for the interchange of data over the Internet, data is often available in XML as the only option—especially with some older APIs.

Let us also see how we can handle another popular online data interchange format besides JSON. The XML API can be used in a similar way, but we must define the desired output format in the endpoint URL: `http://data.consumerfinance.gov/api/views.xml`, as you should be able to see in the following screenshot:

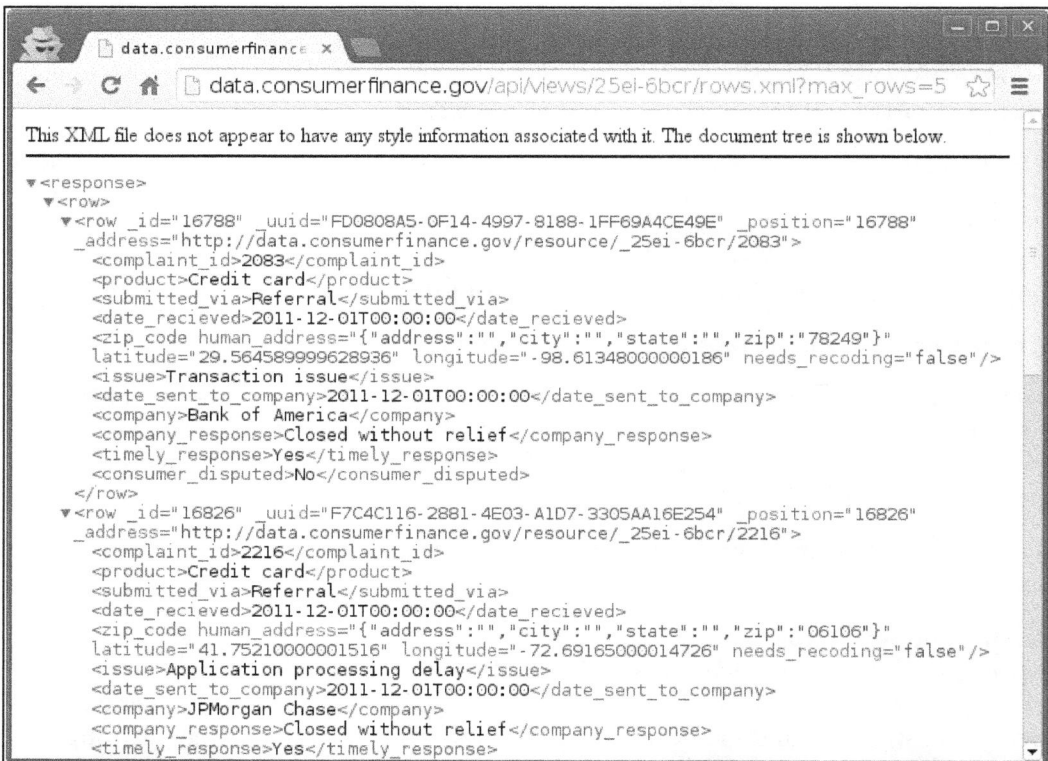

It seems that the XML output of the API differs from what we have seen in the JSON format, and it simply includes the rows that we are interested in. This way, we can simply parse the XML document and extract the rows from the response then transform them to `data.frame`:

```
> library(XML)
> doc <- xmlParse(paste0(u, '/25ei-6bcr/rows.xml?max_rows=5'))
> df  <- xmlToDataFrame(nodes = getNodeSet(doc,"//response/row/row"))
> str(df)
'data.frame':  5 obs. of  11 variables:
 $ complaint_id        : Factor w/ 5 levels "2083","2216",..: 1 2 ...
 $ product             : Factor w/ 1 level "Credit card": 1 1 ...
 $ submitted_via       : Factor w/ 2 levels "Referral","Web": 1 1 ...
 $ date_recieved       : Factor w/ 1 level "2011-12-01T00:00:00" ...
 $ zip_code            : Factor w/ 1 level "": 1 1 ...
 $ issue               : Factor w/ 5 levels  ...
 $ date_sent_to_company: Factor w/ 3 levels "2011-12-01T00:00:00" ...
 $ company             : Factor w/ 5 levels "Amex" ....
 $ company_response    : Factor w/ 1 level "Closed without relief"...
 $ timely_response     : Factor w/ 1 level "Yes": 1 1 ...
 $ consumer_disputed   : Factor w/ 2 levels "No","Yes": 1 1 ...
```

Although we could manually set the desired classes of the variables in the `colClasses` argument passed to `xmlToDataFrame`, just like in `read.tables` we can also fix this issue afterwards with a quick `helper` function:

```
> is.number <- function(x)
+     all(!is.na(suppressWarnings(as.numeric(as.character(x)))))
> for (n in names(df))
+     if (is.number(df[, n]))
+         df[, n] <- as.numeric(as.character(df[, n]))
```

So we tried to guess if a column includes only numbers, and convert those to `numeric` if our helper function returns TRUE. Please note that we first convert the `factor` to `character` before transforming to number, as a direct conversion from `factor` to `numeric` would return the `factor` order instead of the real value. One might also try to resolve this issue with the `type.convert` function, which is used by default in `read.table`.

> To test similar APIs and JSON or XML resources, you may find it interesting to check out the API of Twitter, GitHub, or probably any other online service provider. On the other hand, there is also another open-source service based on R that can return XML, JSON, or CSV files from any R code. Please find more details at http://www.opencpu.org.

So now we can process structured data from various kinds of downloadable data formats but, as there are still some other data source options to master, I promise you it's worth it to keep reading.

Reading data from HTML tables

According to the traditional document formats on the World Wide Web, most texts and data are served in HTML pages. We can often find interesting pieces of information in for example HTML tables, from which it's pretty easy to copy and paste data into an Excel spreadsheet, save that to disk, and load it to R afterwards. But it takes time, it's boring, and can be automated anyway.

Such HTML tables can be easily generated with the help of the aforementioned API of the Customer Compliant Database. If we do not set the required output format for which we used XML or JSON earlier, then the browser returns a HTML table instead, as you should be able to see in the following screenshot:

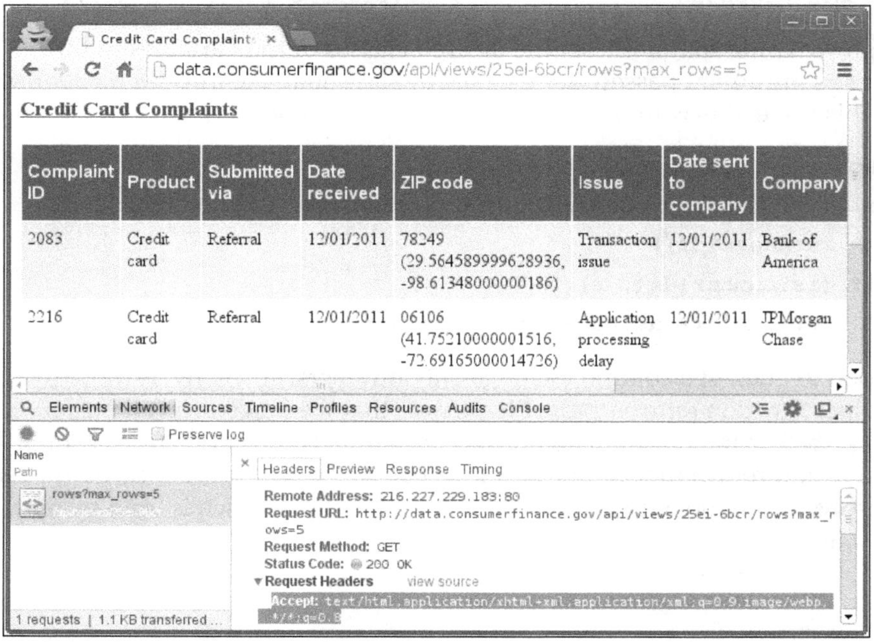

Well, in the R console it's a bit more complicated as the browser sends some non-default HTTP headers while using curl, so the preceding URL would simply return a JSON list. To get HTML, let the server know that we expect HTML output. To do so, simply set the appropriate HTTP header of the query:

```
> doc <- getURL(paste0(u, '/25ei-6bcr/rows?max_rows=5'),
+     httpheader = c(Accept = "text/html"))
```

The XML package provides an extremely easy way to parse all the HTML tables from a document or specific nodes with the help of the readHTMLTable function, which returns a list of data.frames by default:

```
> res <- readHTMLTable(doc)
```

To get only the first table on the page, we can filter res afterwards or pass the which argument to readHTMLTable. The following two R expressions have the very same results:

```
> df <- res[[1]]
> df <- readHTMLTable(doc, which = 1)
```

Reading tabular data from static Web pages

Okay, so far we have seen a bunch of variations on the same theme, but what if we do not find a downloadable dataset in any popular data format? For example, one might be interested in the available R packages hosted at CRAN, whose list is available at http://cran.r-project.org/web/packages/available_packages_by_name.html. How do we scrape that? No need to call RCurl or to specify custom headers, still less do we have to download the file first; it's enough to pass the URL to readHTMLTable:

```
> res <- readHTMLTable('http://cran.r-project.org/Web/packages/available_packages_by_name.html')
```

So readHTMLTable can directly fetch HTML pages, then it extracts all the HTML tables to data.frame R objects, and returns a list of those. In the preceding example, we got a list of only one data.frame with all the package names and descriptions as columns.

Well, this amount of textual information is not really informative with the `str` function. For a quick example of processing and visualizing this type of raw data, and to present the plethora of available features by means of R packages at CRAN, now we can create a word cloud of the package descriptions with some nifty functions from the `wordcloud` and the `tm` packages:

```
> library(wordcloud)
Loading required package: Rcpp
Loading required package: RColorBrewer
> wordcloud(res[[1]][, 2])
Loading required package: tm
```

This short command results in the following screenshot, which shows the most frequent words found in the R package descriptions. The position of the words has no special meaning, but the larger the font size, the higher the frequency. Please see the technical description of the plot following the screenshot:

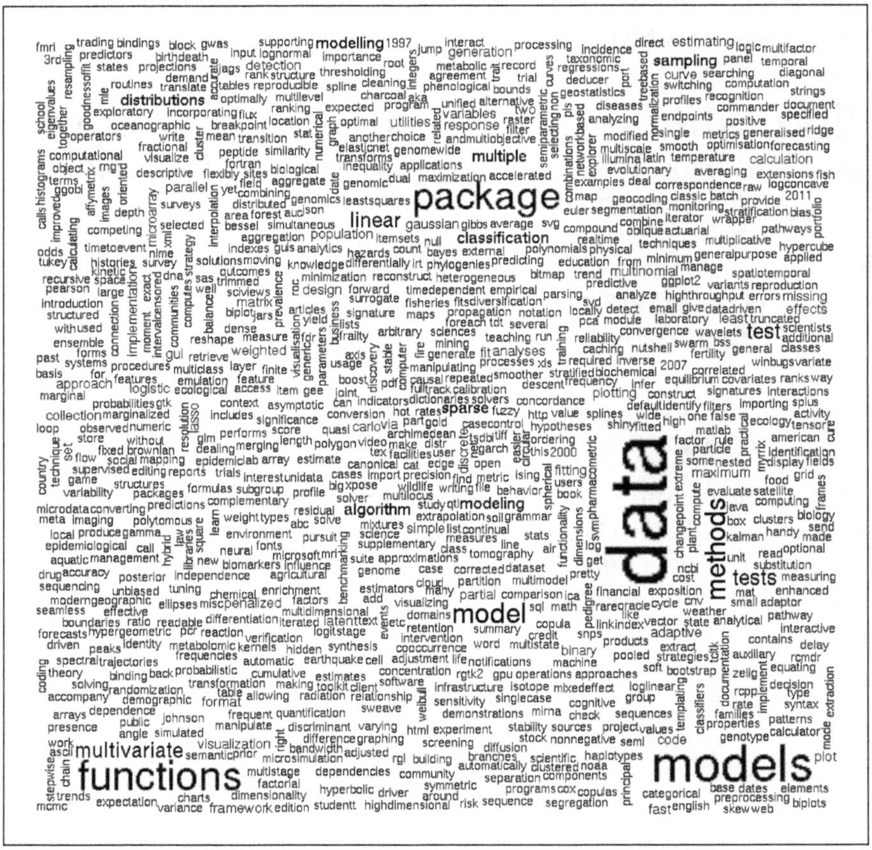

So we simply passed all the strings from the second column of the first `list` element to the `wordcloud` function, which automatically runs a few text-mining scripts from the `tm` package on the text. You can find more details on this topic in *Chapter 7, Unstructured Data*. Then, it renders the words with a relative size weighted by the number of occurrences in the package descriptions. It seems that R packages are indeed primarily targeted at building models and applying multivariate tests on data.

Scraping data from other online sources

Although the `readHTMLTable` function is very useful, sometimes the data is not structured in tables, but rather it's available only as HTML lists. Let's demonstrate such a data format by checking all the R packages listed in the relevant CRAN Task View at `http://cran.r-project.org/web/views/WebTechnologies.html`, as you can see in the following screenshot:

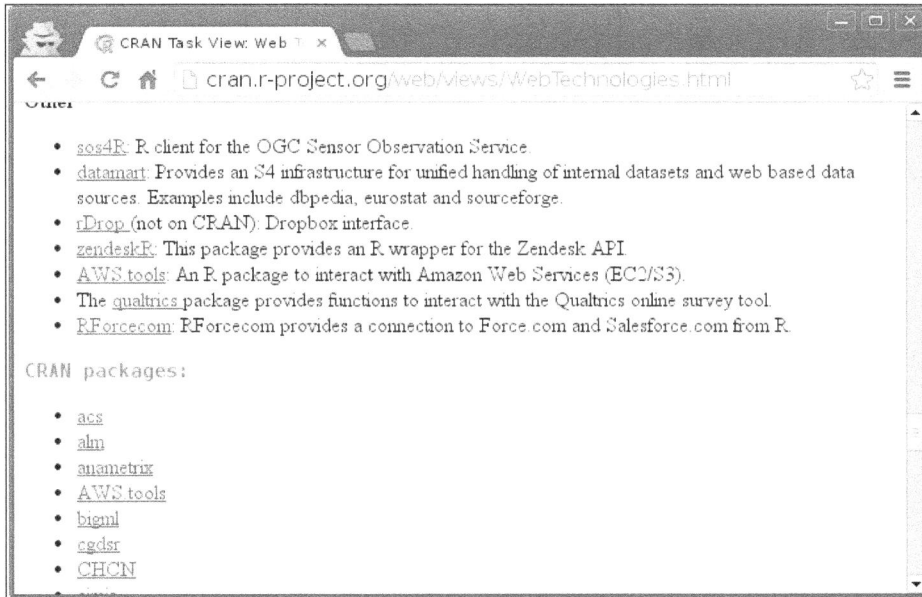

So we see a HTML list of the package names along with a URL pointing to the CRAN, or in some cases to the GitHub repositories. To proceed, first we have to get acquainted a bit with the HTML sources to see how we can parse them. You can do that easily either in Chrome or Firefox: just right-click on the **CRAN** packages heading at the top of the list, and choose **Inspect Element**, as you can see in the following screenshot:

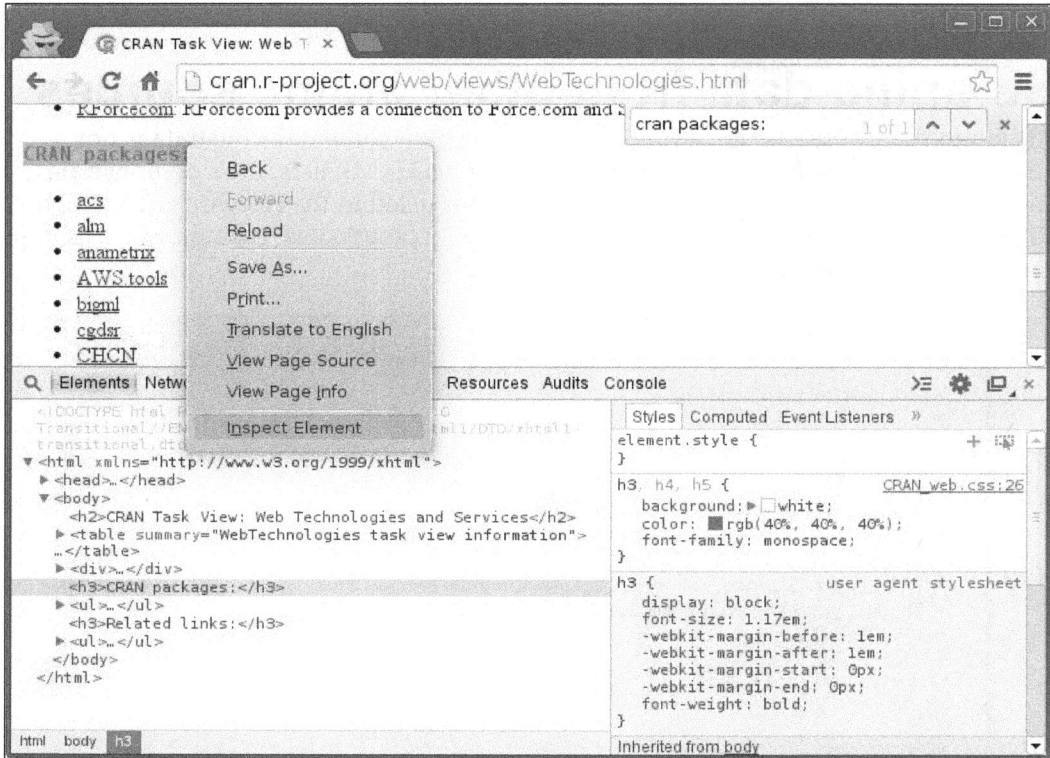

So we have the list of related R packages in an `ul` (unordered list) HTML tag, just after the `h3` (level 3 heading) tag holding the CRAN packages string.

In short:

- We have to parse this HTML file
- Look for the third-level heading holding the search term
- Get all the list elements from the subsequent unordered HTML list

This can be done by, for example, the XML Path Language, which has a special syntax to select nodes in XML/HTML documents via queries.

> For more details and R-driven examples, see *Chapter 4, XPath, XPointer, and XInclude* of the book *XML and Web Technologies for Data Sciences with R* written by Deborah Nolan and Duncan Temple Lang in the Use R! series from Springer. Please see more references in the *Appendix* at the end of the book.

XPath can be rather ugly and complex at first glance. For example, the preceding list can be described with:

```
//h3[text()='CRAN packages:']/following-sibling::ul[1]/li
```

Let me elaborate a bit on this:

1. We are looking for a h3 tag which has CRAN packages as its text, so we are searching for a specific node in the whole document with these attributes.

2. Then the following-siblings expression stands for all the subsequent nodes at the same hierarchy level as the chosen h3 tag.

3. Filter to find only ul HTML tags.

4. As we have several of those, we select only the first of the further siblings with the index (1) between the brackets.

5. Then we simply select all li tags (the list elements) inside that.

Let's try it in R:

```
> page <- htmlParse(file =
+    'http://cran.r-project.org/Web/views/WebTechnologies.html')
> res  <- unlist(xpathApply(doc = page, path =
+    "//h3[text()='CRAN packages:']/following-sibling::ul[1]/li",
+    fun  = xmlValue))
```

And we have the character vector of the related 118 R packages:

```
> str(res)
 chr [1:118] "acs" "alm" "anametrix" "AWS.tools" "bigml" ...
```

XPath is really powerful for selecting and searching for nodes in HTML documents, so is xpathApply. The latter is the R wrapper around most of the XPath functionality in libxml, which makes the process rather quick and efficient. One might rather use the xpathSApply instead, which tries to simplify the returned list of elements, just like sapply does compared to the lapply function. So we can also update our previous code to save the unlist call:

```
> res <- xpathSApply(page, path =
+ "//h3[text()='CRAN packages:']/following-sibling::ul[1]/li",
+    fun = xmlValue)
```

The attentive reader must have noticed that the returned list was a simple character vector, while the original HTML list also included the URLs of the aforementioned packages. Where and why did those vanish?

We can blame xmlValue for this result, which we called instead of the default NULL as the evaluating function to extract the nodes from the original document at the xpathSApply call. This function simply extracts the raw text content of each leaf node without any children, which explains this behavior. What if we are rather interested in the package URLs?

Calling xpathSapply without a specified fun returns all the raw child nodes, which is of no direct help, and we shouldn't try to apply some regular expressions on those. The help page of xmlValue can point us to some similar functions that can be very handy with such tasks. Here we definitely want to use xmlAttrs:

```
> xpathSApply(page,
+    "//h3[text()='CRAN packages:']/following-sibling::ul[1]/li/a",
+    xmlAttrs, 'href')
```

Please note that an updated path was used here, where now we selected all the a tags instead of the li parents. And, instead of the previously introduced xmlValue, now we called xmlAttrs with the 'href' extra argument. This simply extracts all the href arguments of all the related a nodes.

With these primitives, you will be able to fetch any publicly available data from online sources, although sometimes the implementation can end up being rather complex.

On the other hand, please be sure to always consult the terms and conditions and other legal documents of all potential data sources, as fetching data is often prohibited by the copyright owner.

Beside the legal issues, it's also wise to think of fetching and crawling data from the technical point of view of the service provider. If you start to send a plethora of queries to a server without consulting with their administrators beforehand, this action might be construed as a network attack and/or might result in an unwanted load on the servers. To keep it simple, always use a sane delay between your queries. This should be for example, a 2-second pause between queries at a minimum, but it's better to check the *Crawl-delay* directive set in the site's *robot.txt*, which can be found in the root path if available. This file also contains other directives if crawling is allowed or limited. Most of the data provider sites also have some technical documentation on data crawling; please be sure to search for Rate limits and throttling.

And sometimes we are just simply lucky in that someone else has already written the tricky XPath selectors or other interfaces, so we can load data from Web services and homepages with the help of native R packages.

R packages to interact with data source APIs

Although it's great that we can read HTML tables, CSV files and JSON and XML data, and even parse raw HTML documents to store some parts of those in a dataset, there is no sense in spending too much time developing custom tools until we have no other option. First, always start with a quick look on the Web Technologies and Services CRAN Task View; also search R-bloggers, StackOverflow, and GitHub for any possible solution before getting your hands dirty with custom XPath selectors and JSON list magic.

Socrata Open Data API

Let's do this for our previous examples by searching for Socrata, the Open Data Application Program Interface of the Consumer Financial Protection Bureau. Yes, there is a package for that:

```
> library(RSocrata)
Loading required package: httr
Loading required package: RJSONIO
```

```
Attaching package: 'RJSONIO'
```

```
The following objects are masked from 'package:rjson':
```

```
    fromJSON, toJSON
```

As a matter of fact, the RSocrata package uses the same JSON sources (or CSV files), as we did before. Please note the warning message, which says that RSocrata depends on another JSON parser R package rather than the one we used, so some function names are conflicting. It's probably wise to detach('package:rjson') before automatically loading the RJSONIO package.

Loading the Customer Complaint Database by the given URL is pretty easy with RSocrata:

```
> df <- read.socrata(paste0(u, '/25ei-6bcr'))
> str(df)
'data.frame':   18894 obs. of   11 variables:
 $ Complaint.ID        : int  2240 2243 2260 2254 2259 2261 ...
 $ Product             : chr  "Credit card" "Credit card" ...
 $ Submitted.via       : chr  "Web" "Referral" "Referral" ...
 $ Date.received       : chr  "12/01/2011" "12/01/2011" ...
 $ ZIP.code            : chr  ...
 $ Issue               : chr  ...
 $ Date.sent.to.company: POSIXlt, format: "2011-12-19" ...
 $ Company             : chr  "Citibank" "HSBC" ...
 $ Company.response    : chr  "Closed without relief" ...
 $ Timely.response.    : chr  "Yes" "Yes" "No" "Yes" ...
 $ Consumer.disputed.  : chr  "No" "No" "" "No" ...
```

We got numeric values for numbers, and the dates are also automatically processed to POSIXlt!

Similarly, the Web Technologies and Services CRAN Task View contains more than a hundred R packages to interact with data sources on the Web in natural sciences such as ecology, genetics, chemistry, weather, finance, economics, and marketing, but we can also find R packages to fetch texts, bibliography resources, Web analytics, news, and map and social media data besides some other topics. Due to page limitations, here we will only focus on the most-used packages.

Finance APIs

Yahoo! and Google Finance are pretty standard free data sources for all those working in the industry. Fetching for example stock, metal, or foreign exchange prices is extremely easy with the `quantmod` package and the aforementioned service providers. For example, let us see the most recent stock prices for Agilent Technologies with the `A` ticker symbol:

```
> library(quantmod)
Loading required package: Defaults
Loading required package: xts
Loading required package: zoo


Attaching package: 'zoo'


The following objects are masked from 'package:base':

    as.Date, as.Date.numeric


Loading required package: TTR
Version 0.4-0 included new data defaults. See ?getSymbols.
> tail(getSymbols('A', env = NULL))
           A.Open A.High A.Low A.Close A.Volume A.Adjusted
2014-05-09  55.26  55.63 54.81   55.39  1287900      55.39
2014-05-12  55.58  56.62 55.47   56.41  2042100      56.41
2014-05-13  56.63  56.98 56.40   56.83  1465500      56.83
2014-05-14  56.78  56.79 55.70   55.85  2590900      55.85
2014-05-15  54.60  56.15 53.75   54.49  5740200      54.49
2014-05-16  54.39  55.13 53.92   55.03  2405800      55.03
```

By default, `getSymbols` assigns the fetched results to the `parent.frame` (usually the global) environment with the name of the symbols, while specifying `NULL` as the desired environment simply returns the fetched results as an `xts` time-series object, as seen earlier.

Foreign exchange rates can be fetched just as easily:

```
> getFX("USD/EUR")
[1] "USDEUR"
> tail(USDEUR)
```

	USD.EUR
2014-05-13	0.7267
2014-05-14	0.7281
2014-05-15	0.7293
2014-05-16	0.7299
2014-05-17	0.7295
2014-05-18	0.7303

The returned string of `getSymbols` refers to the R variable in which the data was saved inside `.GlobalEnv`. To see all the available data sources, let's query the related S3 methods:

```
> methods(getSymbols)
 [1] getSymbols.csv     getSymbols.FRED    getSymbols.google
 [4] getSymbols.mysql   getSymbols.MySQL   getSymbols.oanda
 [7] getSymbols.rda     getSymbols.RData   getSymbols.SQLite
[10] getSymbols.yahoo
```

So besides some offline data sources, we can query Google, Yahoo!, and OANDA for recent financial information. To see the full list of available symbols, the already loaded `TTR` package might help:

```
> str(stockSymbols())
Fetching AMEX symbols...
Fetching NASDAQ symbols...
Fetching NYSE symbols...
'data.frame':    6557 obs. of  8 variables:
 $ Symbol   : chr  "AAMC" "AA-P" "AAU" "ACU" ...
 $ Name     : chr  "Altisource Asset Management Corp" ...
 $ LastSale : num  841 88.8 1.3 16.4 15.9 ...
 $ MarketCap: num  1.88e+09 0.00 8.39e+07 5.28e+07 2.45e+07 ...
 $ IPOyear  : int  NA NA NA 1988 NA NA NA NA NA NA ...
 $ Sector   : chr  "Finance" "Capital Goods" ...
 $ Industry : chr  "Real Estate" "Metal Fabrications" ...
 $ Exchange : chr  "AMEX" "AMEX" "AMEX" "AMEX" ...
```

[Find more information on how to handle and analyze similar datasets in *Chapter 12, Analyzing Time-series*.]

Fetching time series with Quandl

Quandl provides access to millions of similar time-series data in a standard format, via a custom API, from around 500 data sources. In R, the `Quandl` package provides easy access to all these open data in various industries all around the world. Let us see for example the dividends paid by Agilent Technologies published by the U.S. Securities and Exchange Commission. To do so, simply search for "Agilent Technologies" at the `http://www.quandl.com` homepage, and provide the code of the desired data from the search results to the `Quandl` function:

```
> library(Quandl)
> Quandl('SEC/DIV_A')
        Date Dividend
1 2013-12-27    0.132
2 2013-09-27    0.120
3 2013-06-28    0.120
4 2013-03-28    0.120
5 2012-12-27    0.100
6 2012-09-28    0.100
7 2012-06-29    0.100
8 2012-03-30    0.100
9 2006-11-01    2.057
Warning message:
In Quandl("SEC/DIV_A") :
  It would appear you aren't using an authentication token. Please visit
http://www.quandl.com/help/r or your usage may be limited.
```

As you can see, the API is rather limited without a valid authentication token, which can be redeemed at the `Quandl` homepage for free. To set your token, simply pass that to the `Quandl.auth` function.

This package also lets you:

* Fetch filtered data by time
* Perform some transformations of the data on the server side—such as cumulative sums and the first differential
* Sort the data
* Define the desired class of the returning object—such as `ts`, `zoo`, and `xts`
* Download some meta-information on the data source

The latter is saved as `attributes` of the returning R object. So, for example, to see the frequency of the queried dataset, call:

```
> attr(Quandl('SEC/DIV_A', meta = TRUE), 'meta')$frequency
[1] "quarterly"
```

Google documents and analytics

You might however be more interested in loading your own or custom data from Google Docs, to which end the `RGoogleDocs` package is a great help and is available for download at the `http://www.omegahat.org/` homepage. It provides authenticated access to Google spreadsheets with both read and write access.

Unfortunately, this package is rather outdated and uses some deprecated API functions, so you might be better trying some newer alternatives, such as the recently released `googlesheets` package, which can manage Google Spreadsheets (but not other documents) from R.

Similar packages are also available to interact with Google Analytics or Google Adwords for all those, who would like to analyze page visits or ad performance in R.

Online search trends

On the other hand, we interact with APIs to download public data. Google also provides access to some public data of the World Bank, IMF, US Census Bureau, and so on at `http://www.google.com/publicdata/directory` and also some of their own internal data in the form of search trends at `http://google.com/trends`.

The latter can be queried extremely easily with the `GTrendsR` package, which is not yet available on CRAN, but we can at least practice how to install R packages from other sources. The `GTrendR` code repository can be found on `BitBucket`, from where it's really convenient to install it with the `devtools` package:

> To make sure you install the same version of GTrensR as used in the following, you can specify the `branch`, `commit`, or other reference in the `ref` argument of the `install_bitbucket` (or `install_github`) function. Please see the *References* section in the *Appendix* at the end of the book for the commit hash.

```
> library(devtools)
> install_bitbucket('GTrendsR', 'persican', quiet = TRUE)
Installing bitbucket repo(s) GTrendsR/master from persican
```

```
Downloading master.zip from https://bitbucket.org/persican/gtrendsr/get/
master.zip
```

```
arguments 'minimized' and 'invisible' are for Windows only
```

So installing R packages from BitBucket or GitHub is as easy as providing the name of the code repository and author's username and allowing `devtools` to do the rest: downloading the sources and compiling them.

Windows users should install `Rtools` prior to compiling packages from the source: `http://cran.r-project.org/bin/windows/Rtools/`. We also enabled the quiet mode, to suppress compilation logs and the boring details.

After the package has been installed, we can load it in the traditional way:

```
> library(GTrendsR)
```

First, we have to authenticate with a valid Google username and password before being able to query the Google Trends database. Our search term will be "how to install R":

> Please make sure you provide a valid username and password; otherwise the following query will fail.

```
> conn <- gconnect('some Google username', 'some Google password')
> df   <- gtrends(conn, query = 'how to install R')
> tail(df$trend)
        start        end how.to.install.r
601 2015-07-05 2015-07-11               86
602 2015-07-12 2015-07-18               70
603 2015-07-19 2015-07-25              100
604 2015-07-26 2015-08-01               75
605 2015-08-02 2015-08-08               73
606 2015-08-09 2015-08-15               94
```

The returned dataset includes weekly metrics on the relative amount of search queries on R installation. The data shows that the highest activity was recorded in the middle of July, while only around 75 percent of those search queries were triggered at the beginning of the next month. So Google do not publish raw search query statistics, but rather comparative studies can be done with different search terms and time periods.

Historical weather data

There are also various packages providing access to data sources for all R users in Earth Science. For example, the RNCEP package can download historical weather data from the National Centers for Environmental Prediction for more than one hundred years in six hourly resolutions. The weatherData package provides direct access to http://wunderground.com. For a quick example, let us download the daily temperature averages for the last seven days in London:

```
> library(weatherData)
> getWeatherForDate('London', start_date = Sys.Date()-7, end_date = Sys.
Date())
Retrieving from: http://www.wunderground.com/history/airport/
London/2014/5/12/CustomHistory.html?dayend=19&monthend=5&yearend=2014&r
eq_city=NA&req_state=NA&req_statename=NA&format=1
Checking Summarized Data Availability For London
Found 8 records for 2014-05-12 to 2014-05-19
Data is Available for the interval.
Will be fetching these Columns:
[1] "Date"              "Max_TemperatureC"   "Mean_TemperatureC"
[4] "Min_TemperatureC"
```

	Date	Max_TemperatureC	Mean_TemperatureC	Min_TemperatureC
1	2014-05-12	18	13	9
2	2014-05-13	16	12	8
3	2014-05-14	19	13	6
4	2014-05-15	21	14	8
5	2014-05-16	23	16	9
6	2014-05-17	23	17	11
7	2014-05-18	23	18	12
8	2014-05-19	24	19	13

Please note that an unimportant part of the preceding output was suppressed, but what happened here is quite straightforward: the package fetched the specified URL, which is a CSV file by the way, then parsed that with some additional information. Setting opt_detailed to TRUE would also return intraday data with a 30-minute resolution.

Other online data sources

Of course, this short chapter cannot provide an overview of querying all the available online data sources and R implementations, but please consult the Web Technologies and Services CRAN Task View, R-bloggers, StackOverflow, and the resources in the *References* chapter at the end of the book to look for any already existing R packages or helper functions before creating your own crawler R scripts.

Summary

This chapter focused on how to fetch and process data directly from the Web, including some problems with downloading files, processing XML and JSON formats, parsing HTML tables, applying XPath selectors to extract data from HTML pages, and interacting with RESTful APIs.

Although some examples in this chapter might appear to have been an idle struggle with the Socrata API, it turned out that the `RSocrata` package provides production-ready access to all those data. However, please bear in mind that you will face some situations without ready-made R packages; thus, as a data hacker, you will have to get your hands dirty with all the JSON, HTML and XML sources.

In the next chapter, we will discover how to filter and aggregate the already acquired and loaded data with the top, most-used methods for reshaping and restructuring data.

3
Filtering and Summarizing Data

After loading data from either flat files or databases (as we have seen in *Chapter 1*, *Hello, Data!*), or directly from the web via some APIs (as covered in *Chapter 2*, *Getting Data from the Web*), we often have to aggregate, transform, or filter the original dataset before the actual data analysis could take place.

In this chapter, we will focus on how to:

- Filter rows and columns in data frames
- Summarize and aggregate data
- Improve the performance of such tasks with the `dplyr` and `data.table` packages besides the base R methods

Drop needless data

Although not loading the needless data is the optimal solution (see the *Loading a subset of text files* and *Loading data from databases* sections in *Chapter 1*, *Hello, Data!*), we often have to filter the original dataset inside R. This can be done with the traditional tools and functions from base R, such as `subset`, by using `which` and the `[` or `[[` operator (see the following code), or for example with the SQL-like approach of the `sqldf` package:

```
> library(sqldf)
> sqldf("SELECT * FROM mtcars WHERE am=1 AND vs=1")
   mpg cyl  disp hp drat    wt  qsec vs am gear carb
1 22.8   4 108.0 93 3.85 2.320 18.61  1  1    4    1
2 32.4   4  78.7 66 4.08 2.200 19.47  1  1    4    1
```

3	30.4	4	75.7	52	4.93	1.615	18.52	1	1	4	2
4	33.9	4	71.1	65	4.22	1.835	19.90	1	1	4	1
5	27.3	4	79.0	66	4.08	1.935	18.90	1	1	4	1
6	30.4	4	95.1	113	3.77	1.513	16.90	1	1	5	2
7	21.4	4	121.0	109	4.11	2.780	18.60	1	1	4	2

I am sure that all readers who have a decent SQL background and are just getting in touch with R appreciate this alternative way of filtering data, but I personally prefer the following rather similar, native, and much more concise R version:

```
> subset(mtcars, am == 1 & vs == 1)
```

	mpg	cyl	disp	hp	drat	wt	qsec	vs	am	gear	carb
Datsun 710	22.8	4	108.0	93	3.85	2.320	18.61	1	1	4	1
Fiat 128	32.4	4	78.7	66	4.08	2.200	19.47	1	1	4	1
Honda Civic	30.4	4	75.7	52	4.93	1.615	18.52	1	1	4	2
Toyota Corolla	33.9	4	71.1	65	4.22	1.835	19.90	1	1	4	1
Fiat X1-9	27.3	4	79.0	66	4.08	1.935	18.90	1	1	4	1
Lotus Europa	30.4	4	95.1	113	3.77	1.513	16.90	1	1	5	2
Volvo 142E	21.4	4	121.0	109	4.11	2.780	18.60	1	1	4	2

Please note the slight difference in the results. This is attributed to the fact that the row.names argument of sqldf is FALSE by default, which can of course be overridden to get the exact same results:

```
> identical(
+       sqldf("SELECT * FROM mtcars WHERE am=1 AND vs=1",
+          row.names = TRUE),
+       subset(mtcars, am == 1 & vs == 1)
+       )
[1] TRUE
```

These examples focused on how to drop rows from data.frame, but what if we also want to remove some columns?

The SQL approach is really straightforward; just specify the required columns instead of * in the SELECT statement. On the other hand, subset also supports this approach by the select argument, which can take vectors or an R expression describing, for example, a range of columns:

```
> subset(mtcars, am == 1 & vs == 1, select = hp:wt)
```

	hp	drat	wt
Datsun 710	93	3.85	2.320

```
Fiat 128          66 4.08 2.200
Honda Civic       52 4.93 1.615
Toyota Corolla    65 4.22 1.835
Fiat X1-9         66 4.08 1.935
Lotus Europa     113 3.77 1.513
Volvo 142E       109 4.11 2.780
```

> Pass the unquoted column names as a vector via the c function to select an arbitrary list of columns in the given order, or exclude the specified columns by using the - operator, for example, subset(mtcars, select = -c(hp, wt)).

Let's take this to the next step, and see how we can apply the forementioned filters on some larger datasets, when we face some performance issues with the base functions.

Drop needless data in an efficient way

R works best with datasets that can fit in the actual physical memory, and some R packages provide extremely fast access to this amount of data.

> Some benchmarks (see the *References* section at the end of the book) provide real-life examples of more efficient summarizing R functions than what the current open source (for example, MySQL, PostgreSQL, and Impala) and commercial databases (such as HP Vertica) provide.

Some of the related packages were already mentioned in *Chapter 1, Hello, Data!*, where we benchmarked reading a relatively large amount of data from the hflights package into R.

Let's see how the preceding examples perform on this dataset of a quarter of a million rows:

```
> library(hflights)
> system.time(sqldf("SELECT * FROM hflights WHERE Dest == 'BNA'",
+   row.names = TRUE))
   user  system elapsed
  1.487   0.000   1.493
> system.time(subset(hflights, Dest == 'BNA'))
   user  system elapsed
  0.132   0.000   0.131
```

The `base::subset` function seems to perform pretty well, but can we make it any faster? Well, the second generation of the `plyr` package, called `dplyr` (the relevant details are discussed *High-performance helper functions* section in this chapter and *Chapter 4, Restructuring Data*), provides extremely fast C++ implementations of the most common database manipulation methods in a rather intuitive way:

```
> library(dplyr)
> system.time(filter(hflights, Dest == 'BNA'))
   user  system elapsed
  0.021   0.000   0.022
```

Further, we can extend this solution by dropping some columns from the dataset just like we did with `subset` before, although now, we call the `select` function instead of passing an argument with the same name:

```
> str(select(filter(hflights, Dest == 'BNA'), DepTime:ArrTime))
'data.frame':  3481 obs. of  2 variables:
 $ DepTime: int  1419 1232 1813 900 716 1357 2000 1142 811 1341 ...
 $ ArrTime: int  1553 1402 1948 1032 845 1529 2132 1317 945 1519 ...
```

Therefore, it's like calling the `filter` function instead of `subset`, and we get the results faster than the blink of an eye! The `dplyr` package can work with traditional `data.frame` or `data.table` objects, or can interact directly with the most widely used database engines. Please note that row names are not preserved in `dplyr`, so if you require them, it's worth copying the names to explicit variables before passing them to `dplyr` or directly to `data.table` as follows:

```
> mtcars$rownames <- rownames(mtcars)
> select(filter(mtcars, hp > 300), c(rownames, hp))
       rownames  hp
1 Maserati Bora 335
```

Drop needless data in another efficient way

Let's see a quick example of the `data.table` solution on its own, without `dplyr`.

> The `data.table` package provides an extremely efficient way to handle larger datasets in a column-based, auto-indexed in-memory data structure, with backward compatibility for the traditional `data.frame` methods.

After loading the package, we have to transform the `hflights` traditional `data.frame` to `data.table`. Then, we create a new column, called `rownames`, to which we assign the `rownames` of the original dataset with the help of the := assignment operator specific to `data.table`:

```
> library(data.table)
> hflights_dt <- data.table(hflights)
> hflights_dt[, rownames := rownames(hflights)]
> system.time(hflights_dt[Dest == 'BNA'])
   user  system elapsed
  0.021   0.000   0.020
```

Well, it takes some time to get used to the custom `data.table` syntax and it might even seem a bit strange to the traditional R user at first sight, but it's definitely worth mastering in the long run. You get great performance, and the syntax turns out to be natural and flexible after the relatively steep learning curve of the first few examples.

As a matter of fact, the `data.table` syntax is pretty similar to SQL:

```
DT[i, j, ... , drop = TRUE]
```

This could be described with SQL commands as follows:

```
DT[where, select | update, group by] [having] [order by] [ ]...[ ]
```

Therefore, `[.data.table` (which stands for the `[` operator applied to a `data.table` object) has some different arguments as compared to the traditional `[.data.frame` syntax, as you have already seen in the preceding example.

> Now, we are not dealing with the assignment operator in detail, as this example might be too complex for such an introductory part of the book, and we are probably getting out of our comfort zone. Therefore, please find more details in *Chapter 4*, *Restructuring Data*, or head to `?data.table` for a rather technical overview.

It seems that the first argument (i) of the `[.data.table` operator stands for filtering, or in other words, for the WHERE statement in SQL parlance, while `[.data.frame` expects indices specifying which rows to keep from the original dataset. The real difference between the two arguments is that the former can take any R expression, while the latter traditional method expects mainly integers or logical values.

Anyway, filtering is as easy as passing an R expression to the `i` argument of the `[` operator specific to `data.table`. Further, let's see how we can select the columns in the `data.table` syntax, which should be done in the second argument (`j`) of the call on the basis of the abovementioned general `data.table` syntax:

```
> str(hflights_dt[Dest == 'BNA', list(DepTime, ArrTime)])
Classes 'data.table' and 'data.frame':        3481 obs. of 2 variables:
 $ DepTime: int   1419 1232 1813 900 716 1357 2000 1142 811 1341 ...
 $ ArrTime: int   1553 1402 1948 1032 845 1529 2132 1317 945 1519 ...
 - attr(*, ".internal.selfref")=<externalptr>
```

Okay, so we now have the two expected columns with the 3,481 observations. Note that `list` was used to define the required columns to keep, although the use of `c` (a function from base R to concatenate vector elements) is more traditionally used with `[.data.frame`. The latter is also possible with `[.data.table`, but then, you have to pass the variable names as a character vector and set `with` to `FALSE`:

```
> hflights_dt[Dest == 'BNA', c('DepTime', 'ArrTime'), with = FALSE]
```

> Instead of `list`, you can use a dot as the function name in the style of the `plyr` package; for example: `hflights_dt[, .(DepTime, ArrTime)]`.

Now that we are more or less familiar with our options for filtering data inside a live R session, and we know the overall syntax of the `dplyr` and `data.table` packages, let's see how these can be used to aggregate and summarize data in action.

Aggregation

The most straightforward way of summarizing data is calling the `aggregate` function from the `stats` package, which does exactly what we are looking for: splitting the data into subsets by a grouping variable, then computing summary statistics for them separately. The most basic way to call the `aggregate` function is to pass the numeric vector to be aggregated, and a factor variable to define the splits for the function passed in the `FUN` argument to be applied. Now, let's see the average ratio of diverted flights on each weekday:

```
> aggregate(hflights$Diverted, by = list(hflights$DayOfWeek),
+    FUN = mean)
  Group.1           x
1       1 0.002997672
```

```
2          2 0.002559323
3          3 0.003226211
4          4 0.003065727
5          5 0.002687865
6          6 0.002823121
7          7 0.002589057
```

Well, it took some time to run the preceding script, but please bear in mind that we have just aggregated around a quarter of a million rows to see the daily averages for the number of diverted flights departing from the Houston airport in 2011.

In other words, which also makes sense for all those not into statistics, the percentage of diverted flights per weekday. The results are rather interesting, as it seems that flights are more often diverted in the middle of the week (around 0.3 percent) than over the weekends (around 0.05 percent less), at least from Houston.

An alternative way of calling the preceding function is to supply the arguments inside of the with function, which seems to be a more human-friendly expression after all because it saves us from the repeated mention of the hflights database:

```
> with(hflights, aggregate(Diverted, by = list(DayOfWeek),
+   FUN = mean))
```

The results are not shown here, as they are exactly the same as those shown earlier. The manual for the aggregate function (see ?aggregate) states that it returns the results in a convenient form. Well, checking the column names of the abovementioned returned data does not seem convenient, right? We can overcome this issue by using the formula notation instead of defining the numeric and factor variables separately:

```
> aggregate(Diverted ~ DayOfWeek, data = hflights, FUN = mean)
  DayOfWeek    Diverted
1          1 0.002997672
2          2 0.002559323
3          3 0.003226211
4          4 0.003065727
5          5 0.002687865
6          6 0.002823121
7          7 0.002589057
```

The gain by using the formula notation is at least two-fold:

* There are relatively few characters to type
* The headers and row names are correct in the results
* This version also runs a bit faster than the previous `aggregate` calls; please see the all-out benchmark at the end of this section

The only downside of using the formula notation is that you have to learn it, which might seem a bit awkward at first, but as formulas are highly used in a bunch of R functions and packages, particularly for defining models, it's definitely worth learning how to use them in the long run.

> The formula notation is inherited from the S language with the following general syntax: `response_variable ~ predictor_variable_1 + ... + predictor_variable_n`. The notation also includes some other symbols, such as - for excluding variables and : or * to include the interaction between the variables with or without themselves. See *Chapter 5, Building Models (authored by Renata Nemeth and Gergely Toth)*, and `?formula` in the R console for more details.

Quicker aggregation with base R commands

An alternative solution to aggregate data might be to call the `tapply` or by function, which can apply an R function over a *ragged* array. The latter means that we can provide one or more INDEX variables, which will be coerced to factor, and then, run the provided R function separately on all cells in each subset. The following is a quick example:

```
> tapply(hflights$Diverted, hflights$DayOfWeek, mean)
        1        2        3        4        5        6        7
0.002998 0.002559 0.003226 0.003066 0.002688 0.002823 0.002589
```

Please note that `tapply` returns an `array` object instead of convenient data frames; on the other hand, it runs a lot quicker than the abovementioned aggregate calls. Thus, it might be reasonable to use `tapply` for the computations and then, convert the results to `data.frame` with the appropriate column names.

Convenient helper functions

Such conversions can be done easily and in a very user-friendly way by, for example, using the `plyr` package, a general version of the `dplyr` package, which stands for *plyr specialized for data frames*.

The `plyr` package provides a variety of functions to apply data from `data.frame`, `list`, or `array` objects, and can return the results in any of the mentioned formats. The naming scheme of these functions is easy to remember: the first character of the function name stands for the class of the input data, and the second character represents the output format, all followed by *ply* in all cases. Besides the three abovementioned R classes, there are some special options coded by the characters:

- d stands for `data.frame`
- s stands for `array`
- l stands for `list`
- m is a special input type, which means that we provide multiple arguments in a tabular format for the function
- r input type expects an integer, which specifies the number of times the function will be replicated
- _ is a special output type that does not return anything for the function

Thus, the following most frequently used combinations are available:

- `ddply` takes `data.frame` as input and returns `data.frame`
- `ldply` takes `list` as input but returns `data.frame`
- `l_ply` does not return anything, but it's really useful for example, to iterate through a number of elements instead of a `for` loop; as with a set `.progress` argument, the function can show the current state of iterations, the remaining time

Please find more details, examples, and use cases of `plyr` in *Chapter 4, Restructuring Data*. Here, we will only concentrate on how to summarize data. To this end, we will use `ddply` (not to be confused with the `dplyr` package) in all the following examples: taking `data.frame` as the input argument and returning data with the same class.

So, let's load the package and apply the mean function on the Diverted column over each subset by DayOfWeek:

```
> library(plyr)
> ddply(hflights, .(DayOfWeek), function(x) mean(x$Diverted))
  DayOfWeek          V1
1         1 0.002997672
2         2 0.002559323
3         3 0.003226211
4         4 0.003065727
5         5 0.002687865
6         6 0.002823121
7         7 0.002589057
```

> The . function of the plyr package provides us with a convenient way of referring to a variable (name) as is; otherwise, the content of the DayOfWeek columns would be interpreted by ddply, resulting in an error.

An important thing to note here is that ddply is much quicker than our first attempt with the aggregate function. On the other hand, I am not yet pleased with the results, V1 and such creative column names have always freaked me out. Instead of updating the names of the data.frame post processing let's call the summarise helper function instead of the previously applied anonymous one; here, we can also provide the desired name for our newly computed column:

```
> ddply(hflights, .(DayOfWeek), summarise, Diverted = mean(Diverted))
  DayOfWeek    Diverted
1         1 0.002997672
2         2 0.002559323
3         3 0.003226211
4         4 0.003065727
5         5 0.002687865
6         6 0.002823121
7         7 0.002589057
```

Okay, much better. But, can we do even better?

High-performance helper functions

Hadley Wickham, the author of `ggplot`, `reshape`, and several other R packages, started working on the second generation, or rather a specialized version, of `plyr` in 2008. The basic concept was that `plyr` is most frequently used to transform one `data.frame` to another `data.frame`; therefore, its operation requires extra attention. The `dplyr` package, `plyr` specialized for data frames, provides a faster implementation of the `plyr` functions, written in raw C++, and `dplyr` can also deal with remote databases.

However, the performance improvements also go hand-in-hand with some other changes; for example, the syntax of `dplyr` has changed a lot as compared to `plyr`. Although the previously mentioned `summarise` function does exist in `dplyr`, we do not have the `ddplyr` function any more, as all functions in the package are dedicated to act as some component of `plyr::ddplyr`.

Anyway, to keep the theoretical background short, if we want to summarize the subgroups of a dataset, we have to define the groups before aggregation:

```
> hflights_DayOfWeek <- group_by(hflights, DayOfWeek)
```

The resulting object is the very same `data.frame` that we had previously with one exception: a bunch of metadata was merged to the object by the means of attributes. To keep the following output short, we do not list the whole structure (`str`) of the object, but only the attributes are shown:

```
> str(attributes(hflights_DayOfWeek))
List of 9
 $ names          : chr [1:21] "Year" "Month" "DayofMonth" ...
 $ class          : chr [1:4] "grouped_df" "tbl_df" "tbl" ...
 $ row.names      : int [1:227496] 5424 5425 5426 5427 5428 ...
 $ vars           :List of 1
  ..$ : symbol DayOfWeek
 $ drop           : logi TRUE
 $ indices        :List of 7
  ..$ : int [1:34360] 2 9 16 23 30 33 40 47 54 61 ...
  ..$ : int [1:31649] 3 10 17 24 34 41 48 55 64 70 ...
  ..$ : int [1:31926] 4 11 18 25 35 42 49 56 65 71 ...
  ..$ : int [1:34902] 5 12 19 26 36 43 50 57 66 72 ...
  ..$ : int [1:34972] 6 13 20 27 37 44 51 58 67 73 ...
  ..$ : int [1:27629] 0 7 14 21 28 31 38 45 52 59 ...
```

```
  ..$ : int [1:32058] 1 8 15 22 29 32 39 46 53 60 ...
 $ group_sizes       : int [1:7] 34360 31649 31926 34902 34972 ...
 $ biggest_group_size: int 34972
 $ labels            :'data.frame':  7 obs. of  1 variable:
  ..$ DayOfWeek: int [1:7] 1 2 3 4 5 6 7
  ..- attr(*, "vars")=List of 1
  .. ..$ : symbol DayOfWeek
```

From this metadata, the `indices` attribute is important. It simply lists the IDs of each row for one of the weekdays, so later operations can easily select the subgroups from the whole dataset. So, let's see how the proportion of diverted flights looks like with some performance boost due to using `summarise` from `dplyr` instead of `plyr`:

```
> dplyr::summarise(hflights_DayOfWeek, mean(Diverted))
Source: local data frame [7 x 2]


  DayOfWeek mean(Diverted)
1         1    0.002997672
2         2    0.002559323
3         3    0.003226211
4         4    0.003065727
5         5    0.002687865
6         6    0.002823121
7         7    0.002589057
```

The results are pretty familiar, which is good. However, while running this example, did you measure the execution time? This was close to an instant, which makes `dplyr` even better.

Aggregate with data.table

Do you remember the second argument of `[.data.table`? It's called `j`, which stands for a SELECT or an UPDATE SQL statement, and the most important feature is that it can be any R expression. Thus, we can simply pass a function there and set groups with the help of the `by` argument:

```
> hflights_dt[, mean(Diverted), by = DayOfWeek]
   DayOfWeek            V1
1:         6 0.002823121
```

```
2:          7 0.002589057
3:          1 0.002997672
4:          2 0.002559323
5:          3 0.003226211
6:          4 0.003065727
7:          5 0.002687865
```

I am pretty sure that you are not in the least surprised by how fast the results were returned by `data.table`, as people can get used to great tools very quickly. Further, it was very concise as compared to the previous two-line `dplyr` call, right? The only downside of this solution is that the weekdays are ordered by some hardly intelligible rank. Please see *Chapter 4, Restructuring Data*, for more details on this; for now, let's fix the issue quickly by setting a key, which means that we order `data.table` first by DayOfWeek:

```
> setkey(hflights_dt, 'DayOfWeek')
> hflights_dt[, mean(Diverted), by = DayOfWeek]
   DayOfWeek              V1
1:          1 0.002997672
2:          2 0.002559323
3:          3 0.003226211
4:          4 0.003065727
5:          5 0.002687865
6:          6 0.002823121
7:          7 0.002589057
```

> To specify a name for the second column in the resulting tabular object instead of V1, you can specify the `summary` object as a named list, for example, as `hflights_dt[, list('mean(Diverted)' = mean(Diverted)), by = DayOfWeek]`, where you can use . (dot) instead of `list`, just like in `plyr`.

Besides getting the results in the expected order, summarizing data by an already existing key also runs relatively fast. Let's verify this with some empirical evidence on your machine!

Running benchmarks

As already discussed in the previous chapters, with the help of the microbenchmark package, we can run any number of different functions for a specified number of times on the same machine to get some reproducible results on the performance.

To this end, we have to define the functions that we want to benchmark first. These were compiled from the preceding examples:

```
> AGGR1      <- function() aggregate(hflights$Diverted,
+   by = list(hflights$DayOfWeek), FUN = mean)
> AGGR2      <- function() with(hflights, aggregate(Diverted,
+   by = list(DayOfWeek), FUN = mean))
> AGGR3      <- function() aggregate(Diverted ~ DayOfWeek,
+   data = hflights, FUN = mean)
> TAPPLY     <- function() tapply(X = hflights$Diverted,
+   INDEX = hflights$DayOfWeek, FUN = mean)
> PLYR1      <- function() ddply(hflights, .(DayOfWeek),
+   function(x) mean(x$Diverted))
> PLYR2      <- function() ddply(hflights, .(DayOfWeek), summarise,
+   Diverted = mean(Diverted))
> DPLYR      <- function() dplyr::summarise(hflights_DayOfWeek,
+   mean(Diverted))
```

However, as mentioned before, the summarise function in dplyr needs some prior data restructuring, which also takes time. To this end, let's define another function that also includes the creation of the new data structure along with the real aggregation:

```
> DPLYR_ALL <- function() {
+     hflights_DayOfWeek <- group_by(hflights, DayOfWeek)
+     dplyr::summarise(hflights_DayOfWeek, mean(Diverted))
+ }
```

Similarly, benchmarking data.table also requires some additional variables for the test environment; as hlfights_dt is already sorted by DayOfWeek, let's create a new data.table object for benchmarking:

```
> hflights_dt_nokey <- data.table(hflights)
```

Further, it probably makes sense to verify that it has no keys:

```
> key(hflights_dt_nokey)
NULL
```

Okay, now, we can define the `data.table` test cases along with a function that also includes the transformation to `data.table`, and adding an index just to be fair with `dplyr`:

```
> DT      <- function() hflights_dt_nokey[, mean(FlightNum),
+   by = DayOfWeek]
> DT_KEY <- function() hflights_dt[, mean(FlightNum),
+   by = DayOfWeek]
> DT_ALL <- function() {
+       setkey(hflights_dt_nokey, 'DayOfWeek')
+       hflights_dt[, mean(FlightNum), by = DayOfWeek]
+       setkey(hflights_dt_nokey, NULL)
+ }
```

Now that we have all the described implementations ready for testing, let's load the `microbenchmark` package to do its job:

```
> library(microbenchmark)
> res <- microbenchmark(AGGR1(), AGGR2(), AGGR3(), TAPPLY(), PLYR1(),
+           PLYR2(), DPLYR(), DPLYR_ALL(), DT(), DT_KEY(), DT_ALL())
> print(res, digits = 3)
Unit: milliseconds
        expr     min      lq  median      uq     max neval
     AGGR1() 2279.82 2348.14 2462.02 2597.70 2719.88    10
     AGGR2() 2278.15 2465.09 2528.55 2796.35 2996.98    10
     AGGR3() 2358.71 2528.23 2726.66 2879.89 3177.63    10
    TAPPLY()   19.90   21.05   23.56   29.65   33.88    10
     PLYR1()   56.93   59.16   70.73   82.41  155.88    10
     PLYR2()   58.31   65.71   76.51   98.92  103.48    10
     DPLYR()    1.18    1.21    1.30    1.74    1.84    10
 DPLYR_ALL()    7.40    7.65    7.93    8.25   14.51    10
        DT()    5.45    5.73    5.99    7.75    9.00    10
    DT_KEY()    5.22    5.45    5.63    6.26   13.64    10
    DT_ALL()   31.31   33.26   35.19   38.34   42.83    10
```

The results are pretty spectacular: from more than 2,000 milliseconds, we could improve our tools to provide the very same results in only a bit more than 1 millisecond. The spread can be demonstrated easily on a violin plot with a logarithmic scale:

```
> autoplot(res)
```

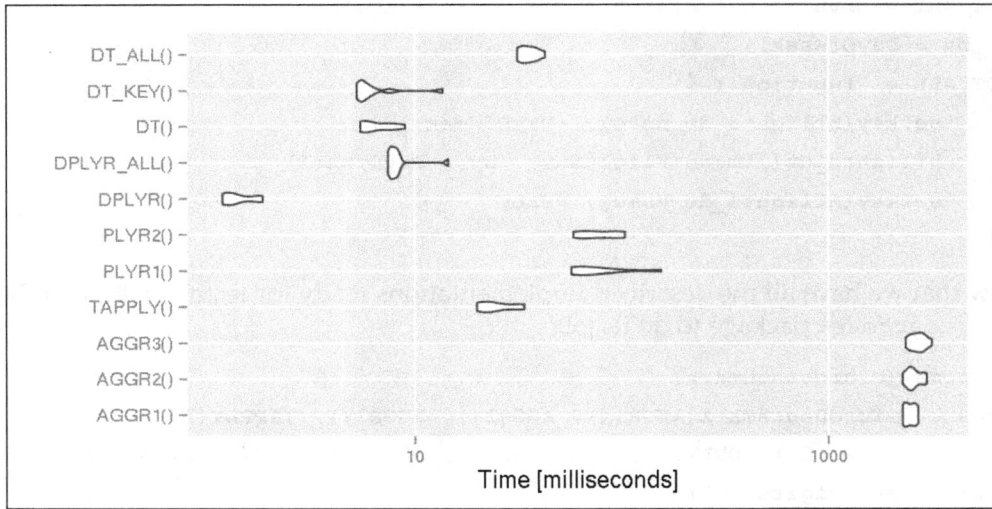

Therefore, `dplyr` seems to be the most efficient solution, although if we also take the extra step (to group `data.frame`) into account, it makes the otherwise clear advantage rather unconvincing. As a matter of fact, if we already have a `data.table` object, and we can save the transformation of a traditional `data.frame` object into `data.table`, then `data.table` performs better than `dplyr`. However, I am pretty sure that you will not really notice the time difference between the two high-performance solutions; both of these do a very good job with even larger datasets.

It's worth mentioning that `dplyr` can work with `data.table` objects as well; therefore, to ensure that you are not locked to either package, it's definitely worth using both if needed. The following is a POC example:

```
> dplyr::summarise(group_by(hflights_dt, DayOfWeek), mean(Diverted))
Source: local data table [7 x 2]


  DayOfWeek mean(Diverted)
1         1     0.002997672
2         2     0.002559323
3         3     0.003226211
```

4	4	0.003065727
5	5	0.002687865
6	6	0.002823121
7	7	0.002589057

Okay, so now we are pretty sure to use either `data.table` or `dplyr` for computing group averages in the future. However, what about more complex operations?

Summary functions

As we have discussed earlier, all aggregating functions can take any valid R functions to apply on the subsets of the data. Some of the R packages make it extremely easy for the users, while a few functions do require you to fully understand the package concept, custom syntax, and options to get the most out of the high-performance opportunities.

For such more advanced topics, please see *Chapter 4*, *Restructuring Data*, and the further readings listed in the *References* section at the end of the book.

Now, we will concentrate on a very simple `summary` function, which is extremely common in any general data analysis project: counting the number of cases per group. This quick example will also highlight some of the differences among the referenced alternatives mentioned in this chapter.

Adding up the number of cases in subgroups

Let's focus on `plyr`, `dplyr` and `data.table` now, as I am pretty sure that you can construct the `aggregate` and `tapply` versions without any serious issues. On the basis of the previous examples, the current task seems fairly easy: instead of the `mean` function, we can simply call the `length` function to return the number of elements in the `Diverted` column:

```
> ddply(hflights, .(DayOfWeek), summarise, n = length(Diverted))
  DayOfWeek     n
1         1 34360
2         2 31649
3         3 31926
4         4 34902
5         5 34972
6         6 27629
7         7 32058
```

Now, we also know that a relatively low number of flights leave Houston on Saturday. However, do we really have to type so much to answer such a simple question? Further, do we really have to name a variable in which we can count the number of cases? You already know the answer:

```
> ddply(hflights, .(DayOfWeek), nrow)
  DayOfWeek    V1
1         1 34360
2         2 31649
3         3 31926
4         4 34902
5         5 34972
6         6 27629
7         7 32058
```

In short, there is no need to choose a variable from data.frame to determine its length, as it's a lot easier (and faster) to simply check the number of rows in the (sub)datasets.

However, we can also return the very same results in a much easier and quicker way. Probably, you have already thought of using the good old table function for such a straightforward task:

```
> table(hflights$DayOfWeek)

    1     2     3     4     5     6     7
34360 31649 31926 34902 34972 27629 32058
```

The only problem with the resulting object is that we have to transform it further, for example, to data.frame in most cases. Well, plyr already has a helper function to do this in one step, with a very intuitive name:

```
> count(hflights, 'DayOfWeek')
  DayOfWeek  freq
1         1 34360
2         2 31649
3         3 31926
4         4 34902
5         5 34972
6         6 27629
7         7 32058
```

Therefore, we end up with some rather simple examples for counting data, but let us also see how to implement summary tables with dplyr. If you simply try to modify our previous dplyr commands, you will soon realize that passing the length or nrow function, as we did in plyr, simply does not work. However, reading the manuals or some related questions on StackOverflow soon points our attention to a handy helper function called n:

```
> dplyr::summarise(hflights_DayOfWeek, n())
Source: local data frame [7 x 2]
```

	DayOfWeek	n()
1	1	34360
2	2	31649
3	3	31926
4	4	34902
5	5	34972
6	6	27629
7	7	32058

However, to be honest, do we really need this relatively complex approach? If you remember the structure of hflights_DayOfWeek, you will soon realize that there is a lot easier and quicker way to find out the overall number of flights on each weekday:

```
> attr(hflights_DayOfWeek, 'group_sizes')
[1] 34360 31649 31926 34902 34972 27629 32058
```

Further, just to make sure that we do not forget the custom (yet pretty) syntax of data.table, let us compute the results with another helper function:

```
> hflights_dt[, .N, by = list(DayOfWeek)]
```

	DayOfWeek	N
1:	1	34360
2:	2	31649
3:	3	31926
4:	4	34902
5:	5	34972
6:	6	27629
7:	7	32058

Summary

In this chapter, we introduced some effective and convenient ways of filtering and summarizing data. We discussed some use cases on filtering the rows and columns of datasets. We also learned how to summarize data for further analysis. After getting familiar with the most popular implementations of such tasks, we compared them with reproducible examples and a benchmarking package.

In the next chapter, we will continue this journey of restructuring datasets and creating new variables.

4
Restructuring Data

We already covered the most basic methods for restructuring data in the *Chapter 3, Filtering and Summarizing Data*, but of course, there are several other, more complex tasks that we will master in the forthcoming pages.

Just to give a quick example on how diversified tools are needed for getting the data in a form that can be used for real data analysis: Hadley Wickham, one of the best known R developers and users, spent one third of his PhD thesis on reshaping data. As he says, "it is unavoidable before doing any exploratory data analysis or visualization."

So now, besides the previous examples of restructuring data, such as the counting of elements in each group, we will focus on some more advanced features, as listed next:

- Transposing matrices
- Splitting, applying, and joining data
- Computing margins of tables
- Merging data frames
- Casting and melting data

Transposing matrices

One of the most used, but often not mentioned, methods for restructuring data is transposing matrices. This simply means switching the columns with rows and vice versa, via the t function:

```
> (m <- matrix(1:9, 3))
     [,1] [,2] [,3]
[1,]    1    4    7
[2,]    2    5    8
[3,]    3    6    9
```

```
> t(m)
      [,1] [,2] [,3]
[1,]    1    2    3
[2,]    4    5    6
[3,]    7    8    9
```

Of course, this S3 method also works with data.frame, and actually, with any tabular object. For more advanced features, such as transposing a multi-dimensional table, take a look at the aperm function from the base package.

Filtering data by string matching

Although some filtering algorithms were already discussed in the previous chapters, the dplyr package contains some magic features that have not yet been covered and are worth mentioning here. As we all know by this time, the subset function in base, or the filter function from dplyr is used for filtering rows, and the select function can be used to choose a subset of columns.

The function filtering rows usually takes an R expression, which returns the IDs of the rows to drop, similar to the which function. On the other hand, providing such R expressions to describe column names is often more problematic for the select function; it's harder if not impossible to evaluate R expressions on column names.

The dplyr package provides some useful functions to select some columns of the data, based on column name patterns. For example, we can keep only the variables ending with the string, delay:

```
> library(dplyr)
> library(hflights)
> str(select(hflights, ends_with("delay")))
'data.frame':  227496 obs. of  2 variables:
 $ ArrDelay: int  -10 -9 -8 3 -3 -7 -1 -16 44 43 ...
 $ DepDelay: int  0 1 -8 3 5 -1 -1 -5 43 43 ...
```

Of course, there is a similar helper function to check the first characters of the column names with starts_with, and both functions can ignore (by default) or take into account the upper or lower case of the characters with the ignore.case parameter. And we have the more general, contains function, looking for substrings in the column names:

```
> str(select(hflights, contains("T", ignore.case = FALSE)))
```

```
'data.frame':  227496 obs. of  7 variables:
 $ DepTime          : int  1400 1401 1352 1403 1405 ...
 $ ArrTime          : int  1500 1501 1502 1513 1507 ...
 $ TailNum          : chr  "N576AA" "N557AA" "N541AA" "N403AA" ...
 $ ActualElapsedTime: int  60 60 70 70 62 64 70 59 71 70 ...
 $ AirTime          : int  40 45 48 39 44 45 43 40 41 45 ...
 $ TaxiIn           : int  7 6 5 9 9 6 12 7 8 6 ...
 $ TaxiOut          : int  13 9 17 22 9 13 15 12 22 19 ...
```

The other option is that we might need a more complex approach with regular expressions, which is another extremely important skill for data scientists. Now, we will provide a regular expression to the matches function, which is to be fitted against all the columns names. Let's select all the columns with a name comprising of 5 or 6 characters:

```
> str(select(hflights, matches("^[[:alpha:]]{5,6}$")))
'data.frame':  227496 obs. of  3 variables:
 $ Month : int  1 1 1 1 1 1 1 1 1 1 ...
 $ Origin: chr  "IAH" "IAH" "IAH" "IAH" ...
 $ TaxiIn: int  7 6 5 9 9 6 12 7 8 6 ...
```

We can keep all column names that do not match a regular expression by using a negative sign before the expression. For example, let's identify the most frequent number of characters in the columns' names:

```
> table(nchar(names(hflights)))

 4  5  6  7  8  9 10 13 16 17
 2  1  2  5  4  3  1  1  1  1
```

And then, let's remove all the columns with 7 or 8 characters from the dataset. Now, we will show the column names from the filtered dataset:

```
> names(select(hflights, -matches("^[[:alpha:]]{7,8}$")))
 [1] "Year"              "Month"             "DayofMonth"
 [4] "DayOfWeek"         "UniqueCarrier"     "FlightNum"
 [7] "ActualElapsedTime" "Origin"            "Dest"
[10] "TaxiIn"            "Cancelled"         "CancellationCode"
```

Rearranging data

Sometimes, we do not want to filter any part of the data (neither the rows, nor the columns), but the data is simply not in the most useful order due to convenience or performance issues, as we have seen, for instance, in *Chapter 3*, *Filtering and Summarizing Data*.

Besides the base `sort` and `order` functions, or providing the order of variables passed to the [operator, we can also use some SQL-like solutions with the `sqldf` package, or query the data in the right format directly from the database. And the previously mentioned `dplyr` package also provides an effective method for ordering data. Let's sort the `hflights` data, based on the actual elapsed time for each of the quarter million flights:

```
> str(arrange(hflights, ActualElapsedTime))
'data.frame':   227496 obs. of   21 variables:
 $ Year            : int  2011 2011 2011 2011 2011 2011 ...
 $ Month           : int  7 7 8 9 1 4 5 6 7 8 ...
 $ DayofMonth      : int  24 25 13 21 3 29 9 21 8 2 ...
 $ DayOfWeek       : int  7 1 6 3 1 5 1 2 5 2 ...
 $ DepTime         : int  2005 2302 1607 1546 1951 2035 ...
 $ ArrTime         : int  2039 2336 1641 1620 2026 2110 ...
 $ UniqueCarrier   : chr  "WN" "XE" "WN" "WN" ...
 $ FlightNum       : int  1493 2408 912 2363 2814 2418 ...
 $ TailNum         : chr  "N385SW" "N12540" "N370SW" "N524SW" ...
 $ ActualElapsedTime: int  34 34 34 34 35 35 35 35 35 35 ...
 $ AirTime         : int  26 26 26 26 23 23 27 26 25 25 ...
 $ ArrDelay        : int  9 -8 -4 15 -19 20 35 -15 86 -9 ...
 $ DepDelay        : int  20 2 7 26 -4 35 45 -8 96 1 ...
 $ Origin          : chr  "HOU" "IAH" "HOU" "HOU" ...
 $ Dest            : chr  "AUS" "AUS" "AUS" "AUS" ...
 $ Distance        : int  148 140 148 148 127 127 148 ...
 $ TaxiIn          : int  3 3 4 3 4 4 5 3 5 4 ...
 $ TaxiOut         : int  5 5 4 5 8 8 3 6 5 6 ...
 $ Cancelled       : int  0 0 0 0 0 0 0 0 0 0 ...
 $ CancellationCode: chr  "" "" "" "" ...
 $ Diverted        : int  0 0 0 0 0 0 0 0 0 0 ...
```

Well, it's pretty straightforward that flights departing to Austin are among the first few records shown. For improved readability, the above three R expressions can be called in a much nicer way with the pipe operator from the automatically imported `magrittr` package, which provides a simple way to pass an R object as the first argument of the subsequent R expression:

```
> hflights %>% arrange(ActualElapsedTime) %>% str
```

So, instead of nesting R functions, we can now start our R command with the core object and pass the results of each evaluated R expression to the next one in the chain. In most cases, this makes the code more convenient to read. Although most hardcore R programmers have already gotten used to reading the nested function calls from inside-out, believe me, it's pretty easy to get used to this nifty feature! Do not let me confuse you with the inspiring painting of René Magritte, which became the slogan, "This is not a pipe," and a symbol of the `magrittr` package:

There is no limit to the number of chainable R expressions and objects one can have. For example, let's also filter a few cases and variables to see how easy it is to follow the data restructuring steps with `dplyr`:

```
> hflights %>%
+       arrange(ActualElapsedTime) %>%
+       select(ActualElapsedTime, Dest) %>%
+       subset(Dest != 'AUS') %>%
+       head %>%
+       str
'data.frame':  6 obs. of  2 variables:
 $ ActualElapsedTime: int   35 35 36 36 37 37
 $ Dest             : chr   "LCH" "LCH" "LCH" "LCH" ...
```

So, now we have filtered the original dataset a few times to see the closest airport after Austin, and the code is indeed easy to read and understand. This is a nice and efficient way to filter data, although some prefer to use nifty one-liners with the `data.table` package:

```
> str(head(data.table(hflights, key = 'ActualElapsedTime')[Dest !=
+    'AUS', c('ActualElapsedTime', 'Dest'), with = FALSE]))
Classes 'data.table' and 'data.frame':  6 obs. of  2 variables:
 $ ActualElapsedTime: int  NA NA NA NA NA NA
 $ Dest             : chr  "MIA" "DFW" "MIA" "SEA" ...
 - attr(*, "sorted")= chr "ActualElapsedTime"
 - attr(*, ".internal.selfref")=<externalptr>
```

Almost perfect! The only problem is that we got different results due to the missing values, which were ordered at the beginning of the dataset while we defined the `data.table` object to be indexed by `ActualElapsedTime`. To overcome this issue, let's drop the NA values, and instead of specifying the column names as strings along with forcing the `with` parameter to be FALSE, let's pass a list of column names:

```
> str(head(na.omit(
+    data.table(hflights, key = 'ActualElapsedTime'))[Dest != 'AUS',
+      list(ActualElapsedTime, Dest)]))
Classes 'data.table' and 'data.frame':  6 obs. of  2 variables:
 $ ActualElapsedTime: int  35 35 36 36 37 37
 $ Dest             : chr  "LCH" "LCH" "LCH" "LCH" ...
 - attr(*, "sorted")= chr "ActualElapsedTime"
 - attr(*, ".internal.selfref")=<externalptr>
```

This is exactly the same results as we have seen before. Please note that in this example, we have omitted the NA values after transforming `data.frame` to `data.table`, indexed by the `ActualElapsedTime` variable, which is a lot faster compared to calling `na.omit` on `hflights` first and then evaluating all the other R expressions:

```
> system.time(str(head(data.table(na.omit(hflights),
+    key = 'ActualElapsedTime')[Dest != 'AUS',
+      c('ActualElapsedTime', 'Dest'), with = FALSE])))
   user  system elapsed
  0.374   0.017   0.390
> system.time(str(head(na.omit(data.table(hflights,
+    key = 'ActualElapsedTime'))[Dest != 'AUS',
```

```
+      c('ActualElapsedTime', 'Dest'), with = FALSE])))
   user   system  elapsed
   0.22    0.00    0.22
```

dplyr versus data.table

You might now be wondering, "which package should we use?"

The dplyr and data.table packages provide a spectacularly different syntax and a slightly less determinative difference in performance. Although data.table seems to be slightly more effective on larger datasets, there is no clear winner in this spectrum — except for doing aggregations on a high number of groups. And to be honest, the syntax of dplyr, provided by the magrittr package, can be also used by the data.table objects if needed.

Also, there is another R package that provides pipes in R, called the pipeR package, which claims to be a lot more effective on larger datasets than magrittr. This performance gain is due to the fact that the pipeR operators do not try to be smart like the F# language's |>-compatible operator in magrittr. Sometimes, this performance overhead is estimated to be 5-15 times more than the ones where no pipes are used at all.

One should take into account the community and support behind an R package before spending a reasonable amount of time learning about its usage. In a nutshell, the data. table package is now mature enough, without doubt, for production usage, as the development was started around 6 years ago by Matt Dowle, who was working for a large hedge fund at that time. The development has been continuous since then. Matt and Arun (co-developer of the package) release new features and performance tweaks from time to time, and they both seem to be keen on providing support on the public R forums and channels, such as mailing lists and StackOverflow.

On the other hand, dplyr is shipped by Hadley Wickham and RStudio, one of the most well-known persons and trending companies in the R community, which translates to an even larger user-base, community, and kind-of-instant support on StackOverflow and GitHub.

In short, I suggest using the packages that fit your needs best, after dedicating some time to discover the power and features they make available. If you are coming from an SQL background, you'll probably find data.table a lot more convenient, while others rather opt for the Hadleyverse (take a look at the R package with this name; it installs a bunch of useful R packages developed by Hadley). You should not mix the two approaches in a single project, as both for readability and performance issues, it's better to stick to only one syntax at a time.

To get a deeper understanding of the pros and cons of the different approaches, I will continue to provide multiple implementations of the same problem in the following few pages as well.

Computing new variables

One of the most trivial actions we usually perform while restructuring a dataset is to create a new variable. For a traditional `data.frame`, it's as simple as assigning a `vector` to a new variable of the R object.

Well, this method also works with `data.table`, but the usage is deprecated due to the fact that there is a much more efficient way of creating one, or even multiple columns in the dataset:

```
> hflights_dt <- data.table(hflights)
> hflights_dt[, DistanceKMs := Distance / 0.62137]
```

We have just computed the distances, in kilometers, between the origin and destination airports with a simple division; although all the hardcore users can head for the `udunits2` package, which includes a bunch of conversion tools based on Unidata's `udunits` library.

And as can be seen previously, data.table uses that special `:=` assignment operator inside of the square brackets, which might seem strange at first glance, but you will love it!

> The `:=` operator can be more than 500 times faster than the traditional `<-` assignment, which is based on the official `data.table` documentation. This speedup is due to not copying the whole dataset into the memory like R used to do before the 3.1 version. Since then, R has used shallow copies, which greatly improved the performance of column updates, but is still beaten by `data.table` powerful in-place updates.

Compare the speed of how the preceding computation was run with the traditional `<-` operator and `data.table`:

```
> system.time(hflights_dt$DistanceKMs <-
+    hflights_dt$Distance / 0.62137)
   user  system elapsed
  0.017   0.000   0.016
> system.time(hflights_dt[, DistanceKMs := Distance / 0.62137])
   user  system elapsed
  0.003   0.000   0.002
```

This is impressive, right? But it's worth double checking what we've just done. The first traditional call, of course, create/updates the `DistanceKMs` variable, but what happens in the second call? The `data.table` syntax did not return anything (visibly), but in the background, the `hflights_dt` R object was updated in-place due to the `:=` operator.

> Please note that the `:=` operator can produce unexpected results when used inside of `knitr`, such as returning the `data.table` visible after the creation of a new variable, or strange rendering of the command when the return is `echo = TRUE`. As a workaround, Matt Dowle suggests increasing the `depthtrigger` option of `data.table`, or one can simply reassign the `data.table` object with the same name. Another solution might be to use my `pander` package over `knitr`. :)

But once again, how was it so fast?

Memory profiling

The magic of the `data.table` package—besides having more than 50 percent of C code in the sources—is copying objects in memory only if it's truly necessary. This means that R often copies objects in memory while updating those, and `data.table` tries to keep these resource-hungry actions at a minimal level. Let's verify this by analyzing the previous example with the help of the `pryr` package, which provides convenient access to some helper functions for memory profiling and understanding R-internals.

First, let's recreate the `data.table` object and let's take a note of the pointer value (location address of the object in the memory), so that we will be able to verify later if the new variable simply updated the same R object, or if it was copied in the memory while the operation took place:

```
> library(pryr)
> hflights_dt <- data.table(hflights)
> address(hflights_dt)
[1] "0x62c88c0"
```

Okay, so `0x62c88c0` refers to the location where `hflights_dt` is stored at the moment. Now, let's check if it changes due to the traditional assignment operator:

```
> hflights_dt$DistanceKMs <- hflights_dt$Distance / 0.62137
> address(hflights_dt)
[1] "0x2c7b3a0"
```

This is definitely a different location, which means that adding a new column to the R object also requires R to copy the whole object in the memory. Just imagine, we now moved 21 columns in memory due to adding another one.

Now, to bring about the usage of `:=` in `data.table`:

```
> hflights_dt <- data.table(hflights)
> address(hflights_dt)
[1] "0x8ca2340"
> hflights_dt[, DistanceKMs := Distance / 0.62137]
> address(hflights_dt)
[1] "0x8ca2340"
```

The location of the R object in the memory did not change! And copying objects in the memory can cost you a lot of resources, thus a lot of time. Take a look at the following example, which is a slightly updated version of the above traditional variable assignment call, but with an added convenience layer of `within`:

```
> system.time(within(hflights_dt, DistanceKMs <- Distance / 0.62137))
   user   system elapsed
  0.027    0.000    0.027
```

Here, using the `within` function probably copies the R object once more in the memory, and hence brings about the relatively serious performance overhead. Although the absolute time difference between the preceding examples might not seem very significant (not in the statistical context), but just imagine how the needless memory updates can affect the processing time of your data analysis with some larger datasets!

Creating multiple variables at a time

One nice feature of `data.table` is the creation of multiple columns with a single command, which can be extremely useful in some cases. For example, we might be interested in the distance of airports in feet as well:

```
> hflights_dt[, c('DistanceKMs', 'DiastanceFeets') :=
+    list(Distance / 0.62137, Distance * 5280)]
```

So, it's as simple as providing a character vector of the desired variable names on the left-hand side and the `list` of appropriate values on the right-hand side of the `:=` operator. This feature can easily be used for some more complex tasks. For example, let's create the dummy variables of the airline carriers:

```
> carriers <- unique(hflights_dt$UniqueCarrier)
> hflights_dt[, paste('carrier', carriers, sep = '_') :=
+    lapply(carriers, function(x) as.numeric(UniqueCarrier == x))]
> str(hflights_dt[, grep('^carrier', names(hflights_dt)),
+    with = FALSE])
Classes 'data.table' and 'data.frame': 227496 obs. of  15 variables:
 $ carrier_AA: num  1 1 1 1 1 1 1 1 1 1 ...
 $ carrier_AS: num  0 0 0 0 0 0 0 0 0 0 ...
 $ carrier_B6: num  0 0 0 0 0 0 0 0 0 0 ...
 $ carrier_CO: num  0 0 0 0 0 0 0 0 0 0 ...
 $ carrier_DL: num  0 0 0 0 0 0 0 0 0 0 ...
 $ carrier_OO: num  0 0 0 0 0 0 0 0 0 0 ...
 $ carrier_UA: num  0 0 0 0 0 0 0 0 0 0 ...
 $ carrier_US: num  0 0 0 0 0 0 0 0 0 0 ...
 $ carrier_WN: num  0 0 0 0 0 0 0 0 0 0 ...
 $ carrier_EV: num  0 0 0 0 0 0 0 0 0 0 ...
 $ carrier_F9: num  0 0 0 0 0 0 0 0 0 0 ...
 $ carrier_FL: num  0 0 0 0 0 0 0 0 0 0 ...
 $ carrier_MQ: num  0 0 0 0 0 0 0 0 0 0 ...
 $ carrier_XE: num  0 0 0 0 0 0 0 0 0 0 ...
 $ carrier_YV: num  0 0 0 0 0 0 0 0 0 0 ...
 - attr(*, ".internal.selfref")=<externalptr>
```

Although it's not a one-liner, and it also introduces a helper variable, it's not that complex to see what we did:

1. First, we saved the `unique` carrier names in a character vector.
2. Then, we defined the new variables' name with the help of that.
3. We iterated our anonymous function over this character vector as well, to return TRUE or FALSE if the carrier name matched the given column.
4. The given column was converted to 0 or 1 through `as.numeric`.
5. And then, we simply checked the structure of all columns whose names start with `carrier`.

This is not perfect, as we usually leave out one label from the dummy variables to reduce redundancy. In the current situation, the last new column is simply the linear combination of the other newly created columns, thus information is duplicated. For this end, it's usually a good practice to leave out, for example, the last category by passing -1 to the n argument in the head function.

Computing new variables with dplyr

The usage of mutate from the dplyr package is identical to that of the base within function, although mutate is a lot quicker than within:

```
> hflights <- hflights %>%
+     mutate(DistanceKMs = Distance / 0.62137)
```

If the analogy of mutate and within has not been made straightforward by the previous example, it's probably also useful to show the same example without using pipes:

```
> hflights <- mutate(hflights, DistanceKMs = Distance / 0.62137)
```

Merging datasets

Besides the previously described elementary actions on a single dataset, joining multiple data sources is one of the most used methods in everyday action. The most often used solution for such a task is to simply call the merge S3 method, which can act as a traditional SQL inner and left/right/full outer joiner of operations—represented in a brief summary by C.L. Moffatt (2008) as follows:

SQL JOINS

SELECT <select_list>
FROM TableA A
LEFT JOIN TableB B
ON A.Key = B.Key

SELECT <select_list>
FROM TableA A
INNER JOIN TableB B
ON A.Key = B.Key

SELECT <select_list>
FROM TableA A
RIGHT JOIN TableB B
ON A.Key = B.Key

SELECT <select_list>
FROM TableA A
LEFT JOIN TableB B
ON A.Key = B.Key
WHERE B.Key IS NULL

SELECT <select_list>
FROM TableA A
RIGHT JOIN TableB B
ON A.Key = B.Key
WHERE A.Key IS NULL

SELECT <select_list>
FROM TableA A
FULL OUTER JOIN TableB B
ON A.Key = B.Key

SELECT <select_list>
FROM TableA A
FULL OUTER JOIN TableB B
ON A.Key = B.Key
WHERE A.Key IS NULL
OR B.Key IS NULL

© C.L. Moffatt, 2008

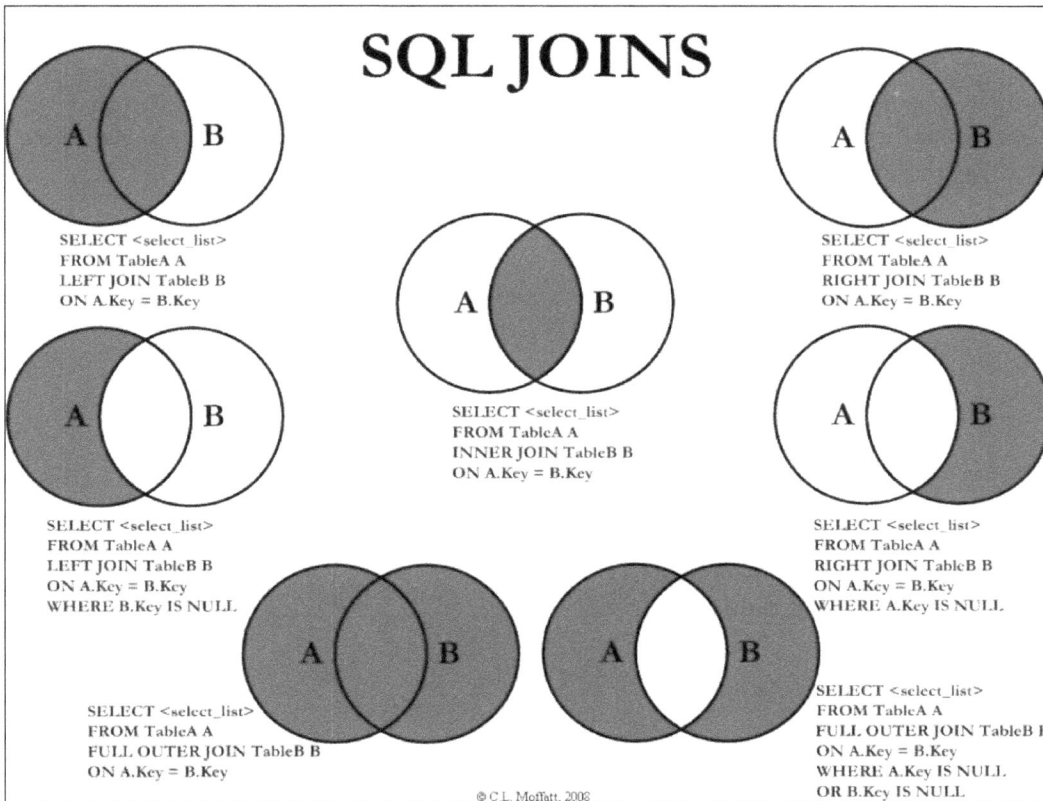

The `dplyr` package provides some easy ways for doing the previously presented join operations right from R, in an easy way:

- `inner_join`: This joins the variables of all the rows, which are found in both datasets
- `left_join`: This includes all the rows from the first dataset and join variables from the other table
- `semi_join`: This includes only those rows from the first dataset that are found in the other one as well
- `anti_join`: This is similar to `semi_join`, but includes only those rows from the first dataset that are not found in the other one

> For more examples, take a look at the *Two-table verbs* `dplyr` vignette, and the Data Wrangling cheat sheet listed in the *References* chapter at the end of the book.

These features are also supported by the `mult` argument of `[` operator of `data.table` call, but for the time being, let's stick to the simpler use cases.

In the following example, we will merge a tiny dataset with the `hflights` data. Let's create the `data.frame` demo by assigning names to the possible values of the `DayOfWeek` variable:

```
> (wdays <- data.frame(
+     DayOfWeek       = 1:7,
+     DayOfWeekString = c("Sunday", "Monday", "Tuesday",
+         "Wednesday", "Thursday", "Friday", "Saturday")
+     ))
  DayOfWeek DayOfWeekString
1         1          Sunday
2         2          Monday
3         3         Tuesday
4         4       Wednesday
5         5        Thursday
6         6          Friday
7         7        Saturday
```

Let's see how we can left-join the previously defined `data.frame` with another `data.frame` and other tabular objects, as `merge` also supports fast operations on, for example, `data.table`:

```
> system.time(merge(hflights, wdays))
   user  system elapsed
  0.700   0.000   0.699
> system.time(merge(hflights_dt, wdays, by = 'DayOfWeek'))
   user  system elapsed
  0.006   0.000   0.009
```

The prior example automatically merged the two tables via the DayOfWeek variable, which was part of both datasets and resulted in an extra variable in the original hflights dataset. However, we had to pass the variable name in the second example, as the by argument of merge.data.table defaults to the key variable of the object, which was missing then. One thing to note is that merging with data.table was a lot faster than the traditional tabular object type.

> Any ideas on how to improve the previous didactical example? Instead of merging, the new variable could be computed as well. See for example, the weekdays function from base R:
> weekdays(as.Date(with(hflights, paste(Year, Month, DayofMonth, sep = '-')))).

A much simpler way of merging datasets is when you simply want to add new rows or columns to the dataset with the same structure. For this end, rbind and cbind, or rBind and cBind for sparse matrices, do a wonderful job.

One of the most often used functions along with these base commands is do.call, which can execute the rbind or cbind call on all elements of a list, thus enabling us, for example, to join a list of data frames. Such lists are usually created by lapply or the related functions from the plyr package. Similarly, rbindlist can be called to merge a list of data.table objects in a much faster way.

Reshaping data in a flexible way

Hadley Wickham has written several R packages to tweak data structures, for example, a major part of his thesis concentrated on how to reshape data frames with his reshape package. Since then, this general aggregation and restructuring package has been renewed to be more efficient with the most commonly used tasks, and it was released with a new version number attached to the name: reshape2 package.

This was a total rewrite of the reshape package, which improves speed at the cost of functionality. Currently, the most important feature of reshape2 is the possibility to convert between the so-called long (narrow) and wide tabular data format. This basically pertains to the columns being stacked below each other, or arranged beside each other.

These features were presented in Hadley's works with the following image on data restructuring, with the related `reshape` functions and simple use cases:

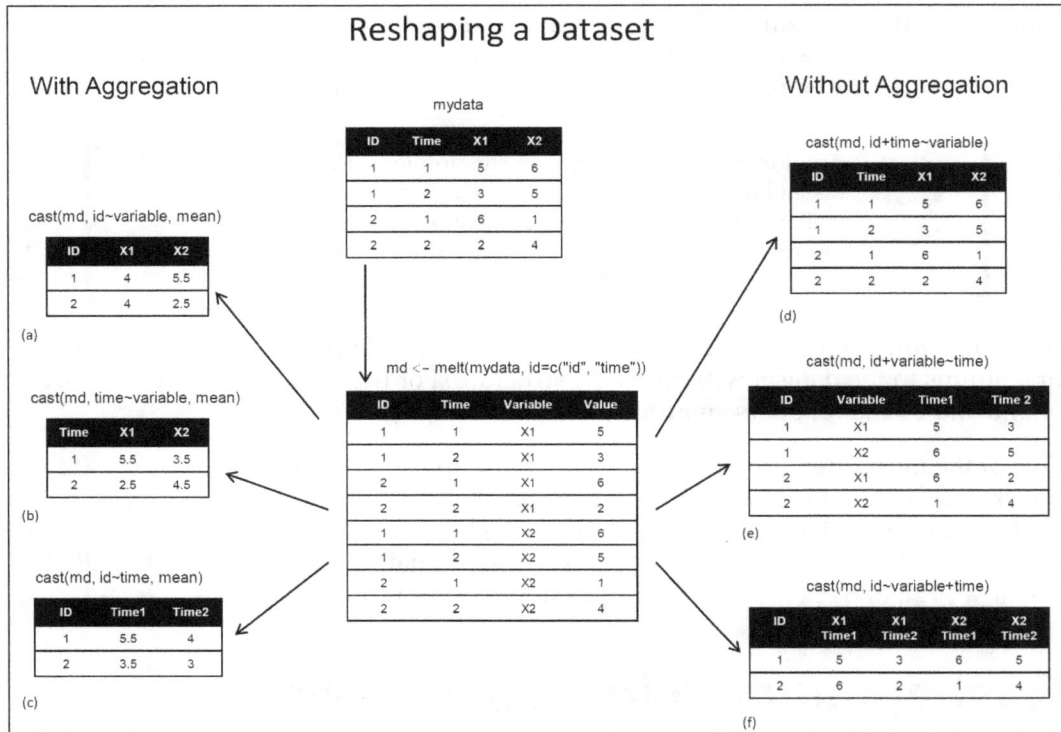

As the `reshape` package is not under active development anymore, and its parts were outsourced to `reshape2`, `plyr`, and most recently to `dplyr`, we will only focus on the commonly used features of `reshape2` in the following pages. This will basically consist of the `melt` and `cast` functions, which provides a smart way of melting the data into a standardized form of measured and identifier variables (long table format), which can later be casted to a new shape for further analysis.

Converting wide tables to the long table format

Melting a data frame means that we transform the tabular data to key-value pairs, based on the given identifier variables. The original column names become the categories of the newly created `variable` column, while all numeric values of those (measured variables) are included in the new `value` column. Here's a quick example:

```
> library(reshape2)
```

```
> head(melt(hflights))
Using UniqueCarrier, TailNum, Origin, Dest, CancellationCode as id
variables
  UniqueCarrier TailNum Origin Dest CancellationCode variable value
1            AA  N576AA    IAH  DFW                      Year  2011
2            AA  N557AA    IAH  DFW                      Year  2011
3            AA  N541AA    IAH  DFW                      Year  2011
4            AA  N403AA    IAH  DFW                      Year  2011
5            AA  N492AA    IAH  DFW                      Year  2011
6            AA  N262AA    IAH  DFW                      Year  2011
```

So, we have just restructured the original data.frame, which had 21 variables and a quarter of a million records, into only 7 columns and more than 3.5 million records. Six out of the seven columns are factor type identifier variables, and the last column stores all the values. But why is it useful? Why should we transform the traditional wide tabular format to the much longer type of data?

For example, we might be interested in comparing the distribution of flight time with the actual elapsed time of the flight, which might not be straightforward to plot with the original data format. Although plotting a scatter plot of the above variables with the ggplot2 package is extremely easy, how would you create two separate boxplots comparing the distributions?

The problem here is that we have two separate variables for the time measurements, while ggplot requires one numeric and one factor variable, from which the latter will be used to provide the labels on the *x*-axis. For this end, let's restructure our dataset with melt by specifying the two numeric variables to treat as measurement variables and dropping all other columns — or in other words, not having any identifier variables:

```
> hflights_melted <- melt(hflights, id.vars = 0,
+    measure.vars = c('ActualElapsedTime', 'AirTime'))
> str(hflights_melted)
'data.frame':  454992 obs. of  2 variables:
 $ variable: Factor w/ 2 levels "ActualElapsedTime",..: 1 1 1 1 1 ...
 $ value   : int   60 60 70 70 62 64 70 59 71 70 ...
```

> In general, it's not a good idea to melt a dataset without identifier variables, as casting it later becomes cumbersome, if not impossible.

Please note that now we have exactly twice as many rows than we had before, and the `variable` column is a factor with only two levels, which represent the two measurement variables. And this resulting `data.frame` is now easy to plot with the two newly created columns:

```
> library(ggplot2)
> ggplot(hflights_melted, aes(x = variable, y = value)) +
+     geom_boxplot()
```

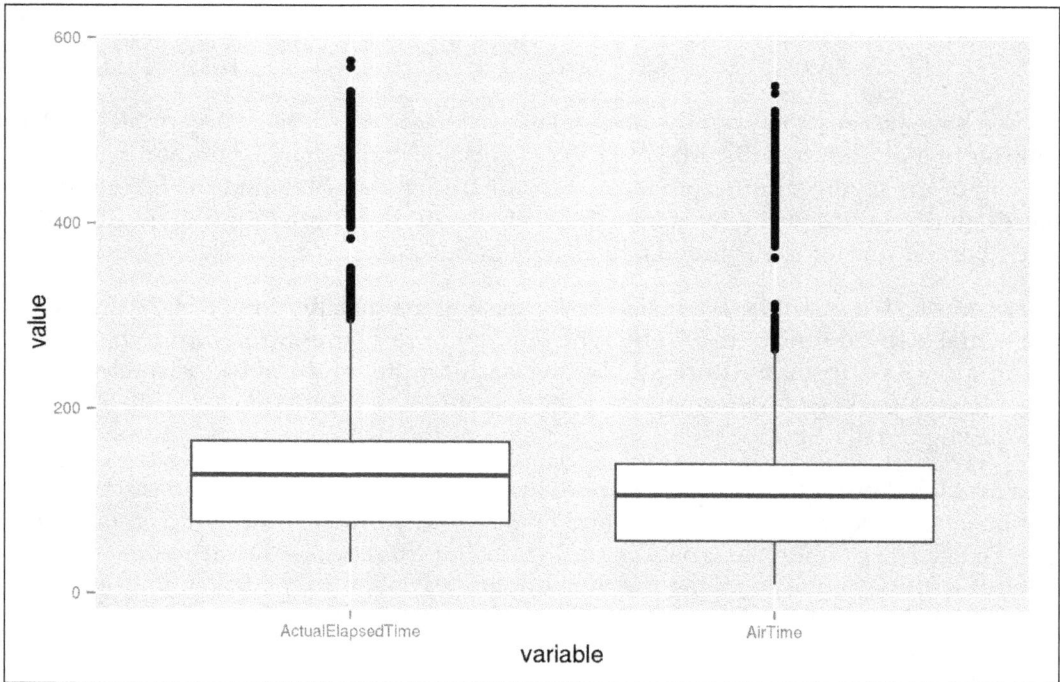

Well, the previous example might not seem mission critical, and to be honest, I first used the `reshape` package when I needed some similar transformation to be able to produce some nifty `ggplot2` charts — as the previous problem simply does not exist if someone is using `base` graphics. For example, you can simply pass the two separate variables of the original dataset to the `boxplot` function.

So, this is kind of entering the world of Hadley Wickham's R packages, and the journey indeed offers some great data analysis practices. Thus, I warmly suggest reading further, for example, on how using `ggplot2` is not easy, if not impossible, without knowing how to reshape datasets efficiently.

Converting long tables to the wide table format

Casting a dataset is the opposite of melting, like turning key-value pairs into a tabular data format. But bear in mind that the key-value pairs can always be combined together in a variety of ways, so this process can result in extremely diversified outputs. Thus, you need a table and a formula to cast, for example:

```
> hflights_melted <- melt(hflights, id.vars = 'Month',
+   measure.vars = c('ActualElapsedTime', 'AirTime'))
> (df <- dcast(hflights_melted, Month ~ variable,
+   fun.aggregate = mean, na.rm = TRUE))
   Month ActualElapsedTime  AirTime
1    1           125.1054 104.1106
2    2           126.5748 105.0597
3    3           129.3440 108.2009
4    4           130.7759 109.2508
5    5           131.6785 110.3382
6    6           130.9182 110.2511
7    7           130.4126 109.2059
8    8           128.6197 108.3067
9    9           128.6702 107.8786
10   10          128.8137 107.9135
11   11          129.7714 107.5924
12   12          130.6788 108.9317
```

This example shows how to aggregate the measured flight times for each month in 2011 with the help of melting and casting the hflights dataset:

1. First, we melted the data.frame with the IDs being the Month, where we only kept two numeric variables for the flight times.
2. Then, we casted the resulting data.frame with a simple formula to show the mean of each month for all measurement variables.

I am pretty sure that now you can quickly restructure this data to be able to plot two separate lines for this basic time-series:

```
> ggplot(melt(df, id.vars = 'Month')) +
+   geom_line(aes(x = Month, y = value, color = variable)) +
```

```
+    scale_x_continuous(breaks = 1:12) +

+    theme_bw()  +

+    theme(legend.position = 'top')
```

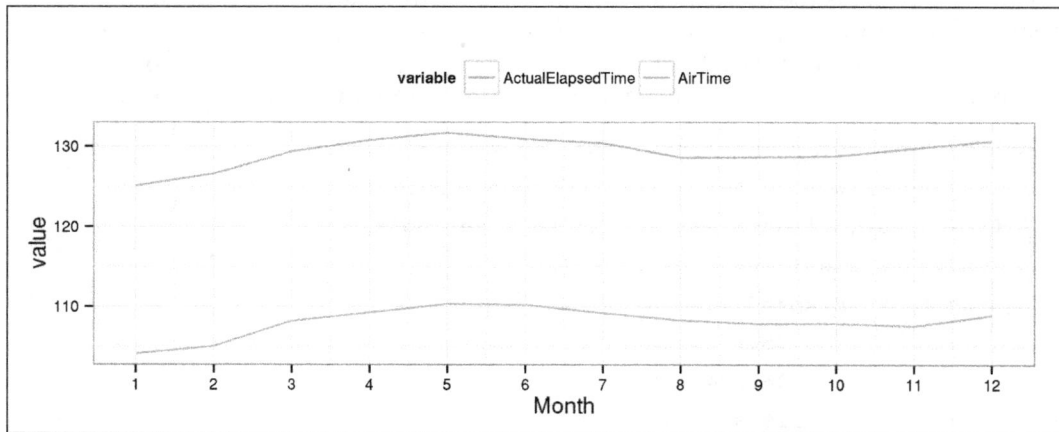

But of course, melting and casting can be used for a variety of things besides aggregating. For example, we can restructure our original database to have a special Month, which includes all the records of the data. This, of course, doubles the number of rows in the dataset, but also lets us easily generate a table on the data with margins. Here's a quick example:

```
> hflights_melted <- melt(add_margins(hflights, 'Month'),

+    id.vars = 'Month',

+    measure.vars = c('ActualElapsedTime', 'AirTime'))

> (df <- dcast(hflights_melted, Month ~ variable,

+    fun.aggregate = mean, na.rm = TRUE))
```

	Month	ActualElapsedTime	AirTime
1	1	125.1054	104.1106
2	2	126.5748	105.0597
3	3	129.3440	108.2009
4	4	130.7759	109.2508
5	5	131.6785	110.3382
6	6	130.9182	110.2511
7	7	130.4126	109.2059
8	8	128.6197	108.3067
9	9	128.6702	107.8786
10	10	128.8137	107.9135

```
11    11              129.7714 107.5924
12    12              130.6788 108.9317
13  (all)             129.3237 108.1423
```

This is very similar to what we have seen previously, but as an intermediate step, we have converted the `Month` variable to be factor with a special level, which resulted in the last line of this table. This row represents the overall arithmetic average of the related measure variables.

Tweaking performance

Some further good news on `reshape2` is that `data.table` has decent support for melting and casting, with highly improved performance. Matt Dowle has published some benchmarks with a 5-10 percent improvement in the processing times of using `cast` and `melt` on `data.table` objects instead of the traditional data frames, which is highly impressive.

To verify these results on your own dataset, simply transform the `data.frame` objects to `data.table` before calling the `reshape2` functions, as the `data.table` package already ships the appropriate S3 methods to extend `reshape2`.

The evolution of the reshape packages

As mentioned before, `reshape2` was a complete rewrite of the `reshape` package, based on around 5 years of experience in using and developing the latter. This update also included some trade-offs, as the original reshape tasks were split among multiple packages. Thus, `reshape2` now offers a lot less compared to the kind of magic features that were supported by `reshape`. Just check, for example `reshape::cast`; especially the `margins` and `add.missing` argument!

But as it turns out, even `reshape2` offers a lot more than simply melting and casting data frames. The birth of the `tidyr` package was inspired by this fact: to have a package in the Hadleyverse that supports easy data cleaning and transformation between the long and wide table formats. In `tidyr` parlance, these operations are called `gather` and `spread`.

Just to give a quick example of this new syntax, let's re-implement the previous examples:

```
> library(tidyr)
> str(gather(hflights[, c('Month', 'ActualElapsedTime', 'AirTime')],
+    variable, value, -Month))
```

```
'data.frame':  454992 obs. of  3 variables:
 $ Month   : int  1 1 1 1 1 1 1 1 1 ...
 $ variable: Factor w/ 2 levels "ActualElapsedTime",..: 1 1 1 1 ...
 $ value   : int  60 60 70 70 62 64 70 59 71 70 ...
```

Summary

In this chapter, we focused on how to transform raw data into an appropriately structured format, before we could run statistical tests. This process is a really important part of our everyday actions, and it takes most of a data scientist's time. But after reading this chapter, you should be confident in how to restructure your data in most cases— so, this is the right time to focus on building some models, which we will do in the next chapter.

5

Building Models

(authored by Renata Nemeth and Gergely Toth)

"All models should be as simple as possible... but no simpler."

– Attributed to Albert Einstein

"All models are wrong... but some are useful."

– George Box

After loading and transforming data, in this chapter, we will focus on how to build statistical models. Models are representations of reality, and, as the preceding citations emphasize, are always simplified representations. Although you can't possibly take everything into account, you should be aware about what to include and exclude in a good model that provides meaningful results.

In this chapter, regression models are discussed on the basis of linear regression models and standard modeling. **Generalized Linear Models** (**GLM**) extend these to allow the response variables to differ in distribution, which will be covered in the *Chapter 6, Beyond the Linear Trend Line (authored by Renata Nemeth and Gergely Toth)*. In all, we will discuss the three most well known regression models:

- **Linear regression** for continuous outcomes (birth weight measured in grams)
- **Logistic regression** for binary outcomes (low birth weight versus normal birth weight)
- **Poisson regression** for count data (number of low birth weight infants per year or per country)

Although there are many other regression models, such as *Cox-regression* which we will not discuss here, the logic in the building of the models and the interpretation are similar. So, after reading this chapter, you will be able to understand those without doubt.

By the end of this chapter, you will learn the most important things about regression models: how to avoid confounding, how to fit, how to interpret, and how to choose the best model among the many different options.

The motivation behind multivariate models

If you would like to measure the strength of association between a response and a predictor, you can choose a simple two-way association measure, such as correlation or the odds ratio, depending on the nature of your data. But, if your aim is to model a complex mechanism by taking into account other predictors as well, you will need regression models.

As Ben Goldacre, the evidence-based columnist for *The Guardian*, tells in his brilliant TED talk that the strong association between olive oil consumption and young looking skin does not imply that olive oil is beneficial to our skin. When modeling a complex association structure, we should also control for other predictors, such as smoking status or physical activity, because those who consume more olive oil are more likely to live a healthy life in general, so it may not be the olive oil itself that prevents skin wrinkles. In short, it seems that the kind of lifestyle is likely to confound the relationship between the variables of interest, making it appear that there might be causality, when in fact there is none.

> A confounder is a third variable that biases (increases or decreases) the association we are interested in. The confounder is always associated with both the response and the predictor.

If we examine the olive oil and skin wrinkles association again by fixing the smoking status, hence building separate models for smokers and non-smokers, the association may vanish. Holding the confounders fixed is the main idea behind controlling confounding via regression models.

Regression models in general are intended to measure associations between a response and a predictor by controlling for others. Potential confounders are entered into the model as predictors, and the regression coefficient of the predictor (the *partial coefficient*) measures the effect adjusted to the confounders.

Linear regression with continuous predictors

Let's start with an actual and illuminating example of confounding. Consider that we would like to predict the amount of air pollution based on the size of the city (measured in population size as thousand of habitants). Air pollution is measured by the sulfur dioxide (SO2) concentration in the air, in milligrams per cubic meter. We will use the US air pollution data set (Hand and others 1994) from the `gamlss.data` package:

```
> library(gamlss.data)
> data(usair)
```

Model interpretation

Let's draw our very first linear regression model by building a formula. The `lm` function from the `stats` package is used to fit linear models, which is an important tool for regression modeling:

```
> model.0 <- lm(y ~ x3, data = usair)
> summary(model.0)


Residuals:
    Min      1Q  Median      3Q     Max
-32.545 -14.456  -4.019  11.019  72.549


Coefficients:
            Estimate Std. Error t value Pr(>|t|)
(Intercept) 17.868316   4.713844   3.791 0.000509 ***
x3           0.020014   0.005644   3.546 0.001035 **
---
Signif. codes:  0 '***' 0.001 '**' 0.01 '*' 0.05 '.' 0.1 ' ' 1


Residual standard error: 20.67 on 39 degrees of freedom
Multiple R-squared:  0.2438,	Adjusted R-squared:  0.2244
F-statistic: 12.57 on 1 and 39 DF,  p-value: 0.001035
```

[✎ Formula notation is one of the best features of R, which lets you define flexible models in a human-friendly way. A typical model has the form of response ~ terms, where response is the continuous response variable, and terms provides one or a series of numeric variables that specifies a linear predictor for the response.]

In the preceding example, the variable, y, denotes air pollution, while x3 stands for the population size. The coefficient of x3 says that a one unit (one thousand) increase in the population size causes a 0.02 unit (0.02 milligram per cubic meter) increase in the sulfur dioxide concentration, and the effect is statistically significant with a p value of 0.001035.

[✎ See more details on the p-value in the *How well does the line fit to the data?* section. To keep it simple for now, we will refer to models as statistically significant when the *p* value is below 0.05.]

The intercept in general is the value of the response variable when each predictor equals to 0, but in this example, there are no cities without inhabitants, so the intercept (17.87) doesn't have a direct interpretation. The two regression coefficients define the regression line:

```
> plot(y ~ x3, data = usair, cex.lab = 1.5)
> abline(model.0, col = "red", lwd = 2.5)
> legend('bottomright', legend = 'y ~ x3', lty = 1, col = 'red',
+     lwd = 2.5, title = 'Regression line')
```

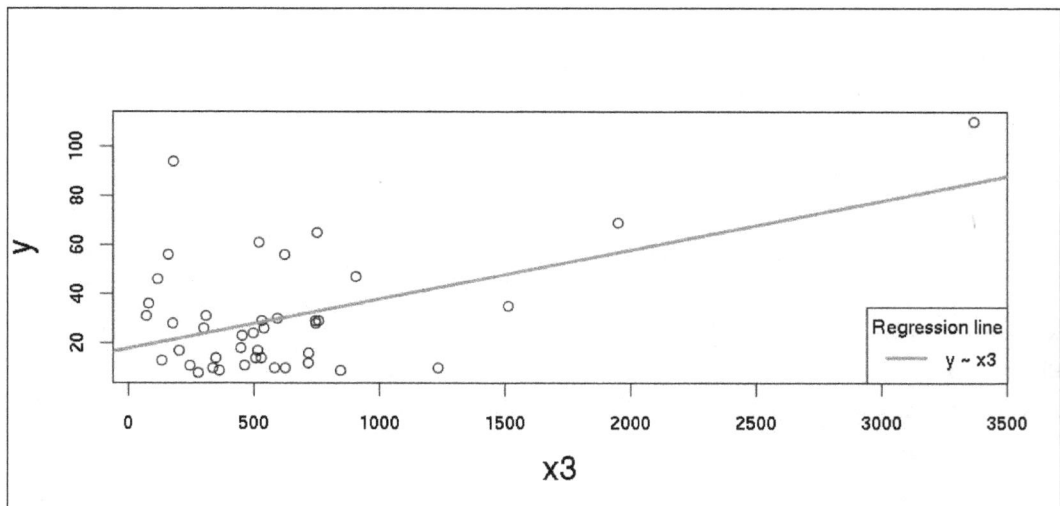

As you can see, the intercept (**17.87**) is the value at which the regression line crosses the y-axis. The other coefficient (**0.02**) is the slope of the regression line: it measures how steep the line is. Here, the function runs uphill because the slope is positive (**y** increases as **x3** increases). Similarly, if the slope is negative, the function runs downhill.

You can easily understand the way the estimates were obtained if you realize how the line was drawn. This is the line that best fits the data points. Here, we refer to the *best fit* as the linear least-squares approach, which is why the model is also known as the **ordinary least squares** (**OLS**) regression.

The least-squares method finds the best fitting line by minimizing the sum of the squares of the residuals, where the residuals represent the error, which is the difference between the observed value (an original dot in the scatterplot) and the fitted or predicted value (a dot on the line with the same *x*-value):

```
> usair$prediction <- predict(model.0)
> usair$residual<- resid(model.0)
> plot(y ~ x3, data = usair, cex.lab = 1.5)
> abline(model.0, col = 'red', lwd = 2.5)
> segments(usair$x3, usair$y, usair$x3, usair$prediction,
+   col = 'blue', lty = 2)
> legend('bottomright', legend = c('y ~ x3', 'residuals'),
+   lty = c(1, 2), col = c('red', 'blue'), lwd = 2.5,
+   title = 'Regression line')
```

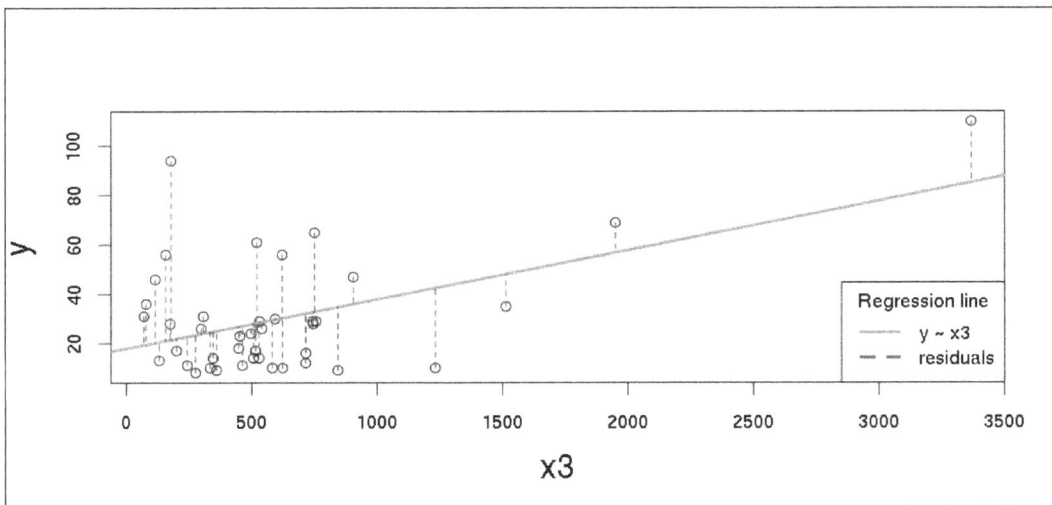

The *linear* term in linear regression refers to the fact that we are interested in a linear relation, which is more natural, easier to understand, and simpler to handle mathematically, as compared to the more complex methods.

Multiple predictors

On the other hand, if we aim to model a more complex mechanism by separating the effect of the population size from the effect of the presence of industries, we have to control for the variable, x2, which describes the number of manufacturers employing more than 20 workers. Now, we can either create a new model by `lm(y ~ x3 + x2, data = usair)`, or use the `update` function to refit the previous model:

```
> model.1 <- update(model.0, . ~ . + x2)
> summary(model.1)


Residuals:
    Min      1Q  Median      3Q     Max
-22.389 -12.831  -1.277   7.609  49.533


Coefficients:
             Estimate Std. Error t value Pr(>|t|)
(Intercept) 26.32508    3.84044   6.855 3.87e-08 ***
x3          -0.05661    0.01430  -3.959 0.000319 ***
x2           0.08243    0.01470   5.609 1.96e-06 ***
---
Signif. codes:  0 '***' 0.001 '**' 0.01 '*' 0.05 '.' 0.1 ' ' 1


Residual standard error: 15.49 on 38 degrees of freedom
Multiple R-squared:  0.5863,    Adjusted R-squared:  0.5645
F-statistic: 26.93 on 2 and 38 DF,  p-value: 5.207e-08
```

Now, the coefficient of x3 is `-0.06`! While the crude association between air pollution and city size was positive in the previous model, after controlling for the number of manufacturers, the association becomes negative. This means that a one thousand increase in the population decreases the SO2 concentration by 0.06 unit, which is a statistically significant effect.

On first sight, this change of sign from positive to negative may be surprising, but it is rather plausible after a closer look; it's definitely not the population size, but rather the level of industrialization that affects the air pollution directly. In the first model, population size showed a positive effect because it implicitly measured industrialization as well. When we hold industrialization fixed, the effect of the population size becomes negative, and growing a city with a fixed industrialization level spreads the air pollution in a wider range.

So, we can conclude that x2 is a confounder here, as it biases the association between y and x3. Although it is beyond the scope of our current research question, we can interpret the coefficient of x2 as well. It says that holding the city size at a constant level, a one unit increase in the number of manufacturers increases the SO2 concentration by 0.08 mgs.

Based on the model, we can predict the expected value of the response for any combination of predictors. For example, we can predict the expected level of sulfur dioxide concentration for a city with 400,000 habitants and 150 manufacturers, each of whom employ more than 20 workers:

```
> as.numeric(predict(model.1, data.frame(x2 = 150, x3 = 400)))
[1] 16.04756
```

You could also calculate the prediction by yourself, multiplying the values with the slopes, and then summing them up with the constant—all these numbers are simply copied and pasted from the previous model summary:

```
> -0.05661 * 400 + 0.08243 * 150 + 26.32508
[1] 16.04558
```

> Prediction outside the range of the data is known as extrapolation. The further the values are from the data, the riskier your prediction becomes. The problem is that you cannot check model assumptions (for example, linearity) outside of your sample data.

If you have two predictors, the regression line is represented by a surface in the three dimensional space, which can be easily shown via the scatterplot3d package:

```
> library(scatterplot3d)
> plot3d <- scatterplot3d(usair$x3, usair$x2, usair$y, pch = 19,
+    type = 'h', highlight.3d = TRUE, main = '3-D Scatterplot')
```

```
> plot3d$plane3d(model.1, lty = 'solid', col = 'red')
```

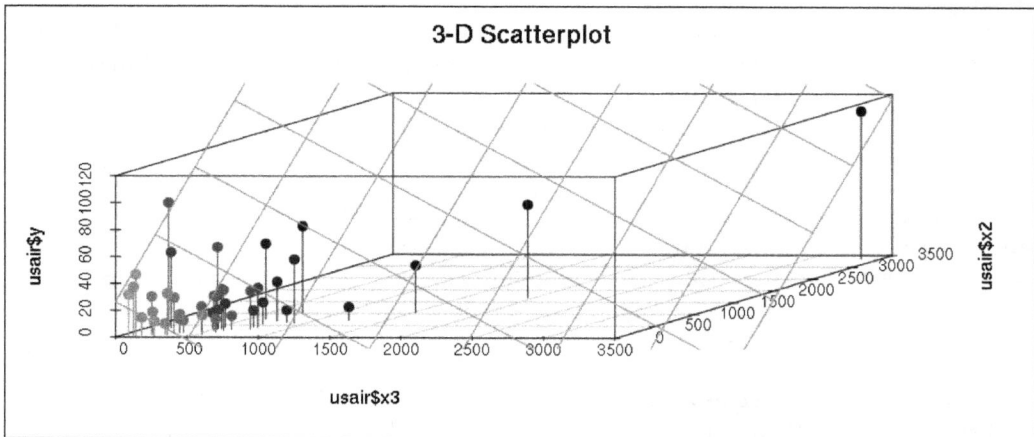

As it's rather hard to interpret this plot, let's draw the 2-dimensional projections of this 3D graph, which might prove to be more informative after all. Here, the value of the third, non-presented variable is held at zero:

```
> par(mfrow = c(1, 2))
> plot(y ~ x3, data = usair, main = '2D projection for x3')
> abline(model.1, col = 'red', lwd = 2.5)
> plot(y ~ x2, data = usair, main = '2D projection for x2')
> abline(lm(y ~ x2 + x3, data = usair), col = 'red', lwd = 2.5)
```

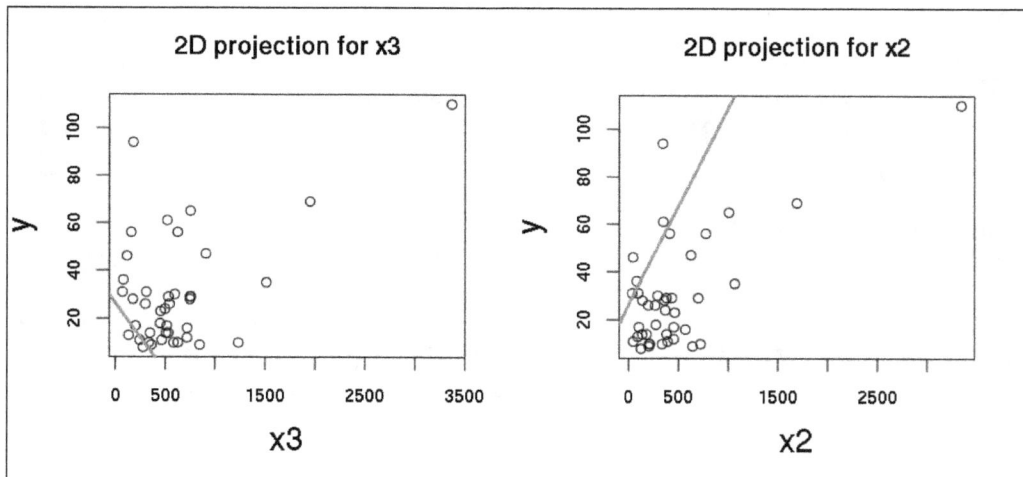

According to the changed sign of the slope, it's well worth mentioning that the *y-x3* regression line has also changed; from uphill, it became downhill.

Model assumptions

Linear regression models with standard estimation techniques make a number of assumptions about the outcome variable, the predictor variables, and also about their relationship:

1. Y is a continuous variable (not binary, nominal, or ordinal)
2. The errors (the residuals) are statistically independent
3. There is a stochastic linear relationship between Y and each X
4. Y has a normal distribution, holding each X fixed
5. Y has the same variance, regardless of the fixed value of the Xs

A violation of assumption **2** occurs in trend analysis, if we use time as the predictor. Since the consecutive years are not independent, the errors will not be independent from each other. For example, if we have a year with high mortality from a specific illness, then we can expect the mortality for the next year to also be high.

A violation of assumption (**3**) says that the relationship is not exactly linear, but there is a deviation from the linear trend line. Assumption **4** and **5** require the conditional distribution of Y to be normal and having the same variance, regardless of the fixed value of Xs. They are needed for inferences of the regression (confidence intervals, F- and t-tests). Assumption **5** is known as the homoscedasticity assumption. If it is violated, heteroscedasticity holds.

The following plot helps in visualizing these assumptions with a simulated dataset:

```
> library(Hmisc)
> library(ggplot2)
> library(gridExtra)
> set.seed(7)
> x   <- sort(rnorm(1000, 10, 100))[26:975]
> y   <- x * 500 + rnorm(950, 5000, 20000)
> df <- data.frame(x = x, y = y, cuts = factor(cut2(x, g = 5)),
+                                resid = resid(lm(y ~ x)))
> scatterPl <- ggplot(df, aes(x = x, y = y)) +
+    geom_point(aes(colour = cuts, fill = cuts), shape = 1,
+  show_guide = FALSE) + geom_smooth(method = lm, level = 0.99)
```

```
> plot_left <- ggplot(df,  aes(x = y, fill = cuts)) +
+    geom_density(alpha = .5) + coord_flip() + scale_y_reverse()
> plot_right <- ggplot(data = df, aes(x = resid, fill = cuts)) +
+    geom_density(alpha = .5) + coord_flip()
> grid.arrange(plot_left, scatterP1, plot_right,
+    ncol=3, nrow=1, widths=c(1, 3, 1))
```

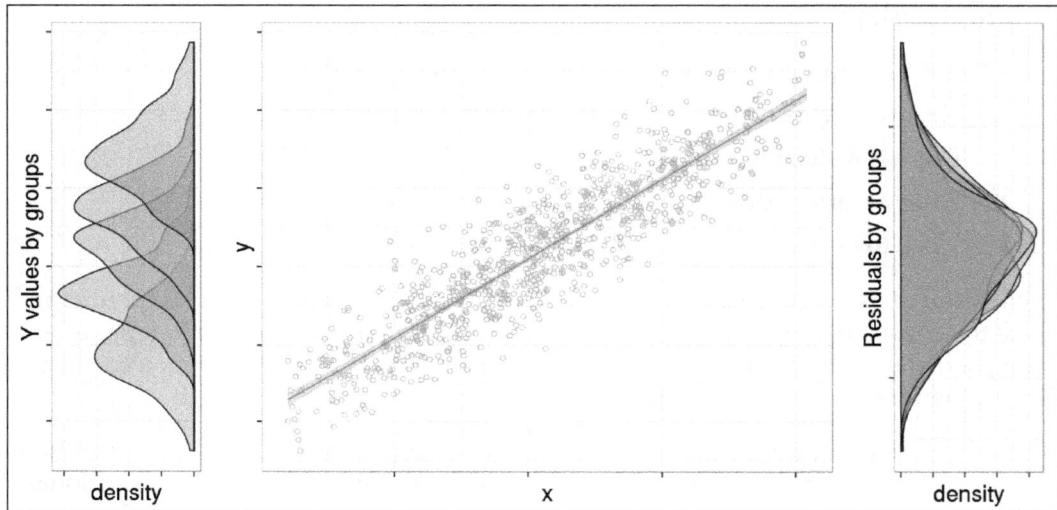

The code bundle, available to be downloaded from the Packt Publishing homepage, includes a slightly longer code chunk for the preceding plot with some tweaks on the plot margins, legends, and titles. The preceding code block focuses on the major parts of the visualization, without wasting too much space in the printed book on the style details.

We will discuss in more detail, how to assess the model assumptions in *Chapter 9, From Big to Smaller Data*. If some of the assumptions fail, a possible solution is to look for outliers. If you have an outlier, do the regression analysis without that observation, and determine how the results differ. Ways of outlier detection will be discussed in more detail in *Chapter 8, Polishing Data*.

The following example illustrates that dropping an outlier (observation number 31) may make the assumptions valid. To quickly verify if a model's assumptions are satisfied, use the gvlma package:

```
> library(gvlma)
> gvlma(model.1)
```

```
Coefficients:
(Intercept)              x3             x2
   26.32508       -0.05661        0.08243


ASSESSMENT OF THE LINEAR MODEL ASSUMPTIONS
USING THE GLOBAL TEST ON 4 DEGREES-OF-FREEDOM:
Level of Significance =   0.05


                     Value  p-value                 Decision
Global Stat         14.1392 0.006864 Assumptions NOT satisfied!
Skewness             7.8439 0.005099 Assumptions NOT satisfied!
Kurtosis             3.9168 0.047805 Assumptions NOT satisfied!
Link Function        0.1092 0.741080    Assumptions acceptable.
Heteroscedasticity   2.2692 0.131964    Assumptions acceptable.
```

It seems that three out of the five assumptions are not satisfied. However, if we build the very same model on the same dataset excluding the 31st observation, we get much better results:

```
> model.2 <- update(model.1, data = usair[-31, ])
> gvlma(model.2)


Coefficients:
(Intercept)              x3             x2
   22.45495       -0.04185        0.06847


ASSESSMENT OF THE LINEAR MODEL ASSUMPTIONS
USING THE GLOBAL TEST ON 4 DEGREES-OF-FREEDOM:
Level of Significance =   0.05


                     Value p-value                 Decision
Global Stat         3.7099  0.4467 Assumptions acceptable.
Skewness            2.3050  0.1290 Assumptions acceptable.
Kurtosis            0.0274  0.8685 Assumptions acceptable.
Link Function       0.2561  0.6128 Assumptions acceptable.
Heteroscedasticity 1.1214  0.2896 Assumptions acceptable.
```

This suggests that we must always exclude the 31st observation from the dataset when building regression models in the future sections.

However, it's important to note that it is not acceptable to drop an observation just because it is an outlier. Before you decide, investigate the particular case. If it turns out that the outlier is due to incorrect data, you should drop it. Otherwise, run the analysis, both with and without it, and state in your research report how the results changed and why you decided on excluding the extreme values.

> You can fit a line for any set of data points; the least squares method will find the optimal solution, and the trend line will be interpretable. The regression coefficients and the R-squared coefficient are also meaningful, even if the model assumptions fail. The assumptions are only needed if you want to interpret the p-values, or if you aim to make good predictions.

How well does the line fit in the data?

Although we know that the trend line is the best fitting among the possible linear trend lines, we don't know how well this fits the actual data. The significance of the regression parameters is obtained by testing the null hypothesis, which states that the given parameter equals to zero. The *F-test* in the output pertains to the hypothesis that each regression parameter is zero. In a nutshell, it tests the significance of the regression in general. A *p-value* below 0.05 can be interpreted as "the regression line is significant." Otherwise, there is not much point in fitting the regression model at all.

However, even if you have a significant F-value, you cannot say too much about the fit of the regression line. We have seen that residuals characterize the error of the fit. The R-squared coefficient summarizes them into a single measure. *R-squared* is the proportion of the variance in the response variable explained by the regression. Mathematically, it is defined as the variance in the predicted Y values, divided by the variance in the observed Y values.

In some cases, despite the significant F-test, the predictors, according to the R-squared, explain only a small proportion (<10 percent) of the total variance. You can interpret this by saying that although the predictors have a statistically significant effect on the response, the response is formed by a mechanism that is much more complex than your model suggests. This phenomenon is common in the area of medicine or biology where complex biological processes are modeled, while it is less common in the area of econometrics, where macro-level, aggregated variables, which usually smooth out small variations in the data.

If we use the population size as the only predictor in our air pollution example, the R-squared equals 0.37, so we can say that 37 percent of the variation in SO2 concentration can be explained by the size of the city:

```
> model.0 <- update(model.0, data = usair[-31, ])
> summary(model.0)[c('r.squared', 'adj.r.squared')]
$r.squared
[1] 0.3728245
$adj.r.squared
[1] 0.3563199
```

After adding the number of manufacturers to the model, the R-squared increases dramatically and almost doubles its previous value:

```
> summary(model.2)[c('r.squared', 'adj.r.squared')]
$r.squared
[1] 0.6433317
$adj.r.squared
[1] 0.6240523
```

It's important to note here that every time you add an extra predictor to your model, the R-squared increases simply because you have more information to predict the response, even if the lastly added predictor doesn't have an important effect. Consequently, a model with more predictors may appear to have a better fit just because it is bigger.

The solution is to use the adjusted R-squared, which takes into account the number of predictors as well. In the previous example, not only the R-squared but also the adjusted R-squared showed a huge advantage in favor of the latter model.

The two previous models are nested, which means that the extended model contains each predictor of the first one. But unfortunately, the adjusted R-squared cannot be used as a base for choosing the best model for non-nested models. If you have non-nested models, you can use the **Akaike Information Criterion (AIC)** measure to select the best model.

AIC is founded on the information theory. It introduces a penalty term for the number of parameters in the model, giving a solution for the problem of bigger models tending to show as better fitted. When using this criterion, you should select the model with the least AIC. As a rule of thumb, two models are essentially indistinguishable if the difference between their AICs is less than 2. In the example that follows, we have two plausible alternative models. Taking the AIC into account, model.4 is better than model.3, as its advantage over model.3 is about 10:

```
> summary(model.3 <- update(model.2, .~. -x2 + x1))$coefficients
              Estimate    Std. Error    t value     Pr(>|t|)
(Intercept) 77.429836 19.463954376   3.978114 3.109597e-04
x3           0.021333  0.004221122   5.053869 1.194154e-05
x1          -1.112417  0.338589453  -3.285444 2.233434e-03

> summary(model.4 <- update(model.2, .~. -x3 + x1))$coefficients
               Estimate    Std. Error    t value     Pr(>|t|)
(Intercept) 64.52477966 17.616612780   3.662723 7.761281e-04
x2           0.02537169  0.003880055   6.539004 1.174780e-07
x1          -0.85678176  0.304807053  -2.810899 7.853266e-03

> AIC(model.3, model.4)
         df      AIC
model.3   4 336.6405
model.4   4 326.9136
```

> Note that AIC can tell nothing about the quality of the model in an absolute sense; your best model may still fit poorly. It does not provide a test for testing model fit either. It is essentially for ranking different models.

Discrete predictors

So far, we have seen only the simple case of both the response and the predictor variables being continuous. Now, let's generalize the model a bit, and enter a discrete predictor into the model. Take the `usair` data and add `x5` (precipitation: average number of wet days per year) as a predictor with three categories (low, middle, and high levels of precipitation), using 30 and 45 as the cut-points. The research question is how these precipitation groups are associated with the SO2 concentration. The association is not necessary linear, as the following plot shows:

```
> plot(y ~ x5, data = usair, cex.lab = 1.5)
> abline(lm(y ~ x5, data = usair), col = 'red', lwd = 2.5, lty = 1)
> abline(lm(y ~ x5, data = usair[usair$x5<=45,]),
+    col = 'red', lwd = 2.5, lty = 3)
> abline(lm(y ~ x5, data = usair[usair$x5 >=30, ]),
+    col = 'red', lwd = 2.5, lty = 2)
> abline(v = c(30, 45), col = 'blue', lwd = 2.5)
> legend('topleft', lty = c(1, 3, 2, 1), lwd = rep(2.5, 4),
+    legend = c('y ~ x5', 'y ~ x5 | x5<=45','y ~ x5 | x5>=30',
+       'Critical zone'), col = c('red', 'red', 'red', 'blue'))
```

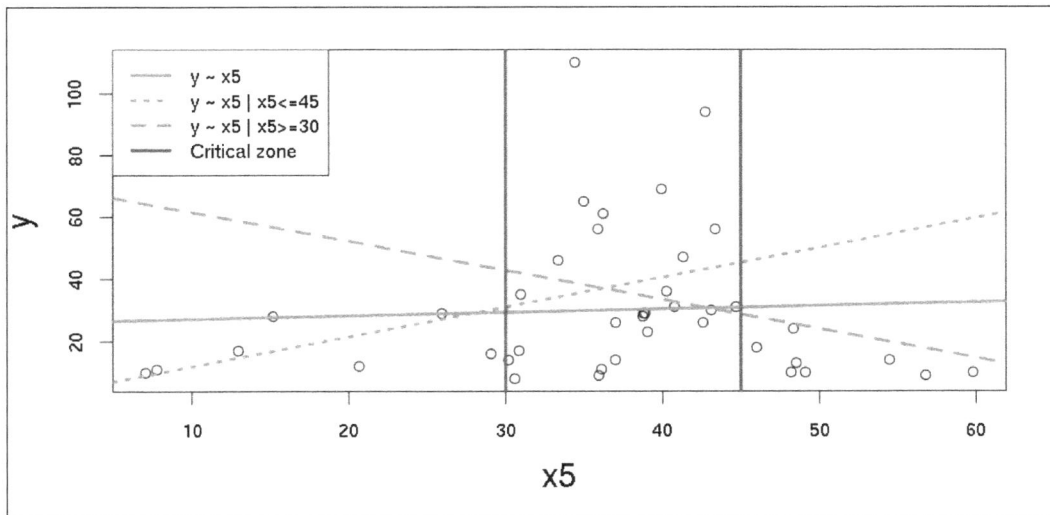

The cut-points 30 and 45 were more or less ad hoc. An advanced way to define optimal cut-points is to use a regression tree. There are various implementations of classification trees in R; a commonly used function is `rpart` from the package with the very same name. The regression tree follows an iterative process that splits the data into partitions, and then continues splitting each partition into smaller groups. In each step, the algorithm selects the best split on the continuous precipitation scale, where the best point minimizes the sum of the squared deviations from the group-level SO2 mean:

```
> library(partykit)
> library(rpart)
> plot(as.party(rpart(y ~ x5, data = usair)))
```

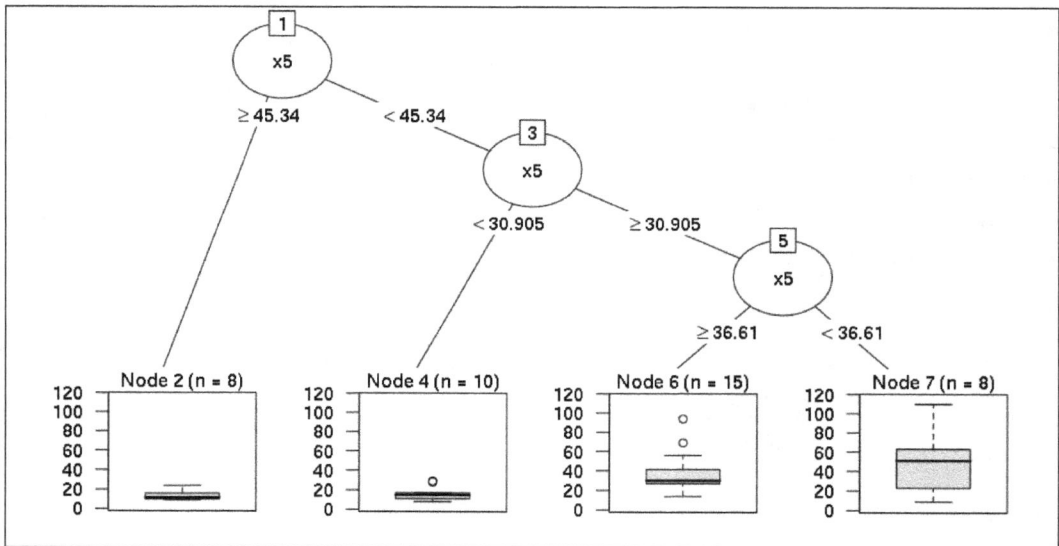

The interpretation of the preceding result is rather straightforward; if we are looking for two groups that differ highly regarding SO2, the optimal cut-point is a precipitation level of 45.34, and if we are looking for three groups, then we will have to split the second group by using the cut-point of 30.91, and so on. The four box-plots describe the SO2 distribution in the four partitions. So, these results confirm our previous assumption, and we have three precipitation groups that strongly differ in their level of SO2 concentration.

> Take a look at *Chapter 10, Classification and Clustering*, for more details and examples on decisions trees.

The following scatterplot also shows that the three groups differ heavily from each other. It seems that the SO2 concentration is highest in the middle group, and the two other groups are very similar:

```
> usair$x5_3 <- cut2(usair$x5, c(30, 45))
> plot(y ~ as.numeric(x5_3), data = usair, cex.lab = 1.5,
+    xlab = 'Categorized annual rainfall(x5)', xaxt = 'n')
> axis(1, at = 1:3, labels = levels(usair$x5_3))
> lines(tapply(usair$y, usair$x5_3, mean), col='red', lwd=2.5, lty=1)
> legend('topright', legend = 'Linear prediction', col = 'red')
```

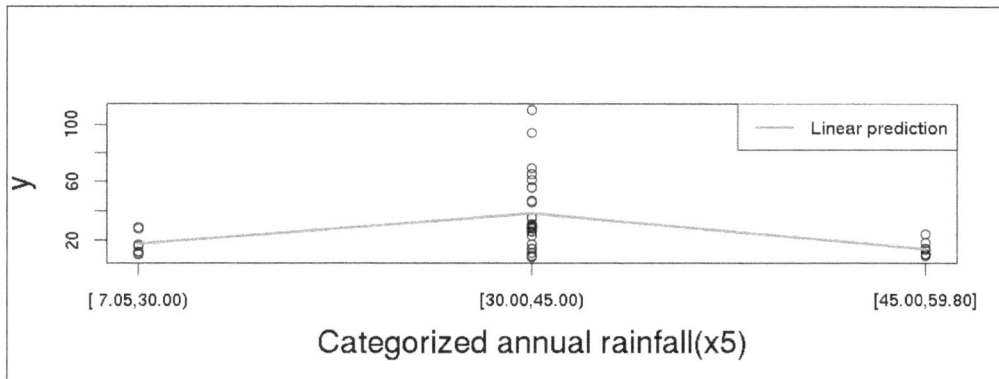

Now, let us refit our linear regression model by adding the three-category precipitation to the predictors. Technically, this goes by adding two dummy variables (learn more about this type of variable in *Chapter 10, Classification and Clustering*) pertaining to the second and third group, as shown in the table that follows:

	Dummy variables	
Categories	first	second
low (0-30)	0	0
middle (30-45)	1	0
high (45+)	0	1

In R, you can run this model using the `glm` (Generalized Linear Models) function, because the classic linear regression doesn't allow non-continuous predictors:

```
> summary(glmmodel.1 <- glm(y ~ x2 + x3 + x5_3, data = usair[-31, ]))
Deviance Residuals:
```

Min	1Q	Median	3Q	Max
-26.926	-4.780	1.543	5.481	31.280

Coefficients:

	Estimate	Std. Error	t value	Pr(>|t|)	
(Intercept)	14.07025	5.01682	2.805	0.00817	**
x2	0.05923	0.01210	4.897	2.19e-05	***
x3	-0.03459	0.01172	-2.952	0.00560	**
x5_3[30.00,45.00)	13.08279	5.10367	2.563	0.01482	*
x5_3[45.00,59.80]	0.09406	6.17024	0.015	0.98792	

Signif. codes: 0 '***' 0.001 '**' 0.01 '*' 0.05 '.' 0.1 ' ' 1

(Dispersion parameter for gaussian family taken to be 139.6349)

    Null deviance: 17845.9 on 39 degrees of freedom
Residual deviance: 4887.2 on 35 degrees of freedom
AIC: 317.74

Number of Fisher Scoring iterations: 2

The second group (wet days between 30 and 45) has a higher average by 15.2 units of SO2, as compared to the first group. This is controlled by the population size and number of manufacturers. The difference is statistically significant.

On the contrary, the third group shows only a slight difference when compared to the first group (0.04 unit lower), which is not significant. The three group mean shows a reversed U-shaped curve. Note that if you used precipitation in its original continuous form, implicitly you would assume a linear relation, so you wouldn't discover this shape. Another important thing to note is that the U-shaped curve here describes the partial association (controlled for x2 and x3), but the crude association, presented on the preceding scatterplot, showed a very similar picture.

The regression coefficients were interpreted as the difference between the group means, and both groups were compared to the omitted category (the first one). This is why the omitted category is usually referred to as the reference category. This way of entering discrete predictors is called reference-category coding. In general, if you have a discrete predictor with *n* categories, you have to define (*n-1*) dummies. Of course, if other contrasts are of interest, you can easily modify the model by entering dummies referring to other (*n-1*) categories.

> If you fit linear regression with discrete predictors, the regression slopes are the differences in the group means. If you also have other predictors, then the group-mean differences will be controlled for these predictors. Remember, the key feature of multivariate regression models is that they model partial two-way associations, holding the other predictors fixed.

You can go further by entering any other types and any number of predictors. If you have an ordinal predictor, it is your decision whether to enter it in its original form, assuming a linear relation, or to form dummies and enter each of them, allowing any type of relation. If you have no background knowledge on how to make this decision, you can try both solutions and compare how the models fit.

Summary

This chapter introduced the concept of how to build and interpret basic models, such as linear regression models. By now, you should be familiar with the motivation behind linear regression models; you should know how to control for confounders, how to enter discrete predictors, how to fit models in R, and how to interpret the results.

In the next chapter, we will extend this knowledge with generalized models, and analyzing the model fit.

6

Beyond the Linear Trend Line
(authored by Renata Nemeth and Gergely Toth)

Linear regression models, which we covered in the previous chapter, can handle continuous responses that have a linear association with the predictors. In this chapter, we will extend these models to allow the response variable to differ in distribution. But, before getting our hands dirty with the generalized linear models, we need to stop for a while and discuss regression models in general.

The modeling workflow

First, some words about the terminology. Statisticians call the Y variable the response, the outcome, or the dependent variable. The X variables are often called the predictors, the explanatory variables, or the independent variables. Some of the predictors are of our main interest, other predictors are added just because they are potential confounders. Continuous predictors are sometimes called covariates.

The GLM is a generalization of linear regression. GLM (also referred to as `glm` in R, from the `stats` package) allows the predictors to be related to the response variable via a link function, and by allowing the magnitude of the variance of each measurement to be a function of its predicted value.

Whatever regression model you use, the main question is, "in what form can we add continuous predictors to the model?" If the relationship between the response and the predictor does not meet the model assumptions, you can transform the variable in some way. For example, a logarithmic or quadratic transformation in a linear regression model is a very common way to solve the problem of non-linear relationships between the independent and dependent variables via linear formulas.

Or, you can transform the continuous predictor into a discrete one by subdividing its range in a proper way. When choosing the classes, one of the best options is to follow some convention, like choosing 18 as a cut-point in the case of age. Or you can follow a more technical way, for example, by categorizing the predictor into quantiles. An advanced way to go about this process would be to use some classification or regression trees, on which you will be able to read more in *Chapter 10, Classification and Clustering*.

Discrete predictors can be added to the model as dummy variables using reference category coding, as we have seen in the previous chapter for linear regression models.

But how do we actually build a model? We have compiled a general workflow to answer this question:

1. First, fit the model with the main predictors and all the relevant confounders, and then reduce the number of confounders by dropping out the non-significant ones. There are some automatic procedures (such as backward elimination) for this.

> The given sample size limits the number of predictors. A rule of thumb for the required sample size is that you should have at least 20 observations per predictor.

2. Decide whether to use the continuous variables in their original or categorized form.

3. Try to achieve a better fit by testing for non-linear relationships, if they are pragmatically relevant.

4. Finally, check the model assumptions.

And how do we find the best model? Is it as simple as the better the fit, the better the model? Unfortunately not. Our aim is to find the best fitting model, but with as few predictors as possible. A good model fit and a low number of independent variables are contradictory to each other.

As we have seen earlier, entering newer predictors into a linear regression model always increases the value of R-squared, and it may result in an over-fitted model. Overfitting means that the model describes the sample with its random noise, instead of the underlying data-generating process. Overfitting occurs, for example, when we have too many predictors in the model for its sample size to accommodate.

Consequently, the best model gives the desired level of fit with as few predictors as possible. AIC is one of those proper measures that takes into account both fit and parsimony. We highly recommend using it when comparing different models, which is very easy via the AIC function from the stats package.

Logistic regression

So far, we have discussed linear regression models, an appropriate method to model continuous response variables. However, non-continuous, binary responses (such as being ill or healthy, being faithful or deciding to switch to a new job, mobile supplier or partner) are also very common. The main difference compared to the continuous case is that now we should rather model probability instead of the expected value of the response variable.

The naive solution would be to use the probability as outcome in a linear model. But the problem with this solution is that the probability should be always between 0 and 1, and this bounded range is not guaranteed at all when using a linear model. A better solution is to fit a logistic regression model, which models not only the probability but also the natural logarithm of the odds, called the **logit**. The logit can be any (positive or negative) number, so the problem of limited range is eliminated.

Let's have a simple example of predicting the probability of the death penalty, using some information on the race of the defendant. This model relates to the much more complicated issue of racism in the infliction of the death penalty, a question with a long history in the USA. We will use the `deathpenalty` dataset from the `catdata` package about the judgment of defendants in cases of multiple murders in Florida between 1976 and 1987. The cases are classified with respect to the death penalty (where 0 refers to no, 1 to yes), the race of the defendant, and the race of the victim (black is referred as 0, white is 1).

First, we expand the frequency table into case form via the `expand.dtf` function from the `vcdExtra` package, then we fit our first generalized model in the dataset:

```
> library(catdata)
> data(deathpenalty)
> library(vcdExtra)
> deathpenalty.expand <- expand.dft(deathpenalty)
> binom.model.0 <- glm(DeathPenalty ~ DefendantRace,
+    data = deathpenalty.expand, family = binomial)
> summary(binom.model.0)
```

```
Deviance Residuals:
    Min       1Q   Median       3Q      Max
-0.4821  -0.4821  -0.4821  -0.4044   2.2558
```

```
Coefficients:
```

```
                Estimate Std. Error z value Pr(>|z|)
(Intercept)      -2.4624      0.2690   -9.155   <2e-16 ***
DefendantRace     0.3689      0.3058    1.206    0.228
---
Signif. codes:   0 '***' 0.001 '**' 0.01 '*' 0.05 '.' 0.1 ' ' 1


(Dispersion parameter for binomial family taken to be 1)

    Null deviance: 440.84  on 673   degrees of freedom
Residual deviance: 439.31  on 672   degrees of freedom
AIC: 443.31


Number of Fisher Scoring iterations: 5
```

The regression coefficient is statistically not significant, so at first sight, we can't see a racial bias in the data. Anyway, for didactic purposes, let's interpret the regression coefficient. It's 0.37, which means that the natural logarithm of the odds of getting a death penalty increases by 0.37 when moving from the black category to the white one. This difference is easily interpretable if you take its exponent, which is the ratio of the odds:

```
> exp(cbind(OR = coef(binom.model.0), confint(binom.model.0)))
                     OR        2.5 %      97.5 %
(Intercept)    0.08522727 0.04818273 0.1393442
DefendantRace  1.44620155 0.81342472 2.7198224
```

The odds ratio pertaining to the race of the defendant is 1.45, which means that white defendants have 45 percent larger odds of getting the death penalty than black defendants.

[Although R produces this, the odds ratio for the intercept is generally not interpreted.]

We can say something more general. We have seen that in linear regression models, the regression coefficient, b, can be interpreted as a one unit increase in X increases Y by b. But, in logistic regression models, a one unit increase in X multiplies the odds of Y by exp(b).

Please note that the preceding predictor was a discrete one, with values of 0 (black) and 1 (white), so it's basically a dummy variable for white, and black is the reference category. We have seen the same solution for entering discrete variables in the case of linear regression models. If you have more than two racial categories, you should define a second dummy for the third race and enter it into the model as well. The exponent of each dummy variables' coefficients equal to the odds ratio, which compares the given category to the reference. If you have a continuous predictor, the exponent of the coefficient equals to the odds ratio pertaining to a one unit increase in the predictor.

Now, let's enter the race of the victim into the examination, since it's a plausible confounder. Let's control for it, and fit the logistic regression model with both the `DefendantRace` and `VictimRace` as predictors:

```
> binom.model.1 <- update(binom.model.0, . ~ . + VictimRace)
> summary(binom.model.1)


Deviance Residuals:
    Min       1Q    Median        3Q       Max
-0.7283  -0.4899   -0.4899   -0.2326    2.6919


Coefficients:
              Estimate Std. Error z value Pr(>|z|)
(Intercept)    -3.5961     0.5069  -7.094 1.30e-12 ***
DefendantRace  -0.8678     0.3671  -2.364   0.0181 *
VictimRace      2.4044     0.6006   4.003 6.25e-05 ***
---
Signif. codes:  0 '***' 0.001 '**' 0.01 '*' 0.05 '.' 0.1 ' ' 1


(Dispersion parameter for binomial family taken to be 1)


    Null deviance: 440.84  on 673   degrees of freedom
Residual deviance: 418.96  on 671   degrees of freedom
AIC: 424.96


Number of Fisher Scoring iterations: 6


> exp(cbind(OR = coef(binom.model.1), confint(binom.model.1)))
```

	OR	2.5 %	97.5 %
(Intercept)	0.02743038	0.008433309	0.06489753
DefendantRace	0.41987565	0.209436976	0.89221877
VictimRace	11.07226549	3.694532608	41.16558028

When controlling for VictimRace, the effect of DefendantRace becomes significant! The odds ratio is 0.42, which means that white defendants' odds of getting the death penalty are only 42 percent of the odds of black defendants, holding the race of the victim fixed. Also, the odds ratio of VictimRace (11.07) shows an extremely strong effect: killers of white victims are 11 times more likely to get a death penalty than killers of black victims.

So, the effect of DefendantRace is exactly the opposite of what we have got in the one-predictor model. The reversed association may seem to be paradoxical, but it can be explained. Let's have a look at the following output:

```
> prop.table(table(factor(deathpenalty.expand$VictimRace,
+              labels = c("VictimRace=0", "VictimRace=1")),
+          factor(deathpenalty.expand$DefendantRace,
+              labels = c("DefendantRace=0", "DefendantRace=1"))), 1)
```

	DefendantRace=0	DefendantRace=1
VictimRace=0	0.89937107	0.10062893
VictimRace=1	0.09320388	0.90679612

The data seems to be homogeneous in some sense: black defendants are more likely to have black victims, and vice versa. If you put these pieces of information together, you start to see that black defendants yield a smaller proportion of death sentences just because they are more likely to have black victims, and those who have black victims are less likely to get a death penalty. The paradox disappears: the crude death penalty and DefendantRace association was confounded by VictimRace.

To sum it up, it seems that taking the available information into account, you can come to the following conclusions:

- Black defendants are more likely to get the death penalty
- Killing a white person is considered to be a more serious crime than killing a black person

Of course, you should draw such conclusions extremely carefully, as the question of racial bias needs a very thorough analysis using all the relevant information regarding the circumstances of the crime, and much more.

Data considerations

Logistic regression models work on the assumption that the observations are totally independent from each other. This assumption is violated, for example, if your observations are consecutive years. The deviance residuals and other diagnostic statistics can help validate the model and detect problems such as the misspecification of the link function. For further reference, see the `LogisticDx` package.

As a general rule of thumb, logistic regression models require at least 10 events per predictors, where an event denotes the observations belonging to the less frequent category in the response. In our death penalty example, death is the less frequent category in the response, and we have 68 death sentences in the database. So, the rule suggests that a maximum of 6-7 predictors are allowed.

The regression coefficients are estimated using the maximum likelihood method. Since there is no closed mathematical form to get these ML estimations, R uses an optimization algorithm instead. In some cases, you may get an error message that the algorithm doesn't reach convergence. In such cases, it is unable to find an appropriate solution. This may occur for a number of reasons, such as having too many predictors, too few events, and so on.

Goodness of model fit

One measure of model fit, to evaluate the performance of the model, is the significance of the overall model. The corresponding likelihood ratio tests whether the given model fits significantly better than a model with just an intercept, which we call the null model.

To obtain the test results, you have to look at the residual deviance in the output. It measures the disagreement between the maxima of the observed and the fitted log likelihood functions.

> Since logistic regression follows the maximal likelihood principle, the goal is to minimize the sum of the deviance residuals. Therefore, this residual is parallel to the raw residual in linear regression, where the goal is to minimize the sum of squared residuals.

The null deviance represents how well the response is predicted by a model with nothing but an intercept. To judge the model, you have to compare the residual deviance to the null deviance; the difference follows a chi-square distribution. The corresponding test is available in the `lmtest` package:

```
> library(lmtest)
> lrtest(binom.model.1)
```

```
Likelihood ratio test

Model 1: DeathPenalty ~ DefendantRace + VictimRace
Model 2: DeathPenalty ~ 1
  #Df  LogLik Df  Chisq Pr(>Chisq)
1   3 -209.48
2   1 -220.42 -2 21.886  1.768e-05 ***
---
Signif. codes:  0 '***' 0.001 '**' 0.01 '*' 0.05 '.' 0.1 ' ' 1
```

The *p* value indicates a highly significant decrease in deviance. This means that the model is significant, and the predictors have a significant effect on the response probability.

You can think of the likelihood ratio as the F-test in the linear regression models. It reveals if the model is significant, but it doesn't tell anything about the goodness-of-fit, which was described by the adjusted R-squared measure in the linear case.

An equivalent statistic for logistic regression models does not exist, but several pseudo R-squared have been developed. These usually range from 0 to 1 with higher values indicating a better fit. We will use the PseudoR2 function from the BaylorEdPsych package to compute this value:

```
> library(BaylorEdPsych)
> PseudoR2(binom.model.1)
         McFadden      Adj.McFadden         Cox.Snell         Nagelkerke
       0.04964600        0.03149893        0.03195036         0.06655297
  McKelvey.Zavoina            Effron             Count          Adj.Count
       0.15176608        0.02918095                NA                 NA
               AIC     Corrected.AIC
      424.95652677      424.99234766
```

But be careful, the pseudo R-squared cannot be interpreted as an OLS R-squared, and there are some documented problems with them as well, but they give us a rough picture. In our case, they say that the explanative power of the model is rather low, which is not surprising if we consider the fact that only two predictors were used in the modeling of such a complex process as judging a crime.

Model comparison

As we have seen in the previous chapter, the adjusted R-squared provides a good base for model comparison when dealing with nested linear regression models. For nested logistic regression models, you can use the likelihood ratio test (such as the `lrtest` function from the `lmtest` library), which compares the difference between the residual deviances.

```
> lrtest(binom.model.0, binom.model.1)
Likelihood ratio test

Model 1: DeathPenalty ~ DefendantRace
Model 2: DeathPenalty ~ DefendantRace + VictimRace
  #Df  LogLik Df Chisq Pr(>Chisq)
1   2 -219.65
2   3 -209.48  1 20.35   6.45e-06 ***
---
Signif. codes:  0 '***' 0.001 '**' 0.01 '*' 0.05 '.' 0.1 ' ' 1
```

> LogLiK, in the preceding output denotes the log-likelihood of the model; you got the residual deviance by multiplying it by 2.

For un-nested models, you can use AIC, just like we did in the case of linear regression models, but in logistic regression models, AIC is part of the standard output, so there is no need to call the AIC function separately. Here, the `binom.model.1` has a lower AIC than `binom.model.0`, and the difference is not negligible since it is greater than 2.

Models for count data

Logistic regression can handle only binary responses. If you have count data, such as the number of deaths or failures in a given period of time, or in a given geographical area, you can use Poisson or negative binomial regression. These data types are particularly common when working with aggregated data, which is provided as a number of events classified in different categories.

Poisson regression

Poisson regression models are generalized linear models with the logarithm as the link function, and they assume that the response has a **Poisson distribution**. The Poisson distribution takes only integer values. It is appropriate for count data, such as events occurring over a fixed period of time, that is, if the events are rather rare, such as a number of hard drive failures per day.

In the following example, we will use the Hard Drive Data Sets for the year of 2013. The dataset was downloaded from https://docs.backblaze.com/public/hard-drive-data/2013_data.zip, but we polished and simplified it a bit. Each record in the original database corresponds to a daily snapshot of one drive. The failure variable, our main point of interest, can be either zero (if the drive is OK), or one (on the last day of the hard drive before failing).

Let's try to determine which factors affect the appearance of a failure. The potential predictive factors are the following:

- model: The manufacturer-assigned model number of the drive
- capacity_bytes: The drive capacity in bytes
- age_month: The drive age in the average month
- temperature: The hard disk drive temperature
- PendingSector: A logical value indicating the occurrence of unstable sectors (waiting for remapping on the given hard drive, on the given day)

We aggregated the original dataset by these variables, where the freq variable denotes the number of records in the given category. It's time to load this final, cleansed, and aggregated dataset:

```
> dfa <- readRDS('SMART_2013.RData')
```

Take a quick look at the number of failures by model:

```
> (ct <- xtabs(~model+failure, data=dfa))
```

| | failure | | | | | | |
model	0	1	2	3	4	5	8
HGST	136	1	0	0	0	0	0
Hitachi	2772	72	6	0	0	0	0
SAMSUNG	125	0	0	0	0	0	0

ST1500DL001	38	0	0	0	0	0	0
ST1500DL003	213	39	6	0	0	0	0
ST1500DM003	84	0	0	0	0	0	0
ST2000DL001	51	4	0	0	0	0	0
ST2000DL003	40	7	0	0	0	0	0
ST2000DM001	98	0	0	0	0	0	0
ST2000VN000	40	0	0	0	0	0	0
ST3000DM001	771	122	34	14	4	2	1
ST31500341AS	1058	75	8	0	0	0	0
ST31500541AS	1010	106	7	1	0	0	0
ST32000542AS	803	12	1	0	0	0	0
ST320005XXXX	209	1	0	0	0	0	0
ST33000651AS	323	12	0	0	0	0	0
ST4000DM000	242	22	10	2	0	0	0
ST4000DX000	197	1	0	0	0	0	0
TOSHIBA	126	2	0	0	0	0	0
WDC	1874	27	1	2	0	0	0

Now, let's get rid of those hard-drive models that didn't have any failure, by removing all rows from the preceding table where there are only zeros beside the first column:

```
> dfa <- dfa[dfa$model %in% names(which(rowSums(ct) - ct[, 1] > 0)),]
```

To get a quick overview on the number of failures, let's plot a histogram on a log scale by model numbers, with the help of the ggplot2 package:

```
> library(ggplot2)
> ggplot(rbind(dfa, data.frame(model='All', dfa[, -1] )),
+    aes(failure)) + ylab("log(count)") +
+    geom_histogram(binwidth = 1, drop=TRUE, origin = -0.5)   +
+    scale_y_log10() + scale_x_continuous(breaks=c(0:10)) +
```

```
+    facet_wrap( ~ model, ncol = 3) +
+    ggtitle("Histograms by manufacturer") + theme_bw()
```

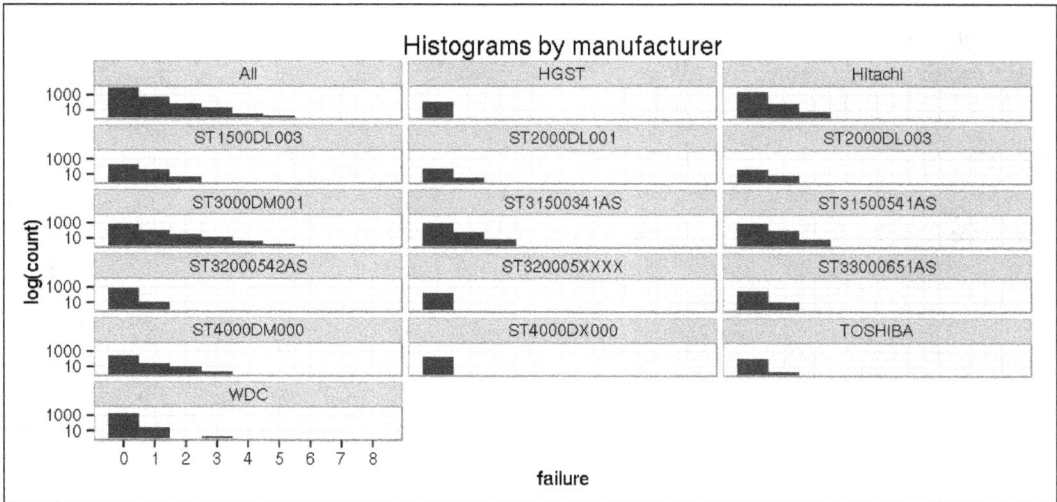

Histograms by manufacturer

Now, it's time to fit a Poisson regression model to the data, using the `model` number as the predictor. The model can be fitted using the `glm` function with the option, `family=poisson`. By default, the expected log count is modeled, so we use the `log` link.

In the database, each observation corresponds to a group with a varying number of hard drives. As we need to handle the different group sizes, we will use the `offset` function:

```
> poiss.base <- glm(failure ~ model, offset(log(freq)),
+    family = 'poisson', data = dfa)
> summary(poiss.base)
```

```
Deviance Residuals:
    Min       1Q    Median       3Q       Max
-2.7337   -0.8052   -0.5160   -0.3291   16.3495
```

Coefficients:

	Estimate	Std. Error	z value	Pr(>|z|)	
(Intercept)	-5.0594	0.5422	-9.331	< 2e-16	***
modelHitachi	1.7666	0.5442	3.246	0.00117	**

```
modelST1500DL003      3.6563        0.5464       6.692 2.20e-11 ***
modelST2000DL001      2.5592        0.6371       4.017 5.90e-05 ***
modelST2000DL003      3.1390        0.6056       5.183 2.18e-07 ***
modelST3000DM001      4.1550        0.5427       7.656 1.92e-14 ***
modelST31500341AS     2.7445        0.5445       5.040 4.65e-07 ***
modelST31500541AS     3.0934        0.5436       5.690 1.27e-08 ***
modelST32000542AS     1.2749        0.5570       2.289  0.02208 *
modelST320005XXXX    -0.4437        0.8988      -0.494  0.62156
modelST33000651AS     1.9533        0.5585       3.497  0.00047 ***
modelST4000DM000      3.8219        0.5448       7.016 2.29e-12 ***
modelST4000DX000    -12.2432      117.6007      -0.104  0.91708
modelTOSHIBA          0.2304        0.7633       0.302  0.76279
modelWDC              1.3096        0.5480       2.390  0.01686 *
---
Signif. codes:  0 '***' 0.001 '**' 0.01 '*' 0.05 '.' 0.1 ' ' 1


(Dispersion parameter for poisson family taken to be 1)


    Null deviance: 22397  on 9858  degrees of freedom
Residual deviance: 17622  on 9844  degrees of freedom
AIC: 24717


Number of Fisher Scoring iterations: 15
```

First, let's interpret the coefficients. The model number is a discrete predictor, so we entered a number of dummy variables to represent it is as a predictor. The reference category is not present in the output by default, but we can query it at any time:

```
> contrasts(dfa$model, sparse = TRUE)
HGST         . . . . . . . . . . . . . .
Hitachi      1 . . . . . . . . . . . . .
ST1500DL003  . 1 . . . . . . . . . . . .
ST2000DL001  . . 1 . . . . . . . . . . .
ST2000DL003  . . . 1 . . . . . . . . . .
ST3000DM001  . . . . 1 . . . . . . . . .
ST31500341AS . . . . . 1 . . . . . . . .
ST31500541AS . . . . . . 1 . . . . . . .
```

```
ST32000542AS  . . . . . . . . 1 . . . . . . .
ST320005XXXX  . . . . . . . . . 1 . . . . . .
ST33000651AS  . . . . . . . . . . 1 . . . . .
ST4000DM000   . . . . . . . . . . . 1 . . . .
ST4000DX000   . . . . . . . . . . . . 1 . .
TOSHIBA       . . . . . . . . . . . . . 1 .
WDC           . . . . . . . . . . . . . . 1
```

So, it turns out that the reference category is HGST, and the dummy variables compare each model with the HGST hard drive. For example, the coefficient of Hitachi is 1.77, so the expected log-count for Hitachi drives is about 1.77 greater than those for HGST drives. Or, you can compute its exponent when speaking about ratios instead of differences:

```
> exp(1.7666)
[1] 5.850926
```

So, the expected number of failures for Hitachi drives is 5.85 times greater than for HGST drives. In general, the interpretation goes as: a one unit increase in X multiplies Y by exp(b).

Similar to logistic regression, let's determine the significance of the model. To do this, we compare the present model to the null model without any predictors, so the difference between the residual deviance and the null deviance can be identified. We expect the difference to be large enough, and the corresponding chi-squared test to be significant:

```
> lrtest(poiss.base)
Likelihood ratio test

Model 1: failure ~ model
Model 2: failure ~ 1
  #Df LogLik  Df  Chisq Pr(>Chisq)
1  15 -12344
2   1 -14732 -14 4775.8  < 2.2e-16 ***
---
Signif. codes:  0 '***' 0.001 '**' 0.01 '*' 0.05 '.' 0.1 ' ' 1
```

And it seems that the model is significant, but we should also try to determine whether any of the model assumptions might fail.

Just like we did with the linear and logistic regression models, we have an independence assumption, where Poisson regression assumes the events to be independent. This means that the occurrence of one failure will not make another more or less likely. In the case of drive failures, this assumption holds. Another important assumption comes from the fact that the response has a Poisson distribution with an equal mean and variance. Our model assumes that the variance and the mean, conditioned on the predictor variables, will be approximately equal.

To decide whether the assumption holds, we can compare the residual deviance to its degree of freedom. For a well-fitting model, their ratio should be close to one. Unfortunately, the reported residual deviance is `17622` on `9844` degrees of freedom, so their ratio is much above one, which suggests that the variance is much greater than the mean. This phenomenon is called **overdispersion**.

Negative binomial regression

In such a case, a negative binomial distribution can be used to model an over-dispersed count response, which is a generalization of the Poisson regression since it has an extra parameter to model the over-dispersion. In other words, Poisson and the negative binomial models are nested models; the former is a subset of the latter one.

In the following output, we use the `glm.nb` function from the `MASS` package to fit a negative binomial regression to our drive failure data:

```
> library(MASS)
> model.negbin.0 <- glm.nb(failure ~ model,
+   offset(log(freq)), data = dfa)
```

To compare this model's performance to the Poisson model, we can use the likelihood ratio test, since the two models are nested. The negative binomial model shows a significantly better fit:

```
> lrtest(poiss.base,model.negbin.0)
Likelihood ratio test

Model 1: failure ~ model
Model 2: failure ~ model
  #Df LogLik Df Chisq Pr(>Chisq)
1  15 -12344
2  16 -11950  1 787.8  < 2.2e-16 ***
---
Signif. codes:  0 '***' 0.001 '**' 0.01 '*' 0.05 '.' 0.1 ' ' 1
```

This result clearly suggests choosing the negative binomial model.

Multivariate non-linear models

So far, the only predictor in our model was the model name, but we have other potentially important information about the drives as well, such as capacity, age, and temperature. Now let's add these to the model, and determine whether the new model is better than the original one.

Furthermore, let's check the importance of `PendingSector` as well. In short, we define a two-step model building procedure with the nested models; hence we can use likelihood ratio statistics to test whether the model fit has significantly increased in both steps:

```
> model.negbin.1 <- update(model.negbin.0, . ~ . + capacity_bytes +
+    age_month + temperature)
> model.negbin.2 <- update(model.negbin.1, . ~ . + PendingSector)
> lrtest(model.negbin.0, model.negbin.1, model.negbin.2)
Likelihood ratio test


Model 1: failure ~ model
Model 2: failure ~ model + capacity_bytes + age_month + temperature
Model 3: failure ~ model + capacity_bytes + age_month + temperature +
    PendingSector
  #Df LogLik Df  Chisq Pr(>Chisq)
1  16 -11950
2  19 -11510  3 878.91  < 2.2e-16 ***
3  20 -11497  1  26.84  2.211e-07 ***
---
Signif. codes:  0 '***' 0.001 '**' 0.01 '*' 0.05 '.' 0.1 ' ' 1
```

Both of these steps are significant, so it was worth adding each predictor to the model. Now, let's interpret the best model:

```
> summary(model.negbin.2)


Deviance Residuals:
    Min      1Q    Median      3Q      Max
-2.7147  -0.7580  -0.4519  -0.2187   9.4018
```

```
Coefficients:
                  Estimate Std. Error z value Pr(>|z|)
(Intercept)      -8.209e+00  6.064e-01 -13.537  < 2e-16 ***
modelHitachi      2.372e+00  5.480e-01   4.328 1.50e-05 ***
modelST1500DL003  6.132e+00  5.677e-01  10.801  < 2e-16 ***
modelST2000DL001  4.783e+00  6.587e-01   7.262 3.81e-13 ***
modelST2000DL003  5.313e+00  6.296e-01   8.440  < 2e-16 ***
modelST3000DM001  4.746e+00  5.470e-01   8.677  < 2e-16 ***
modelST31500341AS 3.849e+00  5.603e-01   6.869 6.49e-12 ***
modelST31500541AS 4.135e+00  5.598e-01   7.387 1.50e-13 ***
modelST32000542AS 2.403e+00  5.676e-01   4.234 2.29e-05 ***
modelST320005XXXX 1.377e-01  9.072e-01   0.152   0.8794
modelST33000651AS 2.470e+00  5.631e-01   4.387 1.15e-05 ***
modelST4000DM000  3.792e+00  5.471e-01   6.931 4.17e-12 ***
modelST4000DX000 -2.039e+01  8.138e+03  -0.003   0.9980
modelTOSHIBA      1.368e+00  7.687e-01   1.780   0.0751 .
modelWDC          2.228e+00  5.563e-01   4.006 6.19e-05 ***
capacity_bytes    1.053e-12  5.807e-14  18.126  < 2e-16 ***
age_month         4.815e-02  2.212e-03  21.767  < 2e-16 ***
temperature      -5.427e-02  3.873e-03 -14.012  < 2e-16 ***
PendingSectoryes  2.240e-01  4.253e-02   5.267 1.39e-07 ***
---
Signif. codes:  0 '***' 0.001 '**' 0.01 '*' 0.05 '.' 0.1 ' ' 1


(Dispersion parameter for Negative Binomial(0.8045) family taken to be 1)


    Null deviance: 17587  on 9858  degrees of freedom
Residual deviance: 12525  on 9840  degrees of freedom
AIC: 23034


Number of Fisher Scoring iterations: 1


          Theta:  0.8045
       Std. Err.:  0.0525


 2 x log-likelihood:  -22993.8850.
```

Each predictor is significant—with a few exceptions of some contrast in model type. For example, Toshiba doesn't differ significantly from the reference category, HGST, when controlling for age, temperature, and so on.

The interpretation of the negative binomial regression parameters is similar to the Poisson model. For example, the coefficient of age_month is 0.048, which shows that a one month increase in age, increases the expected log-count of failures by 0.048. Or, you can opt for using exponentials as well:

```
> exp(data.frame(exp_coef = coef(model.negbin.2)))
                      exp_coef
(Intercept)        2.720600e-04
modelHitachi       1.071430e+01
modelST1500DL003   4.602985e+02
modelST2000DL001   1.194937e+02
modelST2000DL003   2.030135e+02
modelST3000DM001   1.151628e+02
modelST31500341AS  4.692712e+01
modelST31500541AS  6.252061e+01
modelST32000542AS  1.106071e+01
modelST320005XXXX  1.147622e+00
modelST33000651AS  1.182098e+01
modelST4000DM000   4.436067e+01
modelST4000DX000   1.388577e-09
modelTOSHIBA       3.928209e+00
modelWDC           9.283970e+00
capacity_bytes     1.000000e+00
age_month          1.049329e+00
temperature        9.471743e-01
PendingSectoryes   1.251115e+00
```

So, it seems that one month in a lifetime increases the expected number of failures by 4.9 percent, and a larger capacity also increases the number of failures. On the other hand, temperature shows a reversed effect: the exponent of the coefficient is 0.947, which says that one degree of increased warmth decreases the expected number of failures by 5.3 percent.

The effect of the model name can be judged on the basis of comparison to the reference category, which is HGST in our case. One may want to change this reference. For example, for the most common drive: WDC. This can be easily done by changing the order of the factor levels in hard drive models, or simply defining the reference category in the factor via the extremely useful relevel function:

```
> dfa$model <- relevel(dfa$model, 'WDC')
```

Now, let's verify if HGST indeed replaced WDC in the coefficients list, but instead of the lengthy output of summary, we will use the tidy function from the broom package, which can extract the most important features (for the model summary, take a look at the glance function) of different statistical models:

```
> model.negbin.3 <- update(model.negbin.2, data = dfa)
> library(broom)
> format(tidy(model.negbin.3), digits = 4)
```

	term	estimate	std.error	statistic	p.value
1	(Intercept)	-5.981e+00	2.173e-01	-27.52222	9.519e-167
2	modelHGST	-2.228e+00	5.563e-01	-4.00558	6.187e-05
3	modelHitachi	1.433e-01	1.009e-01	1.41945	1.558e-01
4	modelST1500DL003	3.904e+00	1.353e-01	28.84295	6.212e-183
5	modelST2000DL001	2.555e+00	3.663e-01	6.97524	3.054e-12
6	modelST2000DL003	3.085e+00	3.108e-01	9.92496	3.242e-23
7	modelST3000DM001	2.518e+00	9.351e-02	26.92818	1.028e-159
8	modelST31500341AS	1.620e+00	1.069e-01	15.16126	6.383e-52
9	modelST31500541AS	1.907e+00	1.016e-01	18.77560	1.196e-78
10	modelST32000542AS	1.751e-01	1.533e-01	1.14260	2.532e-01
11	modelST320005XXXX	-2.091e+00	7.243e-01	-2.88627	3.898e-03
12	modelST33000651AS	2.416e-01	1.652e-01	1.46245	1.436e-01
13	modelST4000DM000	1.564e+00	1.320e-01	11.84645	2.245e-32
14	modelST4000DX000	-1.862e+01	1.101e+03	-0.01691	9.865e-01
15	modelTOSHIBA	-8.601e-01	5.483e-01	-1.56881	1.167e-01
16	capacity_bytes	1.053e-12	5.807e-14	18.12597	1.988e-73
17	age_month	4.815e-02	2.212e-03	21.76714	4.754e-105
18	temperature	-5.427e-02	3.873e-03	-14.01175	1.321e-44
19	PendingSectoryes	2.240e-01	4.253e-02	5.26709	1.386e-07

> Use the broom package to extract model coefficients, compare model fit, and other metrics to be passed to, for example, ggplot2.

The effect of temperature suggests that the higher the temperature, the lower the number of hard drive failures. However, everyday experiences show a very different picture, for example, as described at `https://www.backblaze.com/blog/hard-drive-temperature-does-it-matter`. Google engineers found that temperature was not a good predictor of failure, while Microsoft and the University of Virginia found that it had a significant effect. Disk drive manufacturers suggest keeping disks at cooler temperatures.

So, let's take a closer look at this interesting question, and we will have the `temperature` as a predictor of drive failure. First, let's classify temperature into six equal categories, and then we will draw a bar plot presenting the mean number of failures per categories. Note that we have to take into account the different groups' sizes, so we will weight by `freq`, and as we are doing some data aggregation, it's the right time to convert our dataset into a `data.table` object:

```
> library(data.table)
> dfa <- data.table(dfa)
> dfa[, temp6 := cut2(temperature, g = 6)]
> temperature.weighted.mean <- dfa[, .(wfailure =
+     weighted.mean(failure, freq)), by = temp6]
> ggplot(temperature.weighted.mean, aes(x = temp6, y = wfailure)) +
+     geom_bar(stat = 'identity') + xlab('Categorized temperature') +
+     ylab('Weighted mean of disk faults') + theme_bw()
```

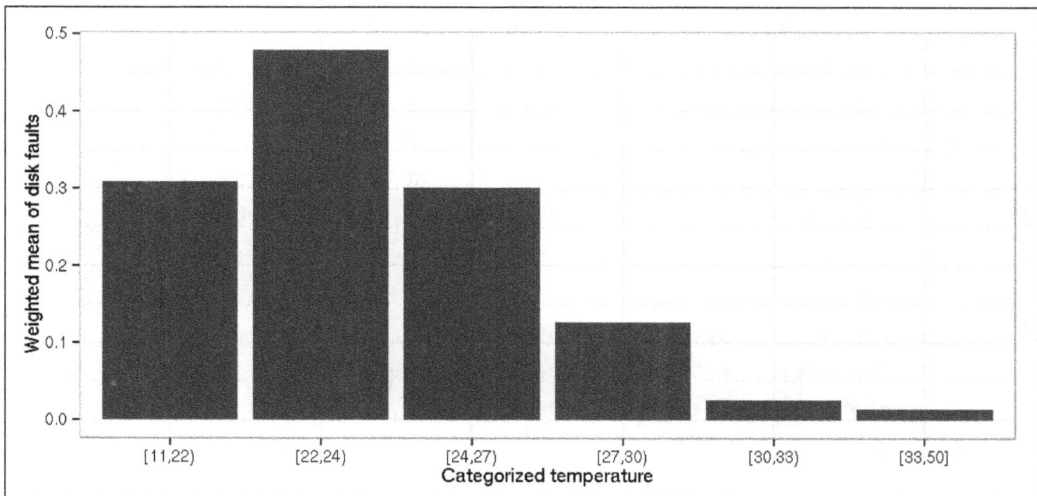

The assumption of linear relation is clearly not supported. The bar plot suggests using the temperature in this classified form, instead of the original continuous variable when entering the model. To actually see which model is better, let's compare those! Since they are not nested, we have to use the AIC, which strongly supports the categorized version:

```
> model.negbin.4 <- update(model.negbin.0, .~. + capacity_bytes +
+   age_month + temp6 + PendingSector, data = dfa)
> AIC(model.negbin.3,model.negbin.4)
                df     AIC
model.negbin.3 20 23033.88
model.negbin.4 24 22282.47
```

Well, it was really worth categorizing temperature! Now, let's check the other two continuous predictors as well. Again, we will use `freq` as a weighting factor:

```
> weighted.means <- rbind(
+     dfa[, .(l = 'capacity', f = weighted.mean(failure, freq)),
+         by = .(v = capacity_bytes)],
+     dfa[, .(l = 'age', f = weighted.mean(failure, freq)),
+         by = .(v = age_month)])
```

As in the previous plots, we will use `ggplot2` to plot the distribution of these discrete variables, but instead of a bar plot, we will use a stair-line chart to overcome the issue of the fixed width of bar charts:

```
> ggplot(weighted.means, aes(x = l, y = f)) + geom_step() +
+     facet_grid(. ~ v, scales = 'free_x') + theme_bw() +
+     ylab('Weighted mean of disk faults') + xlab('')
```

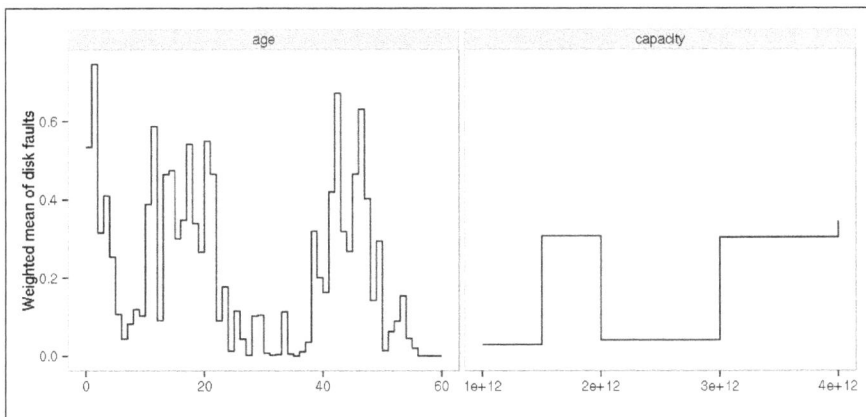

The relations are again, clearly not linear. The case of age is particularly interesting; there seems to be highly risky periods in the hard drives' lifetime. Now, let's force R to use capacity as a nominal variable (it has only five values, so there is no real need to categorize it), and let's classify age into 8 equally sized categories:

```
> dfa[, capacity_bytes := as.factor(capacity_bytes)]
> dfa[, age8 := cut2(age_month, g = 8)]
> model.negbin.5 <- update(model.negbin.0, .~. + capacity_bytes +
+    age8 + temp6 + PendingSector, data = dfa)
```

According to the AIC, the last model with the categorized age and capacity is much better, and is the best fitting model so far:

```
> AIC(model.negbin.5, model.negbin.4)
                df      AIC
model.negbin.5 33 22079.47
model.negbin.4 24 22282.47
```

If you look at the parameter estimates, you can see that the first dummy variable on capacity significantly differ from the reference:

```
> format(tidy(model.negbin.5), digits = 3)
```

	term	estimate	std.error	statistic	p.value
1	(Intercept)	-6.1648	1.84e-01	-3.34e+01	2.69e-245
2	modelHGST	-2.4747	5.63e-01	-4.40e+00	1.10e-05
3	modelHitachi	-0.1119	1.21e-01	-9.25e-01	3.55e-01
4	modelST1500DL003	31.7680	7.05e+05	4.51e-05	1.00e+00
5	modelST2000DL001	1.5216	3.81e-01	3.99e+00	6.47e-05
6	modelST2000DL003	2.1055	3.28e-01	6.43e+00	1.29e-10
7	modelST3000DM001	2.4799	9.54e-02	2.60e+01	5.40e-149
8	modelST31500341AS	29.4626	7.05e+05	4.18e-05	1.00e+00
9	modelST31500541AS	29.7597	7.05e+05	4.22e-05	1.00e+00
10	modelST32000542AS	-0.5419	1.93e-01	-2.81e+00	5.02e-03
11	modelST320005XXXX	-2.8404	7.33e-01	-3.88e+00	1.07e-04
12	modelST33000651AS	0.0518	1.66e-01	3.11e-01	7.56e-01
13	modelST4000DM000	1.2243	1.62e-01	7.54e+00	4.72e-14
14	modelST4000DX000	-29.6729	2.55e+05	-1.16e-04	1.00e+00
15	modelTOSHIBA	-1.1658	5.48e-01	-2.13e+00	3.33e-02
16	capacity_bytes1500301910016	-27.1391	7.05e+05	-3.85e-05	1.00e+00

17 capacity_bytes2000398934016	1.8165	2.08e-01	8.73e+00	2.65e-18
18 capacity_bytes3000592982016	2.3515	1.88e-01	1.25e+01	8.14e-36
19 capacity_bytes4000787030016	3.6023	2.25e-01	1.60e+01	6.29e-58
20 age8[5, 9)	-0.5417	7.55e-02	-7.18e+00	7.15e-13
21 age8[9,14)	-0.0683	7.48e-02	-9.12e-01	3.62e-01
22 age8[14,19)	0.3499	7.24e-02	4.83e+00	1.34e-06
23 age8[19,25)	0.7383	7.33e-02	1.01e+01	7.22e-24
24 age8[25,33)	0.5896	1.14e-01	5.18e+00	2.27e-07
25 age8[33,43)	1.5698	1.05e-01	1.49e+01	1.61e-50
26 age8[43,60]	1.9105	1.06e-01	1.81e+01	3.59e-73
27 temp6[22,24)	0.7582	5.01e-02	1.51e+01	8.37e-52
28 temp6[24,27)	0.5005	4.78e-02	1.05e+01	1.28e-25
29 temp6[27,30)	0.0883	5.40e-02	1.64e+00	1.02e-01
30 temp6[30,33)	-1.0627	9.20e-02	-1.15e+01	7.49e-31
31 temp6[33,50]	-1.5259	1.37e-01	-1.11e+01	1.23e-28
32 PendingSectoryes	0.1301	4.12e-02	3.16e+00	1.58e-03

The next three capacities are more likely to cause failures, but the trend is not linear. The effect of age also does not seem to be linear. In general, aging increases the number of failures, but there are some exceptions. For example, drives are significantly more likely to have a failure in the first (reference) age group than in the second one. This finding is plausible since drives have a higher failure rate at the beginning of their operation. The effect of temperature suggests that the middle temperature (22-30 degrees Celsius) is more likely to cause failures than low or high temperatures. Remember that each effect is controlled for every other predictor.

It would also be important to judge the effect-size of different predictors, comparing them to each other. As a picture is worth a thousand words, let's summarize the coefficients with the confidence intervals in one plot.

First, we have to extract the significant terms from the model:

```
> tmnb5 <- tidy(model.negbin.5)
> str(terms <- tmnb5$term[tmnb5$p.value < 0.05][-1])
 chr [1:22] "modelHGST" "modelST2000DL001" "modelST2000DL003" ...
```

Then, let's identify the confidence intervals of the coefficients using the confint function and the good old plyr package:

```
> library(plyr)
> ci <- ldply(terms, function(t) confint(model.negbin.5, t))
```

Unfortunately, this resulting data frame is not yet complete. We need to add the term names, and also, let's extract the grouping variables via a simple, regular expression:

```
> names(ci) <- c('min', 'max')
> ci$term <- terms
> ci$variable <- sub('[A-Z0-9\\]\\[,() ]*$', '', terms, perl = TRUE)
```

And now we have the confidence intervals of the coefficients in a nicely formatted dataset, which can be easily plotted by `ggplot`:

```
> ggplot(ci, aes(x = factor(term), color = variable)) +
+      geom_errorbar(ymin = min, ymax = max) + xlab('') +
+      ylab('Coefficients (95% conf.int)') + theme_bw() +
+      theme(axis.text.x = element_text(angle = 90, hjust = 1),
+          legend.position = 'top')
```

It can be easily seen that although each predictor is significant, the size of their effects strongly differ. For example, `PendingSector` has just a slight effect on the number of failures, but `age`, `capacity`, and `temperature` have a much stronger effect, and the hard drive model is the predictor that best differentiates the number of failures.

As we have mentioned in the *Logistic regression* section, different pseudo R-squared measures are available for nonlinear models as well. We again warn you to use these metrics with reservation. Anyway, in our case, they uniformly suggest the model's explanative power to be pretty good:

```
> PseudoR2(model.negbin.6 )
       McFadden      Adj.McFadden       Cox.Snell       Nagelkerke
      0.3352654         0.3318286       0.4606953        0.5474952
McKelvey.Zavoina            Effron           Count        Adj.Count
             NA         0.1497521       0.9310444       -0.1943522
            AIC     Corrected.AIC
  12829.5012999     12829.7044941
```

Summary

This chapter introduced three well known nonlinear regression models: the logistic, Poisson, and negative binomial models, and you became familiar with the general logic of modeling. It was also shown how the same concepts, such as effect of predictors, goodness of fit, explanative power, model comparison for nested and non-nested models, and model building are applied in different contexts. Now, having spent some time on mastering the data analysis skills, in the next chapter, we will get back to some hardcore data science problems, such as the cleansing and structuring of data.

7
Unstructured Data

In the previous chapter, we looked at different ways of building and fitting models on structured data. Unfortunately, these otherwise extremely useful methods are of no use (yet) when dealing with, for example, a pile of PDF documents. Hence, the following pages will focus on methods to deal with non-tabular data, such as:

- Extracting metrics from a collection of text documents
- Filtering and parsing **natural language texts** (**NLP**)
- Visualizing unstructured data in a structured way

Text mining is the process of analyzing natural language text; in most cases from online content, such as emails and social media streams (Twitter or Facebook). In this chapter, we are going to cover the most used methods of the `tm` package—although, there is a variety of further types of unstructured data, such as text, image, audio, video, non-digital contents, and so on, which we cannot discuss for the time being.

Importing the corpus

A corpus is basically a collection of text documents that you want to include in the analytics. Use the `getSources` function to see the available options to import a corpus with the `tm` package:

```
> library(tm)
> getSources()
[1] "DataframeSource" "DirSource"   "ReutersSource"   "URISource"
[2] "VectorSource"
```

So, we can import text documents from a `data.frame`, a `vector`, or directly from a uniform resource identifier with the `URISource` function. The latter stands for a collection of hyperlinks or file paths, although this is somewhat easier to handle with `DirSource`, which imports all the textual documents found in the referenced directory on our hard drive. By calling the `getReaders` function in the R console, you can see the supported text file formats:

```
> getReaders()
[1]  "readDOC"                "readPDF"
[3]  "readPlain"              "readRCV1"
[5]  "readRCV1asPlain"        "readReut21578XML"
[7]  "readReut21578XMLasPlain" "readTabular"
[9]  "readXML"
```

So, there are some nifty functions to read and parse MS Word, PDFs, plain text, or XML files among a few other file formats. The previous `Reut` reader stands for the Reuters demo corpus that is bundled with the `tm` package.

But let's not stick to some factory default demo files! You can see the package examples in the vignette or reference manual. As we have already fetched some textual data in *Chapter 2, Getting Data from the Web*, let's see how we can process and analyze that content:

```
> res <- XML::readHTMLTable(paste0('http://cran.r-project.org/',
+                    'web/packages/available_packages_by_name.html'),
+              which = 1)
```

> The preceding command requires a live Internet connection and could take 15-120 seconds to download and parse the referenced HTML page. Please note that the content of the downloaded HTML file might be different from what is shown in this chapter, so please be prepared for slightly different outputs in your R session, as compared to what we published in this book.

So, now we have a `data.frame` with more than 5,000 R package names and short descriptions. Let's build a corpus from the vector source of package descriptions, so that we can parse those further and see the most important trends in package development:

```
> v <- Corpus(VectorSource(res$V2))
```

We have just created a `VCorpus` (in-memory) object, which currently holds 5,880 package descriptions:

```
> v
<<VCorpus (documents: 5880, metadata (corpus/indexed): 0/0)>>
```

As the default `print` method (see the preceding output) shows a concise overview on the corpus, we will need to use another function to inspect the actual content:

```
> inspect(head(v, 3))
<<VCorpus (documents: 3, metadata (corpus/indexed): 0/0)>>

[[1]]
<<PlainTextDocument (metadata: 7)>>
A3: Accurate, Adaptable, and Accessible Error Metrics for
Predictive Models

[[2]]
<<PlainTextDocument (metadata: 7)>>
Tools for Approximate Bayesian Computation (ABC)

[[3]]
<<PlainTextDocument (metadata: 7)>>
ABCDE_FBA: A-Biologist-Can-Do-Everything of Flux Balance
Analysis with this package
```

Here, we can see the first three documents in the corpus, along with some metadata. Until now, we have not done much more than when in the *Chapter 2*, *Getting Data from the Web*, we visualized a wordcloud of the expression used in the package descriptions. But that's exactly where the journey begins with text mining!

Cleaning the corpus

One of the nicest features of the `tm` package is the variety of bundled transformations to be applied on corpora (corpuses). The `tm_map` function provides a convenient way of running the transformations on the corpus to filter out all the data that is irrelevant in the actual research. To see the list of available transformation methods, simply call the `getTransformations` function:

```
> getTransformations()
[1] "as.PlainTextDocument" "removeNumbers"
```

```
[3]  "removePunctuation"     "removeWords"
[5]  "stemDocument"          "stripWhitespace"
```

We should usually start with removing the most frequently used, so called stopwords from the corpus. These are the most common, short function terms, which usually carry less important meanings than the other expressions in the corpus, especially the keywords. The package already includes such lists of words in different languages:

```
> stopwords("english")
  [1] "i"         "me"          "my"            "myself"        "we"
  [6] "our"       "ours"        "ourselves"     "you"           "your"
 [11] "yours"     "yourself"    "yourselves"    "he"            "him"
 [16] "his"       "himself"     "she"           "her"           "hers"
 [21] "herself"   "it"          "its"           "itself"        "they"
 [26] "them"      "their"       "theirs"        "themselves"    "what"
 [31] "which"     "who"         "whom"          "this"          "that"
 [36] "these"     "those"       "am"            "is"            "are"
 [41] "was"       "were"        "be"            "been"          "being"
 [46] "have"      "has"         "had"           "having"        "do"
 [51] "does"      "did"         "doing"         "would"         "should"
 [56] "could"     "ought"       "i'm"           "you're"        "he's"
 [61] "she's"     "it's"        "we're"         "they're"       "i've"
 [66] "you've"    "we've"       "they've"       "i'd"           "you'd"
 [71] "he'd"      "she'd"       "we'd"          "they'd"        "i'll"
 [76] "you'll"    "he'll"       "she'll"        "we'll"         "they'll"
 [81] "isn't"     "aren't"      "wasn't"        "weren't"       "hasn't"
 [86] "haven't"   "hadn't"      "doesn't"       "don't"         "didn't"
 [91] "won't"     "wouldn't"    "shan't"        "shouldn't"     "can't"
 [96] "cannot"    "couldn't"    "mustn't"       "let's"         "that's"
[101] "who's"     "what's"      "here's"        "there's"       "when's"
[106] "where's"   "why's"       "how's"         "a"             "an"
[111] "the"       "and"         "but"           "if"            "or"
[116] "because"   "as"          "until"         "while"         "of"
[121] "at"        "by"          "for"           "with"          "about"
[126] "against"   "between"     "into"          "through"       "during"
[131] "before"    "after"       "above"         "below"         "to"
[136] "from"      "up"          "down"          "in"            "out"
[141] "on"        "off"         "over"          "under"         "again"
```

[146]	"further"	"then"	"once"	"here"	"there"
[151]	"when"	"where"	"why"	"how"	"all"
[156]	"any"	"both"	"each"	"few"	"more"
[161]	"most"	"other"	"some"	"such"	"no"
[166]	"nor"	"not"	"only"	"own"	"same"
[171]	"so"	"than"	"too"	"very"	

Skimming through this list verifies that removing these rather unimportant words will not really modify the meaning of the R package descriptions. Although there are some rare cases in which removing the stopwords is not a good idea at all! Carefully examine the output of the following R command:

```
> removeWords('to be or not to be', stopwords("english"))
[1] "      "
```

This does not suggest that the memorable quote from Shakespeare is meaningless, or that we can ignore any of the stopwords in all cases. Sometimes, these words have a very important role in the context, where replacing the words with a space is not useful, but rather deteriorative. Although I would suggest, that in most cases, removing the stopwords is highly practical for keeping the number of words to process at a low level.

To iteratively apply the previous call on each document in our corpus, the tm_map function is extremely useful:

```
> v <- tm_map(v, removeWords, stopwords("english"))
```

Simply pass the corpus and the transformation function, along with its parameters, to tm_map, which takes and returns a corpus of any number of documents:

```
> inspect(head(v, 3))
<<VCorpus (documents: 3, metadata (corpus/indexed): 0/0)>>


[[1]]
<<PlainTextDocument (metadata: 7)>>
A3 Accurate Adaptable Accessible Error Metrics Predictive Models


[[2]]
<<PlainTextDocument (metadata: 7)>>
Tools Approximate Bayesian Computation ABC


[[3]]
```

```
<<PlainTextDocument (metadata: 7)>>
ABCDEFBA ABiologistCanDoEverything Flux Balance Analysis package
```

We can see that the most common function words and a few special characters are now gone from the package descriptions. But what happens if someone starts the description with uppercase stopwords? This is shown in the following example:

```
> removeWords('To be or not to be.', stopwords("english"))
[1] "To      ."
```

It's clear that the uppercase version of the `to` common word was not removed from the sentence, and the trailing dot was also preserved. For this end, usually, we should simply transform the uppercase letters to lowercase, and replace the punctuations with a space to keep the clutter among the keywords at a minimal level:

```
> v <- tm_map(v, content_transformer(tolower))
> v <- tm_map(v, removePunctuation)
> v <- tm_map(v, stripWhitespace)
> inspect(head(v, 3))
<<VCorpus (documents: 3, metadata (corpus/indexed): 0/0)>>

[[1]]
[1] a3 accurate adaptable accessible error metrics predictive models

[[2]]
[1] tools approximate bayesian computation abc

[[3]]
[1] abcdefba abiologistcandoeverything flux balance analysis package
```

So, we first called the `tolower` function from the `base` package to transform all characters from upper to lower case. Please note that we had to wrap the `tolower` function in the `content_transformer` function, so that our transformation really complies with the `tm` package's object structure. This is usually required when using a transformation function outside of the `tm` package.

Then, we removed all the punctuation marks from the text with the help of the `removePunctutation` function. The punctuations marks are the ones referred to as [:punct:] in regular expressions, including the following characters: ! " # $ % & ' ( ) * + , - . / : ; < = > ? @ [\ ] ^ _ ` { | } ~'. Usually, it's safe to remove these separators, especially when we analyze the words on their own and not their relations.

And we also removed the multiple whitespace characters from the document, so that we find only one space between the filtered words.

Visualizing the most frequent words in the corpus

Now that we have cleared up our corpus a bit, we can generate a much more useful wordcloud, as compared to the proof-of-concept demo we generated in *Chapter 2, Getting Data from the Web*:

```
> wordcloud::wordcloud(v)
```

Further cleanup

There are still some small disturbing glitches in the wordlist. Maybe, we do not really want to keep numbers in the package descriptions at all (or we might want to replace all numbers with a placeholder text, such as NUM), and there are some frequent technical words that can be ignored as well, for example, package. Showing the plural version of nouns is also redundant. Let's improve our corpus with some further tweaks, step by step!

Removing the numbers from the package descriptions is fairly straightforward, as based on the previous examples:

```
> v <- tm_map(v, removeNumbers)
```

To remove some frequent domain-specific words with less important meanings, let's see the most common words in the documents. For this end, first we have to compute the TermDocumentMatrix function that can be passed later to the findFreqTerms function to identify the most popular terms in the corpus, based on frequency:

```
> tdm <- TermDocumentMatrix(v)
```

This object is basically a matrix which includes the words in the rows and the documents in the columns, where the cells show the number of occurrences. For example, let's take a look at the first 5 words' occurrences in the first 20 documents:

```
> inspect(tdm[1:5, 1:20])
<<TermDocumentMatrix (terms: 5, documents: 20)>>
Non-/sparse entries: 5/95
Sparsity          : 95%
Maximal term length: 14
Weighting          : term frequency (tf)
```

	Docs																			
Terms	1	2	3	4	5	6	7	8	9	10	11	12	13	14	15	16	17	18	19	20
aalenjohansson	0	0	0	0	0	0	0	0	0	0	0	0	0	0	0	0	0	0	0	0
abc	0	1	0	1	1	0	1	0	0	0	0	0	0	0	0	0	0	0	0	0
abcdefba	0	0	1	0	0	0	0	0	0	0	0	0	0	0	0	0	0	0	0	0
abcsmc	0	0	0	0	0	0	0	0	0	0	0	0	0	0	0	0	0	0	0	0
aberrations	0	0	0	0	0	0	0	0	0	0	0	0	0	0	0	0	0	0	0	0

Extracting the overall number of occurrences for each word is fairly easy. In theory, we could compute the `rowSums` function of this sparse matrix. But let's simply call the `findFreqTerms` function, which does exactly what we were up to. Let's show all those terms that show up in the descriptions at least a 100 times:

```
> findFreqTerms(tdm, lowfreq = 100)
 [1] "analysis"     "based"        "bayesian"     "data"
 [5] "estimation"   "functions"    "generalized"  "inference"
 [9] "interface"    "linear"       "methods"      "model"
[13] "models"       "multivariate" "package"      "regression"
[17] "series"       "statistical"  "test"         "tests"
[21] "time"         "tools"        "using"
```

Manually reviewing this list suggests ignoring the `based` and `using` words, besides the previously suggested `package` term:

```
> myStopwords <- c('package', 'based', 'using')
> v <- tm_map(v, removeWords, myStopwords)
```

Stemming words

Now, let's get rid of the plural forms of the nouns, which also occur in the preceding top 20 lists of the most common words! This is not as easy as it sounds. We might apply some regular expressions to cut the trailing s from the words, but this method has many drawbacks, such as not taking into account some evident English grammar rules.

But we can, instead, use some stemming algorithms, especially Porter's stemming algorithm, which is available in the `SnowballC` package. The `wordStem` function supports 16 languages (take a look at the `getStemLanguages` for details), and can identify the stem of a character vector as easily as calling the function:

```
> library(SnowballC)
> wordStem(c('cats', 'mastering', 'modelling', 'models', 'model'))
[1] "cat"    "master" "model"  "model"  "model"
```

The only penalty here is the fact that Porter's algorithm does not provide real English words in all cases:

```
> wordStem(c('are', 'analyst', 'analyze', 'analysis'))
[1] "ar"     "analyst" "analyz" "analysi"
```

So later, we will have to tweak the results further; to reconstruct the words with the help of a language lexicon database. The easiest way to construct such a database is copying the words of the already existing corpus:

```
> d <- v
```

Then, let's stem all the words in the documents:

```
> v <- tm_map(v, stemDocument, language = "english")
```

Now, we called the `stemDocument` function, which is a wrapper around the `SnowballC` package's `wordStem` function. We specified only one parameter, which sets the language of the stemming algorithm. And now, let's call the `stemCompletion` function on our previously defined directory, and let's formulate each stem to the shortest relevant word found in the database.

Unfortunately, it's not as straightforward as the previous examples, as the `stemCompletion` function takes a character vector of words instead of documents that we have in our corpus. So thus, we have to write our own transformation function with the previously used `content_transformer` helper. The basic idea is to split each documents into words by a space, apply the `stemCompletion` function, and then concatenate the words into sentences again:

```
> v <- tm_map(v, content_transformer(function(x, d) {
+         paste(stemCompletion(
+                 strsplit(stemDocument(x), ' ')[[1]],
+                 d),
+         collapse = ' ')
+       }), d)
```

> The preceding example is rather resource hungry, so please be prepared for high CPU usage for around 30 to 60 minutes on a standard PC. As you can (technically) run the forthcoming code samples without actually performing this step, you may feel free to skip to the next code chunk, if in a hurry.

It took some time, huh? Well, we had to iterate through all the words in each document found in the corpus , but it's well worth the trouble! Let's see the top used terms in the cleaned corpus:

```
> tdm <- TermDocumentMatrix(v)
> findFreqTerms(tdm, lowfreq = 100)
  [1] "algorithm"     "analysing"     "bayesian"      "calculate"
  [5] "cluster"       "computation"   "data"          "distributed"
```

[9] "estimate"	"fit"	"function"	"general"
[13] "interface"	"linear"	"method"	"model"
[17] "multivariable"	"network"	"plot"	"random"
[21] "regression"	"sample"	"selected"	"serial"
[25] "set"	"simulate"	"statistic"	"test"
[29] "time"	"tool"	"variable"	

While previously the very same command returned 23 terms, out of which we removed 3, now we see more than 30 words occurring more than 100 times in the corpus. We got rid of the plural versions of the nouns and a few other similar variations of the same terms, so the density of the document term matrix also increased:

```
> tdm
<<TermDocumentMatrix (terms: 4776, documents: 5880)>>
Non-/sparse entries: 27946/28054934
Sparsity           : 100%
Maximal term length: 35
Weighting          : term frequency (tf)
```

We not only decreased the number of different words to be indexed in the next steps, but we also identified a few new terms that are to be ignored in our further analysis, for example, set does not seem to be an important word in the package descriptions.

Lemmatisation

While stemming terms, we started to remove characters from the end of words in the hope of finding the stem, which is a heuristic process often resulting in not-existing words, as we have seen previously. We tried to overcome this issue by completing these stems to the shortest meaningful words by using a dictionary, which might result in derivation in the meaning of the term, for example, removing the ness suffix.

Another way to reduce the number of inflectional forms of different terms, instead of deconstructing and then trying to rebuild the words, is morphological analysis with the help of a dictionary. This process is called lemmatisation, which looks for lemma (the canonical form of a word) instead of stems.

The Stanford NLP Group created and maintains a Java-based NLP tool called Stanford CoreNLP, which supports lemmatization besides many other NLP algorithms such as tokenization, sentence splitting, POS tagging, and syntactic parsing.

You can use CoreNLP from R via the `rJava` package, or you might install the `coreNLP` package, which includes some wrapper functions around the `CoreNLP` Java library, which are meant for providing easy access to, for example, lammatisation. Please note that after installing the R package, you have to use the `downloadCoreNLP` function to actually install and make accessible the features of the Java library.

Analyzing the associations among terms

The previously computed `TermDocumentMatrix`, can also be used to identify the association between the cleaned terms found in the corpus. This simply suggests the correlation coefficient computed on the joint occurrence of term-pairs in the same document, which can be queried easily with the `findAssocs` function.

Let's see which words are associated with `data`:

```
> findAssocs(tdm, 'data', 0.1)
                data
set             0.17
analyzing       0.13
longitudinal 0.11
big             0.10
```

Only four terms seem to have a higher correlation coefficient than 0.1, and it's not surprising at all that `analyzing` is among the top associated words. Probably, we can ignore the `set` term, but it seems that `longitudinal` and `big` data are pretty frequent idioms in package descriptions. So, what other `big` terms do we have?

```
> findAssocs(tdm, 'big', 0.1)
                big
mpi             0.38
pbd             0.33
program         0.32
unidata         0.19
demonstration 0.17
netcdf          0.15
forest          0.13
```

```
packaged        0.13
base            0.12
data            0.10
```

Checking the original corpus reveals that there are several R packages starting with **pbd**, which stands for **Programming with Big Data**. The pbd packages are usually tied to Open MPI, which pretty well explains the high association between these terms.

Some other metrics

And, of course, we can use the standard data analysis tools as well after quantifying our package descriptions a bit. Let's see, for example, the length of the documents in the corpus:

```
> vnchar <- sapply(v, function(x) nchar(x$content))
> summary(vnchar)
   Min. 1st Qu.  Median   Mean 3rd Qu.    Max.
   2.00   27.00   37.00  39.85   50.00  168.00
```

So, the average package description consists of around 40 characters, while there is a package with only two characters in the description. Well, two characters after removing numbers, punctuations, and the common words. To see which package has this very short description, we might simply call the which.min function:

```
> (vm <- which.min(vnchar))
[1] 221
```

And this is what's strange about it:

```
> v[[vm]]
<<PlainTextDocument (metadata: 7)>>
NA
> res[vm, ]
    V1    V2
221    <NA>
```

So, this is not a real package after all, but rather an empty row in the original table. Let's visually inspect the overall number of characters in the package descriptions:

```
> hist(vnchar, main = 'Length of R package descriptions',
+      xlab = 'Number of characters')
```

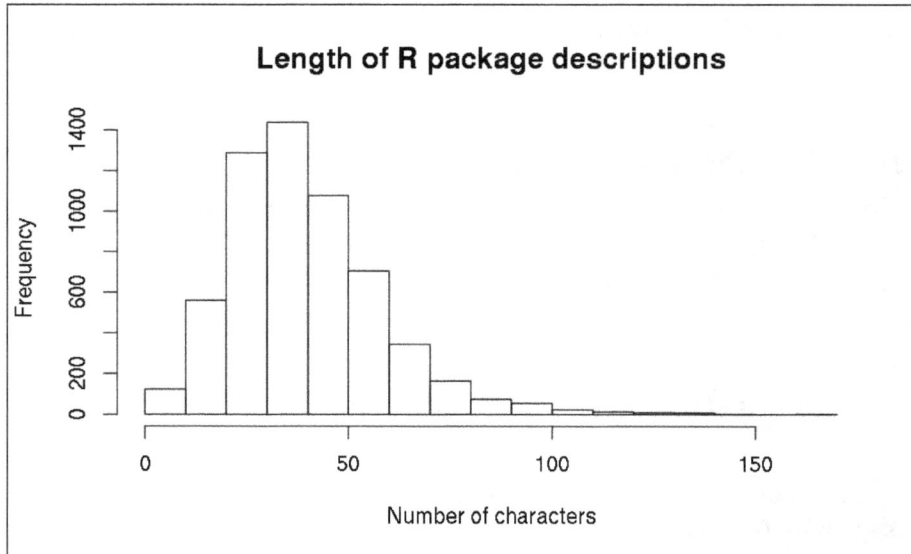

The histogram suggests that most packages have a rather short description with no more than one sentence, based on the fact that an average English sentence includes around 15-20 words with 75-100 characters.

The segmentation of documents

To identify the different groups of cleaned terms, based on the frequency and association of the terms in the documents of the corpus, one might directly use our `tdm` matrix to run, for example, the classic hierarchical cluster algorithm.

On the other hand, if you would rather like to cluster the R packages based on their description, we should compute a new matrix with `DocumentTermMatrix`, instead of the previously used `TermDocumentMatrix`. Then, calling the clustering algorithm on this matrix would result in the segmentation of the packages.

For more details on the available methods, algorithms, and guidance on choosing the appropriate functions for clustering, please see *Chapter 10, Classification and Clustering*. For now, we will fall back to the traditional `hclust` function, which provides a built-in way of running hierarchical clustering on distance matrices. For a quick demo, let's demonstrate this on the so-called `Hadleyverse`, which describes a useful collection of R packages developed by Hadley Wickham:

```
> hadleyverse <- c('ggplot2', 'dplyr', 'reshape2', 'lubridate',
+    'stringr', 'devtools', 'roxygen2', 'tidyr')
```

Now, let's identify which elements of the v corpus hold the cleaned terms of the previously listed packages:

```
> (w <- which(res$V1 %in% hadleyverse))
[1] 1104 1230 1922 2772 4421 4658 5409 5596
```

And then, we can simply compute the (dis)similarity matrix of the used terms:

```
> plot(hclust(dist(DocumentTermMatrix(v[w]))),
+    xlab = 'Hadleyverse packages')
```

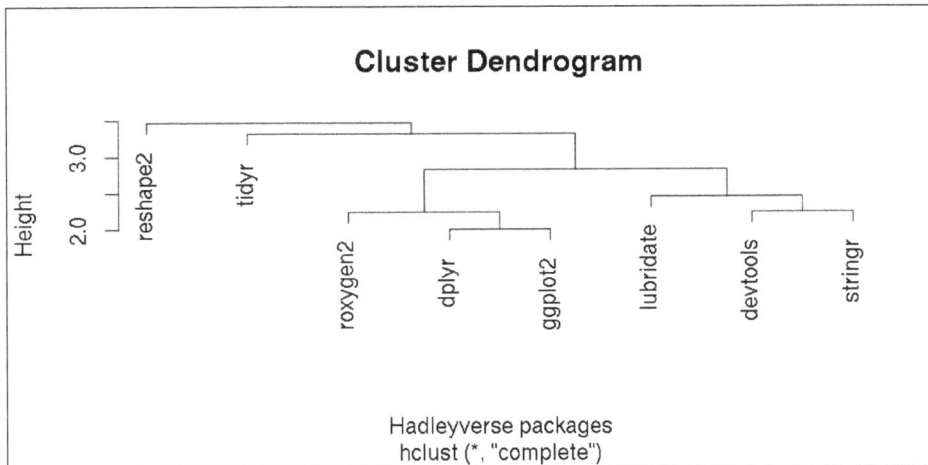

Besides the `reshape2` and `tidyr` packages that we covered in *Chapter 4, Restructuring Data*, we can see two separate clusters in the previous plot (the highlighted terms in the following list are copied from the package descriptions):

- Packages that *make* things a bit *easier*
- Others dealing with the language, *documentation* and *grammar*

To verify this, you might be interested in the cleansed terms for each package:

```
> sapply(v[w], function(x) structure(content(x),
+     .Names = meta(x, 'id')))
                                          devtools
      "tools make developing r code easier"
                                             dplyr
            "a grammar data manipulation"
                                           ggplot2
      "an implementation grammar graphics"
                                         lubridate
         "make dealing dates little easier"
                                          reshape2
   "flexibly reshape data reboot reshape "
                                          roxygen2
              "insource documentation r"
                                           stringr
              "make easier work strings"
                                             tidyr
"easily tidy data spread gather functions"
```

An alternative and probably more appropriate, long-term approach for clustering documents based on NLP algorithms, would be fitting topic models, for example, via the `topicmodels` package. This R package comes with a detailed and very useful vignette, which includes some theoretical background as well as some hands-on examples. But for a quick start, you might simply try to run the LDA or CTM functions on our previously created `DocumentTermMatrix`, and specify the number of topics for the models to be built. A good start, based on our previous clustering example, might be `k=3`.

Summary

The preceding examples and quick theoretical background introduced text mining algorithms to structure plain English texts into numbers for further analysis. In the next chapter, we will concentrate on some similarly important methods in the process of data analysis, such as how to polish this kind of data in the means of identifying outliers, extreme values, and how to handle missing data.

8
Polishing Data

When working with data, you will usually find that it may not always be perfect or clean in the means of missing values, outliers and similar anomalies. Handling and cleaning imperfect or so-called dirty data is part of every data scientist's daily life, and even more, it can take up to 80 percent of the time we actually deal with the data!

Dataset errors are often due to the inadequate data acquisition methods, but instead of repeating and tweaking the data collection process, it is usually better (in the means of saving money, time and other resources) or unavoidable to polish the data by a few simple functions and algorithms. In this chapter, we will cover:

- Different use cases of the `na.rm` argument of various functions
- The `na.action` and related functions to get rid of missing data
- Several packages that offer a user-friendly way of data imputation
- The `outliers` package with several statistical tests for extreme values
- How to implement Lund's outlier test on our own as a brain teaser
- Referring to some robust methods

The types and origins of missing data

First, we have to take a quick look at the possible different sources of missing data to identify why and how we usually get missing values. There are quite a few different reasons for data loss, which can be categorized into 3 different types.

For example, the main cause of missing data might be a malfunctioning device or the human factor of incorrectly entering data. **Missing Completely at Random (MCAR)** means that every value in the dataset has the same probability of being missed, so no systematic error or distortion is to be expected due to missing data, and nor can we explain the pattern of missing values. This is the best situation if we have NA (meaning: no answer, not applicable or not available) values in our data set.

But a much more frequent and unfortunate type of missing data is **Missing at Random (MAR)** compared to MCAR. In the case of MAR, the pattern of missing values is known or at least can be identified, although it has nothing to do with the actual missing values. For example, one might think of a population where males are more loners or lazier compared to females, thus they prefer not to answer all the questions in a survey – regardless of the actual question. So it's not that the males are not giving away their salary due to the fact that they make more or less compared to females, but they tend to skip a few questions in the questionnaire at random.

> This classification and typology of missing data was first proposed by Donald B. Rubin in 1976 in his *Inference and Missing Data*, published in *Biometrika 63(3): 581 – 592*, later reviewed and extended in a book jointly written by *Roderick J. A. Little* (2002): *Statistical Analysis with Missing Data*, *Wiley* – which is well worth of reading for further details.

And the worst scenario would be **Missing Not at Random (MNAR)**, where data is missing for a specific reason that is highly related to the actual question, which classifies missing values as nonignorable non-response.

This happens pretty often in surveys with sensitive questions or due to design flaws in the research preparation. In such cases, data is missing due to some latent process going on in the background, which is often the thing we wanted to come to know better with the help of the research – which can turn out to be a rather cumbersome situation.

So how can we resolve these problems? Sometimes it's relatively easy. For example, if we have lot of observations, MCAR is not a real problem at all due to the law of large numbers, as the probability of having missing value(s) is the same for each observation. We basically have two options to deal with unknown or missing data:

- Removing missing values and/or observations
- Replacing missing values with some estimates

Identifying missing data

The easiest way of dealing with missing values, especially with MCAR data, is simply removing all the observations with any missing values. If we want to exclude every row of a `matrix` or `data.frame` object which has at least one missing value, we can use the `complete.cases` function from the `stats` package to identify those.

For a quick start, let's see how many rows have at least one missing value:

```
> library(hflights)
> table(complete.cases(hflights))
```

```
FALSE    TRUE
 3622 223874
```

This is around 1.5 percent of the quarter million rows:

```
> prop.table(table(complete.cases(hflights))) * 100
     FALSE        TRUE
  1.592116  98.407884
```

Let's see what the distribution of NA looks like within different columns:

```
> sort(sapply(hflights, function(x) sum(is.na(x))))
            Year            Month       DayofMonth
               0                0                0
       DayOfWeek    UniqueCarrier        FlightNum
               0                0                0
         TailNum           Origin             Dest
               0                0                0
        Distance        Cancelled CancellationCode
               0                0                0
        Diverted          DepTime         DepDelay
               0             2905             2905
         TaxiOut          ArrTime           TaxiIn
            2947             3066             3066
ActualElapsedTime          AirTime         ArrDelay
            3622             3622             3622
```

By-passing missing values

So it seems that missing data relatively frequently occurs with the time-related variables, but we have no missing values among the flight identifiers and dates. On the other hand, if one value is missing for a flight, the chances are rather high that some other variables are missing as well – out of the overall number of 3,622 cases with at least one missing value:

```
> mean(cor(apply(hflights, 2, function(x)
+    as.numeric(is.na(x)))), na.rm = TRUE)
[1] 0.9589153
```

```
Warning message:
In cor(apply(hflights, 2, function(x) as.numeric(is.na(x)))) :
  the standard deviation is zero
```

Okay, let's see what we have done here! First, we have called the `apply` function to transform the values of `data.frame` to `0` or `1`, where `0` stands for an observed, while `1` means a missing value. Then we computed the correlation coefficients of this newly created matrix, which of course returned a lot of missing values due to fact that some columns had only one unique value without any variability, as shown in the warning message. For this, we had to specify the `na.rm` parameter to be `TRUE`, so that the `mean` function would return a real value instead of an `NA`, by removing the missing values among the correlation coefficients returned by the `cor` function.

So one option is the heavy use of the `na.rm` argument, which is supported by most functions that are sensitive to missing data — to name a few from the `base` and `stats` packages: `mean`, `median`, `sum`, `max` and `min`.

To compile the complete list of functions that have the `na.rm` argument in the base package, we can follow the steps described in a very interesting SO answer located at `http://stackoverflow.com/a/17423072/564164`. I found this answer motivating because I truly believe in the power of analyzing the tools we use for analysis, or in other words, spending some time on understanding how R works in the background.

First, let's make a list of all the functions found in `baseenv` (the environment of the `base` package) along with the complete function arguments and body:

```
> Funs <- Filter(is.function, sapply(ls(baseenv()), get, baseenv()))
```

Then we can `Filter` all those functions from the returned list, which have `na.rm` among the formal arguments via the following:

```
> names(Filter(function(x)
+     any(names(formals(args(x))) %in% 'na.rm'), Funs))
 [1] "all"                "any"
 [3] "colMeans"           "colSums"
 [5] "is.unsorted"        "max"
 [7] "mean.default"       "min"
 [9] "pmax"               "pmax.int"
[11] "pmin"               "pmin.int"
[13] "prod"               "range"
[15] "range.default"      "rowMeans"
[17] "rowsum.data.frame"  "rowsum.default"
[19] "rowSums"            "sum"
```

```
[21]  "Summary.data.frame"        "Summary.Date"
[23]  "Summary.difftime"          "Summary.factor"
[25]  "Summary.numeric_version"   "Summary.ordered"
[27]  "Summary.POSIXct"           "Summary.POSIXlt"
```

This can be easily applied to any R package by changing the environment variable to for example `'package:stats'` in the case of the `stats` package:

```
> names(Filter(function(x)
+   any(names(formals(args(x))) %in% 'na.rm'),
+     Filter(is.function,
+       sapply(ls('package:stats'), get, 'package:stats'))))
 [1] "density.default" "fivenum"        "heatmap"
 [4] "IQR"             "mad"            "median"
 [7] "median.default" "medpolish"      "sd"
[10] "var"
```

So these are the functions that have the `na.rm` argument in the `base` and the `stats` packages, where we have seen that the fastest and easiest way of ignoring missing values in single function calls (without actually removing the `NA` values from the dataset) is setting `na.rm` to `TRUE`. But why doesn't `na.rm` default to `TRUE`?

Overriding the default arguments of a function

If you are annoyed by the fact that most functions return `NA` if your R object includes missing values, then you can override those by using some custom wrapper functions, such as:

```
> myMean <- function(...) mean(..., na.rm = TRUE)
> mean(c(1:5, NA))
[1] NA
> myMean(c(1:5, NA))
[1] 3
```

Another option might be to write a custom package which would override the factory defaults of the `base` and `stats` function, like in the `rapportools` package, which includes miscellaneous helper functions with sane defaults for reporting:

```
> library(rapportools)
```

```
Loading required package: reshape

Attaching package: 'rapportools'

The following objects are masked from 'package:stats':

    IQR, median, sd, var

The following objects are masked from 'package:base':

    max, mean, min, range, sum


> mean(c(1:5, NA))
[1] 3
```

The problem with this approach is that you've just permanently overridden those functions listed, so you'll need to restart your R session or detach the rapportools package to reset to the standard arguments, like:

```
> detach('package:rapportools')
> mean(c(1:5, NA))
[1] NA
```

A more general solution to override the default arguments of a function is to rely on some nifty features of the Defaults package, which is although not under active maintenance, but it does the job:

```
> library(Defaults)
> setDefaults(mean.default, na.rm = TRUE)
> mean(c(1:5, NA))
[1] 3
```

Please note that here we had to update the default argument value of mean.default instead of simply trying to tweak mean, as that latter would result in an error:

```
> setDefaults(mean, na.rm = TRUE)
Warning message:
In setDefaults(mean, na.rm = TRUE) :
  'na.rm' was not set, possibly not a formal arg for 'mean'
```

This is due to the fact that mean is an S3 method without any formal arguments:

```
> mean
function (x, ...)
{
    if (exists(".importDefaults"))
        .importDefaults(calling.fun = "mean")
    UseMethod("mean")
}
<environment: namespace:base>
> formals(mean)
$x


$...
```

Either methods you prefer, you can automatically call those functions when R starts by adding a few lines of code in your Rprofile file.

> You can customize the R environment via a global or user-specific Rprofile file. This is a normal R script which is usually placed in the user's home directory with a leading dot in the file name, which is run every time a new R session is started. There you can call any R functions wrapped in the .First or .Last functions to be run at the start or at the end of the R session. Such useful additions might be loading some R packages, printing custom greetings or KPI metrics from a database, or for example installing the most recent versions of all R packages.

But it's probably better not to tweak your R environment in such a non-standard way, as you might soon experience some esoteric and unexpected errors or silent malfunctions in your analysis.

For example, I've got used to working in a temporary directory at all times by specifying setwd('/tmp') in my Rprofile, which is very useful if you start R sessions frequently for some quick jobs. On the other hand, it's really frustrating to spend 15 minutes of your life debugging why some random R function does not seem to do its job, and why it's returning some file not found error messages instead.

So please be warned: if you update the factory default arguments of R functions, do not ever think of ranting about some new bugs you have found in some major functions of base R on the R mailing lists, before trying to reproduce those errors in a vanilla R session with starting R with the --vanilla command line option.

Getting rid of missing data

An alternative way of using the na.rm argument in R functions is removing NA from the dataset before passing that to the analysis functions. This means that we are removing the missing values from the dataset permanently, so that they won't cause any problems at later stages in the analysis. For this, we could use either the na.omit or the na.exclude functions:

```
> na.omit(c(1:5, NA))
[1] 1 2 3 4 5
attr(,"na.action")
[1] 6
attr(,"class")
[1] "omit"
> na.exclude(c(1:5, NA))
[1] 1 2 3 4 5
attr(,"na.action")
[1] 6
attr(,"class")
[1] "exclude"
```

The only difference between these two functions is the class of the na.action attribute of the returned R object, which are omit and exclude respectively. This minor difference is only important when modelling. The na.exclude function returns NA for residuals and predictions, while na.omit suppresses those elements of the vector:

```
> x <- rnorm(10); y <- rnorm(10)
> x[1] <- NA; y[2] <- NA
> exclude <- lm(y ~ x, na.action = "na.exclude")
> omit <- lm(y ~ x, na.action = "na.omit")
> residuals(exclude)
    1     2     3     4     5     6     7     8     9    10
   NA    NA -0.89 -0.98  1.45 -0.23  3.11 -0.23 -1.04 -1.20

> residuals(omit)
    3     4     5     6     7     8     9    10
-0.89 -0.98  1.45 -0.23  3.11 -0.23 -1.04 -1.20
```

Important thing to note in case of tabular data, like a `matrix` or `data.frame`, these functions remove the whole row if it contains at least one missing value. For a quick demo, let's create a matrix with 3 columns and 3 rows with values incrementing from 1 to 9, but replacing all values divisible by 4 with `NA`:

```
> m <- matrix(1:9, 3)
> m[which(m %% 4 == 0, arr.ind = TRUE)] <- NA
> m
     [,1] [,2] [,3]
[1,]    1   NA    7
[2,]    2    5   NA
[3,]    3    6    9
> na.omit(m)
     [,1] [,2] [,3]
[1,]    3    6    9
attr(,"na.action")
[1] 1 2
attr(,"class")
[1] "omit"
```

As seen here, we can find the row numbers of the removed cases in the `na.action` attribute.

Filtering missing data before or during the actual analysis

Let's suppose we want to calculate the `mean` of the actual length of flights:

```
> mean(hflights$ActualElapsedTime)
[1] NA
```

The result is `NA` of course, because as identified previously, this variable contains missing values, and almost every R operation with `NA` results in `NA`. So let's overcome this issue as follows:

```
> mean(hflights$ActualElapsedTime, na.rm = TRUE)
[1] 129.3237
> mean(na.omit(hflights$ActualElapsedTime))
[1] 129.3237
```

Any performance issues there? Or other means of deciding which method to use?

```
> library(microbenchmark)
> NA.RM    <- function()
+                 mean(hflights$ActualElapsedTime, na.rm = TRUE)
> NA.OMIT <- function()
+                 mean(na.omit(hflights$ActualElapsedTime))
> microbenchmark(NA.RM(), NA.OMIT())
Unit: milliseconds
      expr       min        lq     median        uq       max neval
    NA.RM()  7.105485  7.231737  7.500382  8.002941  9.850411   100
  NA.OMIT() 12.268637 12.471294 12.905777 13.376717 16.008637   100
```

The first glance at the performance of these options computed with the help of the `microbenchmark` package (please see the *Loading text files of reasonable size* section in the *Chapter 1, Hello Data* for more details) suggests that using `na.rm` is the better solution in case of a single function call.

On the other hand, if we want to reuse the data at some later phase in the analysis, it is more viable and effective to omit the missing values and observations only once from the dataset, instead of always specifying `na.rm` to be `TRUE`.

Data imputation

And sometimes omitting missing values is not reasonable or possible at all, for example due to the low number of observations or if it seems that missing data is not random. Data imputation is a real alternative in such situations, and this method can replace NA with some real values based on various algorithms, such as filling empty cells with:

- A known scalar
- The previous value appearing in the column (hot-deck)
- A random element from the same column
- The most frequent value in the column
- Different values from the same column with given probability
- Predicted values based on regression or machine learning models

The hot-deck method is often used while joining multiple datasets together. In such a situation, the `roll` argument of `data.table` can be very useful and efficient, otherwise be sure to check out the `hotdeck` function in the `VIM` package, which offers some really useful ways of visualizing missing data. But when dealing with an already given column of a dataset, we have some other simple options as well.

For instance, imputing a known scalar is a pretty simple situation, where we know that all missing values are for example due to some research design patterns. Let's think of a database that stores the time you arrived to and left the office every weekday, and by computing the difference between those two, we can analyze the number of work hours spent in the office from day to day. If this variable returns NA for a time period, actually it means that we were outside of the office all day, so thus the computed value should be zero instead of NA.

And not just in theory, but this is pretty easy to implement in R as well (example is continued from the previous demo code where we defined m with two missing values):

```
> m[which(is.na(m), arr.ind = TRUE)] <- 0
> m
     [,1] [,2] [,3]
[1,]    1    0    7
[2,]    2    5    0
[3,]    3    6    9
```

Similarly, replacing missing values with a random number, a `sample` of other values or with the `mean` of a variable can be done relatively easily:

```
> ActualElapsedTime <- hflights$ActualElapsedTime
> mean(ActualElapsedTime, na.rm = TRUE)
[1] 129.3237
> ActualElapsedTime[which(is.na(ActualElapsedTime))] <-
+   mean(ActualElapsedTime, na.rm = TRUE)
> mean(ActualElapsedTime)
[1] 129.3237
```

Which can be even easier with the `impute` function from the `Hmisc` package:

```
> library(Hmisc)
> mean(impute(hflights$ActualElapsedTime, mean))
[1] 129.3237
```

It seems that we have preserved the value of the arithmetic mean of course, but you should be aware of some very serious side-effects:

```
> sd(hflights$ActualElapsedTime, na.rm = TRUE)
[1] 59.28584
> sd(ActualElapsedTime)
[1] 58.81199
```

When replacing missing values with the mean, the variance of the transformed variable will be naturally lower compared to the original distribution. This can be extremely problematic in some situations, where some more sophisticated methods are needed.

Modeling missing values

Besides the previous mentioned univariate methods, you may also fit models on the complete cases in the dataset, rather than fitting those models on the remaining rows to estimate the missing values. Or in a nutshell, we are replacing the missing values with multivariate predictions.

There are a plethora of related functions and packages, for example you might be interested in checking the transcan function in the Hmisc package, or the imputeR package, which includes a wide variety of models for imputing categorical and continuous variables as well.

Most of the imputation methods and models are for one type of variable: either continuous or categorical. In case of mixed-type dataset, we typically use different algorithms to handle the different types of missing data. The problem with this approach is that some of the possible relations between different types of data might be ignored, resulting in some partial models.

To overcome this issue, and to save a few pages in the book on the description of the traditional regression and other related methods for data imputation (although you can find some related methods in the *Chapter 5, Buildings Models (authored by Renata Nemeth and Gergely Toth)* and the *Chapter 6, Beyond the Linear Trend Line (authored by Renata Nemeth and Gergely Toth))*, we will concentrate on a non-parametric method that can handle categorical and continuous variables at the same time via a very user-friendly interface in the missForest package.

This iterative procedure fits a random forest model on the available data in order to predict the missing values. As our hflights data is relatively large for such a process and running the sample code would takes ages, we will rather use the standard iris dataset in the next examples.

First let's see the original structure of the dataset, which does not include any missing values:

```
> summary(iris)
  Sepal.Length     Sepal.Width     Petal.Length     Petal.Width
 Min.   :4.300    Min.   :2.000    Min.   :1.000    Min.   :0.100
 1st Qu.:5.100    1st Qu.:2.800    1st Qu.:1.600    1st Qu.:0.300
 Median :5.800    Median :3.000    Median :4.350    Median :1.300
 Mean   :5.843    Mean   :3.057    Mean   :3.758    Mean   :1.199
 3rd Qu.:6.400    3rd Qu.:3.300    3rd Qu.:5.100    3rd Qu.:1.800
 Max.   :7.900    Max.   :4.400    Max.   :6.900    Max.   :2.500
       Species
 setosa    :50
 versicolor:50
 virginica :50
```

Now let's load the package and add some missing values (completely at random) to the dataset in the means of producing a reproducible minimal example for the forthcoming models:

```
> library(missForest)
> set.seed(81)
> miris <- prodNA(iris, noNA = 0.2)
> summary(miris)
  Sepal.Length     Sepal.Width     Petal.Length     Petal.Width
 Min.   :4.300    Min.   :2.000    Min.   :1.100    Min.   :0.100
 1st Qu.:5.200    1st Qu.:2.800    1st Qu.:1.600    1st Qu.:0.300
 Median :5.800    Median :3.000    Median :4.450    Median :1.300
 Mean   :5.878    Mean   :3.062    Mean   :3.905    Mean   :1.222
 3rd Qu.:6.475    3rd Qu.:3.300    3rd Qu.:5.100    3rd Qu.:1.900
 Max.   :7.900    Max.   :4.400    Max.   :6.900    Max.   :2.500
 NA's   :28       NA's   :29       NA's   :32       NA's   :33
       Species
 setosa    :40
 versicolor:38
 virginica :44
 NA's      :28
```

So now we have around 20 percent of missing values in each column, which is also stated in the bottom row of the preceding summary. The number of completely random missing values is between 28 and 33 cases per variable.

The next step should be building the random forest models to replace the missing values with real numbers and factor levels. As we also have the original dataset, we can use that complete matrix to test the performance of the method via the `xtrue` argument, which computes and returns the error rate when we call the function with `verbose`. This is useful in such didactical examples to show how the model and predictions improves from iteration to iteration:

```
> iiris <- missForest(miris, xtrue = iris, verbose = TRUE)
  missForest iteration 1 in progress...done!
    error(s): 0.1512033 0.03571429
    estimated error(s): 0.1541084 0.04098361
    difference(s): 0.01449533 0.1533333
    time: 0.124 seconds

  missForest iteration 2 in progress...done!
    error(s): 0.1482248 0.03571429
    estimated error(s): 0.1402145 0.03278689
    difference(s): 9.387853e-05 0
    time: 0.114 seconds

  missForest iteration 3 in progress...done!
    error(s): 0.1567693 0.03571429
    estimated error(s): 0.1384038 0.04098361
    difference(s): 6.271654e-05 0
    time: 0.152 seconds

  missForest iteration 4 in progress...done!
    error(s): 0.1586195 0.03571429
    estimated error(s): 0.1419132 0.04918033
    difference(s): 3.02275e-05 0
    time: 0.116 seconds

  missForest iteration 5 in progress...done!
    error(s): 0.1574789 0.03571429
```

```
estimated error(s): 0.1397179 0.04098361

difference(s): 4.508345e-05 0

time: 0.114 seconds
```

The algorithm ran for 5 iterations before stopping, when it seemed that the error rate was not improving any further. The returned `missForest` object includes a few other values besides the imputed dataset:

```
> str(iiris)
List of 3
 $ ximp    :'data.frame':  150 obs. of  5 variables:
  ..$ Sepal.Length: num [1:150] 5.1 4.9 4.7 4.6 5 ...
  ..$ Sepal.Width : num [1:150] 3.5 3.3 3.2 3.29 3.6 ...
  ..$ Petal.Length: num [1:150] 1.4 1.4 1.3 1.42 1.4 ...
  ..$ Petal.Width : num [1:150] 0.2 0.218 0.2 0.2 0.2 ...
  ..$ Species     : Factor w/ 3 levels "setosa","versicolor",..: ...
 $ OOBerror: Named num [1:2] 0.1419 0.0492
  ..- attr(*, "names")= chr [1:2] "NRMSE" "PFC"
 $ error   : Named num [1:2] 0.1586 0.0357
  ..- attr(*, "names")= chr [1:2] "NRMSE" "PFC"
 - attr(*, "class")= chr "missForest"
```

The Out of Box error is an estimate on how good our model was based on the **normalized root mean squared error computed (NRMSE)** for numeric values and the **proportion of falsely classified (PFC)** entries for factors. And as we also provided the complete dataset for the previously run model, we also get the true imputation error ratio – which is pretty close to the above estimates.

> Please find more details on random forests and related machine learning topics in the *Chapter 10, Classification and Clustering*.

But how does this approach compare to a much simpler imputation method, like replacing missing values with the mean?

Comparing different imputation methods

In the comparison, only the first four columns of the `iris` dataset will be used, thus it is not dealing with the factor variable at the moment. Let's prepare this demo dataset:

```
> miris <- miris[, 1:4]
```

In `iris_mean`, we replace all the missing values to the mean of the actual columns:

```
> iris_mean <- impute(miris, fun = mean)
```

And in `iris_forest`, we predict the missing values by fitting random forest model:

```
> iris_forest <- missForest(miris)
  missForest iteration 1 in progress...done!
  missForest iteration 2 in progress...done!
  missForest iteration 3 in progress...done!
  missForest iteration 4 in progress...done!
  missForest iteration 5 in progress...done!
```

Now let's simply check the accuracy of the two models by comparing the correlations of `iris_mean` and `iris_forest` with the complete `iris` dataset. For `iris_forest`, we will extract the actual imputed dataset from the `ximp` attribute, and we will silently ignore the factor variable of the original `iris` table:

```
> diag(cor(iris[, -5], iris_mean))
Sepal.Length  Sepal.Width Petal.Length  Petal.Width
   0.6633507    0.8140169    0.8924061    0.4763395
> diag(cor(iris[, -5], iris_forest$ximp))
Sepal.Length  Sepal.Width Petal.Length  Petal.Width
   0.9850253    0.9320711    0.9911754    0.9868851
```

These results suggest that the nonparametric random forest model did a lot better job compared to the simple univariate solution of replacing missing values with the mean.

Not imputing missing values

Please note that these methods have their drawbacks likewise. Replacing the missing values with a predicted one often lacks any error term and residual variance with most models.

This also means that we are lowering the variability, and overestimating some association in the dataset at the same time, which can seriously affect the results of our data analysis. For this, some simulation techniques were introduced in the past to overcome the problem of distorting the dataset and our hypothesis tests with some arbitrary models.

Multiple imputation

The basic idea behind multiple imputation is to fit models several times in a row on the missing values. This Monte Carlo method usually creates some (like 3 to 10) parallel versions of the simulated complete dataset, each of these is analyzed separately, and then we combine the results to produce the actual estimates and confidence intervals. See for example the `aregImpute` function from the `Hmisc` package for more details.

On the other hand, do we really have to remove or impute missing values in all cases? For more details on this question, please see the last section of this chapter. But before that, let's get to know some other requirements for polishing data.

Extreme values and outliers

An outlier or extreme value is defined as a data point that deviates so far from the other observations, that it becomes suspicious to be generated by a totally different mechanism or simply by error. Identifying outliers is important because those extreme values can:

- Increase error variance
- Influence estimates
- Decrease normality

Or in other words, let's say your raw dataset is a piece of rounded stone to be used as a perfect ball in some game, which has to be cleaned and polished before actually using it. The stone has some small holes on its surface, like missing values in the data, which should be filled – with data imputation.

On the other hand, the stone does not only has holes on its surface, but some mud also covers some parts of the item, which is to be removed. But how can we distinguish mud from the real stone? In this section, we will focus on what the `outliers` package and some related methods have to offer for identifying extreme values.

As this package has some conflicting function names with the `randomForest` package (automatically loaded by the `missForest` package), it's wise to detach the latter before heading to the following examples:

```
> detach('package:missForest')
> detach('package:randomForest')
```

The outlier function returns the value with the largest difference from the mean, which, contrary to its name, not necessarily have to be an outlier. Instead, the function can be used to give the analyst an idea about which values can be outliers:

```
> library(outliers)
> outlier(hflights$DepDelay)
[1] 981
```

So there was a flight with more than 16 hours of delay before actually taking off! This is impressive, isn't it? Let's see if it's normal to be so late:

```
> summary(hflights$DepDelay)
   Min. 1st Qu.  Median    Mean 3rd Qu.    Max.    NA's
-33.000  -3.000   0.000   9.445   9.000 981.000    2905
```

Well, mean is around 10 minutes, but as it's even larger than the third quarter and the median is zero, it's not that hard to guess that the relatively large mean is due to some extreme values:

```
> library(lattice)
> bwplot(hflights$DepDelay)
```

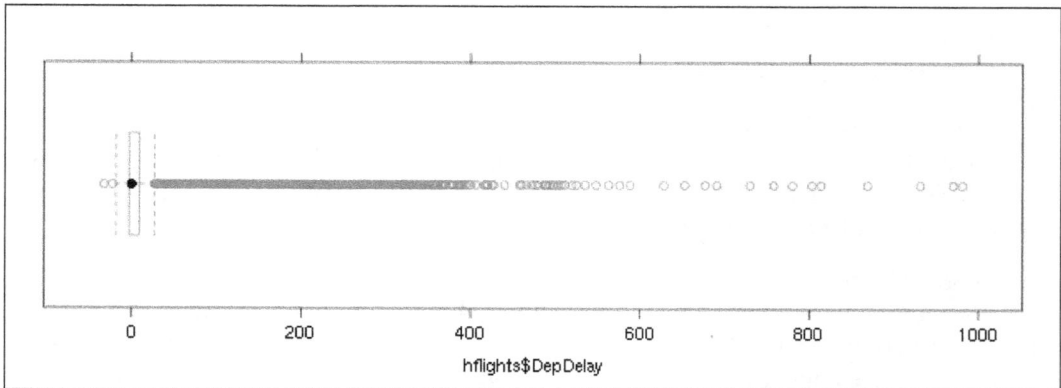

The preceding boxplot clearly shows that most flights were delayed by only a few minutes, and the interquartile range is around 10 minutes:

```
> IQR(hflights$DepDelay, na.rm = TRUE)
[1] 12
```

All the blue circles in the preceding image are the whiskers are possible extreme values, as being higher than the 1.5 IQR of the upper quartile. But how can we (statistically) test a value?

Testing extreme values

The `outliers` package comes with several bundled extreme value detection algorithms, like:

- Dixon's Q test (`dixon.test`)
- Grubb's test (`grubbs.test`)
- Outlying and inlying variance (`cochran.test`)
- Chi-squared test (`chisq.out.test`)

These functions are extremely easy to use. Just pass a vector to the statistical tests and the returning p-value of the significance test will clearly indicate if the data has any outliers. For example, let's test 10 random numbers between 0 and 1 against a relatively large number to verify it's an extreme value in this small sample:

```
> set.seed(83)
> dixon.test(c(runif(10), pi))


    Dixon test for outliers


data:  c(runif(10), pi)
Q = 0.7795, p-value < 2.2e-16
alternative hypothesis: highest value 3.14159265358979 is an outlier
```

But unfortunately, we cannot use these convenient functions in our live dataset, as the methods assume normal distribution, which is definitely not true in our cases as we all know from experience: flights tend to be late more often than arriving a lot sooner to their destinations.

For this, we should use some more robust methods, such as the `mvoutlier` package, or some very simple approaches like Lund suggested around 40 years ago. This test basically computes the distance of each value from the mean with the help of a very simple linear regression:

```
> model <- lm(hflights$DepDelay ~ 1)
```

Just to verify we are now indeed measuring the distance from the mean:

```
> model$coefficients
(Intercept)
   9.444951
> mean(hflights$DepDelay, na.rm = TRUE)
[1] 9.444951
```

Now let's compute the critical value based on the F distribution and two helper variables (where a stands for the alpha value and n represents the number of cases):

```
> a <- 0.1
> (n <- length(hflights$DepDelay))
[1] 227496
> (F <- qf(1 - (a/n), 1, n-2, lower.tail = TRUE))
[1] 25.5138
```

Which can be passed to Lund's formula:

```
> (L <- ((n - 1) * F / (n - 2 + F))^0.5)
[1] 5.050847
```

Now let's see how many values have a higher standardized residual than this computed critical value:

```
> sum(abs(rstandard(model)) > L)
[1] 1684
```

But do we really have to remove these outliers from our data? Aren't extreme values normal? Sometimes these artificial edits in the raw data, like imputing missing values or removing outliers, makes more trouble than it's worth.

Using robust methods

Fortunately, there are some robust methods for analyzing datasets, which are generally less sensitive to extreme values. These robust statistical methods have been developed since 1960, but there are some well-known related methods from even earlier, like using the median instead of the mean as a central tendency. Robust methods are often used when the underlying distribution of our data is not considered to follow the Gaussian curve, so most good old regression models do not work (see more details in the *Chapter 5, Buildings Models (authored by Renata Nemeth and Gergely Toth)* and the *Chapter 6, Beyond the Linear Trend Line (authored by Renata Nemeth and Gergely Toth)*).

Let's take the traditional linear regression example of predicting the sepal length of iris flowers based on the petal length with some missing data. For this, we will use the previously defined `miris` dataset:

```
> summary(lm(Sepal.Length ~ Petal.Length, data = miris))

Call:
```

```
lm(formula = Sepal.Length ~ Petal.Length, data = miris)


Residuals:
     Min       1Q     Median       3Q       Max
-1.26216  -0.36157   0.01461   0.35293   1.01933


Coefficients:
             Estimate  Std. Error  t value  Pr(>|t|)
(Intercept)   4.27831     0.11721    36.50   <2e-16 ***
Petal.Length  0.41863     0.02683    15.61   <2e-16 ***
---
Signif. codes:  0 '***' 0.001 '**' 0.01 '*' 0.05 '.' 0.1 ' ' 1


Residual standard error: 0.4597 on 92 degrees of freedom
  (56 observations deleted due to missingness)
Multiple R-squared:  0.7258,   Adjusted R-squared:  0.7228
F-statistic: 243.5 on 1 and 92 DF,  p-value: < 2.2e-16
```

So it seems that our estimate for the sepal and petal length ratio is around 0.42, which is not too far from the real value by the way:

```
> lm(Sepal.Length ~ Petal.Length, data = iris)$coefficients
 (Intercept) Petal.Length
   4.3066034    0.4089223
```

The difference between the estimated and real coefficients is due to the artificially introduced missing values in a previous section. Can we produce even better estimates? We might impute the missing data with any of the previously mentioned methods, or instead we should rather fit a robust linear regression from the MASS package predicting Sepal.Length with the Petal.Length variable:

```
> library(MASS)
> summary(rlm(Sepal.Length ~ Petal.Length, data = miris))


Call: rlm(formula = Sepal.Length ~ Petal.Length, data = miris)
Residuals:
     Min       1Q     Median       3Q       Max
-1.26184  -0.36098   0.01574   0.35253   1.02262


Coefficients:
```

```
                 Value    Std. Error  t value
(Intercept)      4.2739   0.1205      35.4801
Petal.Length     0.4195   0.0276      15.2167


Residual standard error: 0.5393 on 92 degrees of freedom
  (56 observations deleted due to missingness)
```

Now let's compare the coefficients of the models run against the original (full) and the simulated data (with missing values):

```
> f <- formula(Sepal.Length ~ Petal.Length)
> cbind(
+      orig = lm(f, data = iris)$coefficients,
+      lm   = lm(f, data = miris)$coefficients,
+      rlm  = rlm(f, data = miris)$coefficients)
                   orig         lm         rlm
(Intercept)   4.3066034  4.2783066  4.2739350
Petal.Length  0.4089223  0.4186347  0.4195341
```

To be honest, there's not much difference between the standard linear regression and the robust version. Surprised? Well, the dataset included missing values completely at random, but what happens if the dataset includes other types of missing values or an outlier? Let's verify this by simulating some dirtier data issues (with updating the sepal length of the first observation from 1.4 to 14 – let's say due to a data input error) and rebuilding the models:

```
> miris$Sepal.Length[1] <- 14
> cbind(
+      orig = lm(f, data = iris)$coefficients,
+      lm   = lm(f, data = miris)$coefficients,
+      rlm  = rlm(f, data = miris)$coefficients)
                   orig         lm         rlm
(Intercept)   4.3066034  4.6873973  4.2989589
Petal.Length  0.4089223  0.3399485  0.4147676
```

It seems that the lm model's performance decreased a lot, while the coefficients of the robust model are almost identical to the original model regardless of the outlier in the data. We can conclude that robust methods are pretty impressive and powerful tools when it comes to extreme values! For more information on the related methods already implemented in R, visit the related CRAN Task View at http://cran.r-project.org/web/views/Robust.html.

Summary

This chapter focused on some of the hardest challenges in data analysis in the means of cleansing data, and we covered the most important topics on missing and extreme values. Depending on your field of interest or industry you are working for, dirty data can be a rare or major issue (for example I've seen some projects in the past when regular expressions were applied to a JSON file to make that valid), but I am sure you will find the next chapter interesting and useful despite your background – where we will learn about multivariate statistical techniques.

9
From Big to Small Data

Now that we have some cleansed data ready for analysis, let's first see how we can find our way around the high number of variables in our dataset. This chapter will introduce some statistical techniques to reduce the number of variables by dimension reduction and feature extraction, such as:

- **Principal Component Analysis (PCA)**
- **Factor Analysis (FA)**
- **Multidimensional Scaling (MDS)** and a few other techniques

> Most dimension reduction methods require that two or more numeric variables in the dataset are highly associated or correlated, so the columns in our matrix are not totally independent of each other. In such a situation, the goal of dimension reduction is to decrease the number of columns in the dataset to the actual matrix rank; or, in other words, the number of variables can be decreased whilst most of the information content can be retained. In linear algebra, the matrix rank refers to the dimensions of the vector space generated by the matrix — or, in simpler terms, the number of independent columns and rows in a quadratic matrix. Probably it's easier to understand rank by a quick example: imagine a dataset on students where we know the gender, the age, and the date of birth of respondents. This data is redundant as the age can be computed (via a linear transformation) from the date of birth. Similarly, the year variable is static (without any variability) in the `hflights` dataset, and the elapsed time can be also computed by the departure and arrival times.

These transformations basically concentrate on the common variance identified among the variables and exclude the remaining total (unique) variance. This results in a dataset with fewer columns, which is probably easier to maintain and process, but at the cost of some information loss and the creation of artificial variables, which are usually harder to comprehend compared to the original columns.

In the case of perfect dependence, all but one of the perfectly correlated variables can be omitted, as the rest provide no additional information about the dataset. Although it does not happen often, in most cases it's still totally acceptable to keep only one or a few components extracted from a set of questions, for example in a survey for further analysis.

Adequacy tests

The first thing you want to do, when thinking about reducing the number of dimensions or looking for latent variables in the dataset with multivariate statistical analysis, is to check whether the variables are correlated and the data is normally distributed.

Normality

The latter is often not a strict requirement. For example, the results of a PCA can be still valid and interpreted if we do not have multivariate normality; on the other hand, maximum likelihood factor analysis does have this strong assumption.

[
You should always use the appropriate methods to achieve your data analysis goals, based on the characteristics of your data.
]

Anyway, you can use (for example) qqplot to do a pair-wise comparison of variables, and qqnorm to do univariate normality tests of your variables. First, let's demonstrate this with a subset of hflights:

```
> library(hlfights)
> JFK <- hflights[which(hflights$Dest == 'JFK'),
+                 c('TaxiIn', 'TaxiOut')]
```

So we filter our dataset to only those flights heading to the John F. Kennedy International Airport and we are interested in only two variables describing how long the taxiing in and out times were in minutes. The preceding command with the traditional [indexing can be refactored with subset for much more readable source code:

```
> JFK <- subset(hflights, Dest == 'JFK', select = c(TaxiIn, TaxiOut))
```

Please note that now there's no need to quote variable names or refer to the
`data.frame` name inside the `subset` call. For more details on this, please see
Chapter 3, Filtering and Summarizing Data. And now let's see how the values of
these two columns are distributed:

```
> par(mfrow = c(1, 2))
> qqnorm(JFK$TaxiIn, ylab = 'TaxiIn')
> qqline(JFK$TaxiIn)
> qqnorm(JFK$TaxiOut, ylab = 'TaxiOut')
> qqline(JFK$TaxiOut)
```

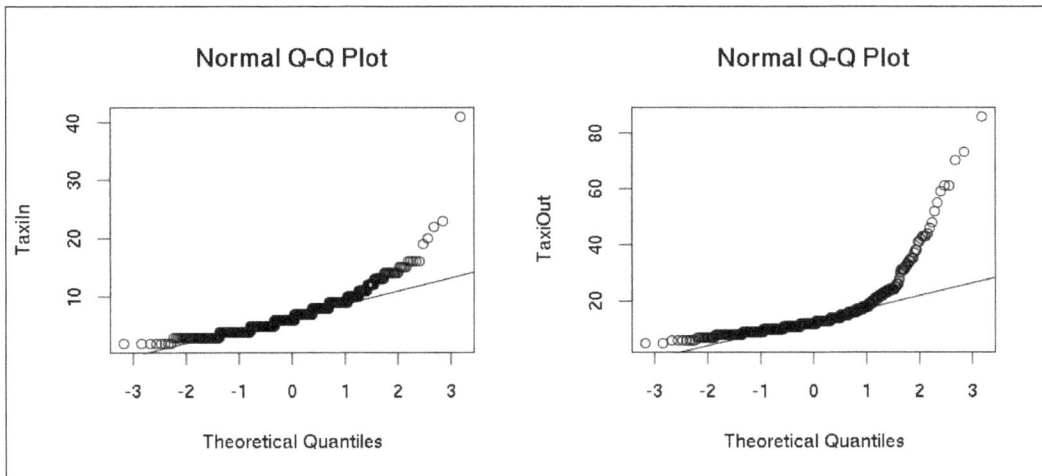

To render the preceding plot, we created a new graphical device (with `par` to hold
two plots in a row), then called `qqnorm`, to show the quantiles of the empirical
variables against the normal distribution, and also added a line for the latter with
`qqline` for easier comparison. If the data was scaled previously, `qqline` would
render a 45-degree line.

Checking the QQ-plots suggest that the data does not fit the normal distribution
very well, which can be also verified by an analytical test such as the Shapiro-Wilk
normality test:

```
> shapiro.test(JFK$TaxiIn)

    Shapiro-Wilk normality test

data:   JFK$TaxiIn
W = 0.8387, p-value < 2.2e-16
```

The `p-value` is really small, so the null hypothesis (stating that the data is normally distributed) is rejected. But how can we test normality for a bunch of variables without and beyond separate statistical tests?

Multivariate normality

Similar statistical tests exist for multiple variables as well; these methods provide different ways to check if the data fits the multivariate normal distribution. To this end, we will use the MVN package, but similar methods can be also found in the `mvnormtest` package. The latter includes the multivariate version of the previously discussed Shapiro-Wilk test as well.

But Mardia's test is more often used to check multivariate normality and, even better, it does not limit the sample size to below 5,000. After loading the MVN package, calling the appropriate R function is pretty straightforward with a very intuitive interpretation—after getting rid of the missing values in our dataset:

```
> JFK <- na.omit(JFK)
> library(MVN)
> mardiaTest(JFK)
   Mardia's Multivariate Normality Test
---------------------------------------------
   data : JFK

   g1p              : 20.84452
   chi.skew         : 2351.957
   p.value.skew     : 0

   g2p              : 46.33207
   z.kurtosis       : 124.6713
   p.value.kurt     : 0

   chi.small.skew   : 2369.368
   p.value.small    : 0

   Result           : Data is not multivariate normal.
---------------------------------------------
```

For more details on handling and filtering missing values, please see *Chapter 8, Polishing Data*.

Out of the three p values, the third one refers to cases when the sample size is extremely small (<20), so now we only concentrate on the first two values, both below 0.05. This means that the data does not seem to be multivariate normal. Unfortunately, Mardia's test fails to perform well in some cases, so more robust methods might be more appropriate to use.

The MVN package can run the Henze-Zirkler's and Royston's Multivariate Normality Test as well. Both return user-friendly and easy to interpret results:

```
> hzTest(JFK)
  Henze-Zirkler's Multivariate Normality Test
---------------------------------------------
  data : JFK

  HZ      : 42.26252
  p-value : 0

  Result  : Data is not multivariate normal.
---------------------------------------------


> roystonTest(JFK)
  Royston's Multivariate Normality Test
---------------------------------------------
  data : JFK

  H       : 264.1686
  p-value : 4.330916e-58

  Result  : Data is not multivariate normal.
---------------------------------------------
```

A more visual method to test multivariate normality is to render similar QQ plots to those we used before. But, instead of comparing only one variable with the theoretical normal distribution, let's first compute the squared Mahalanobis distance between our variables, which should follow a chi-square distribution with the degrees of freedom being the number of our variables. The MVN package can automatically compute all the required values and render those with any of the preceding normality test R functions; just set the `qqplot` argument to be TRUE:

```
> mvt <- roystonTest(JFK, qqplot = TRUE)
```

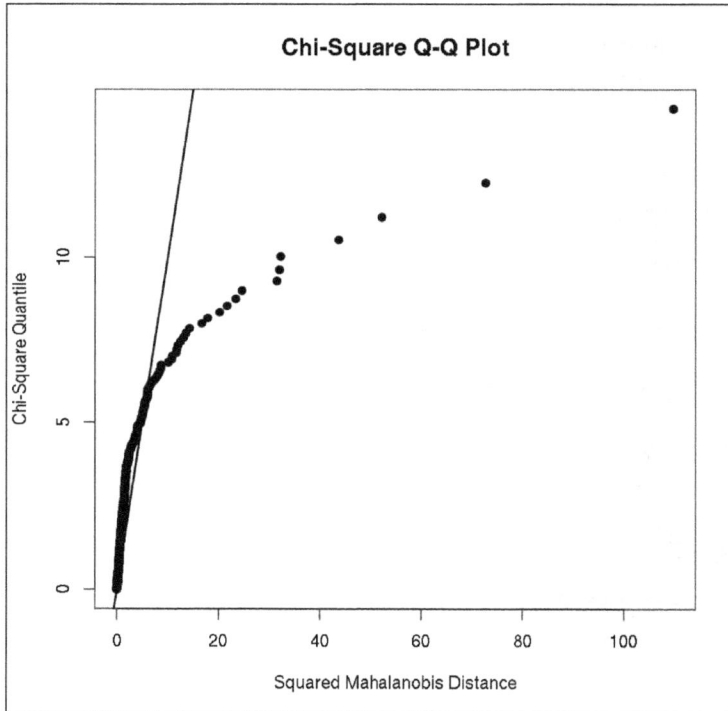

If the dataset was normally distributed, the points shown in the preceding graphs should fit the straight line. Other alternative graphical methods can produce more visual and user-friendly plots with the previously created `mvt` R object. The MVN package ships the `mvnPlot` function, which can render perspective and contour plots for two variables and thus provides a nice way to test bivariate normality:

```
> par(mfrow = c(1, 2))
> mvnPlot(mvt, type = "contour", default = TRUE)
> mvnPlot(mvt, type = "persp", default = TRUE)
```

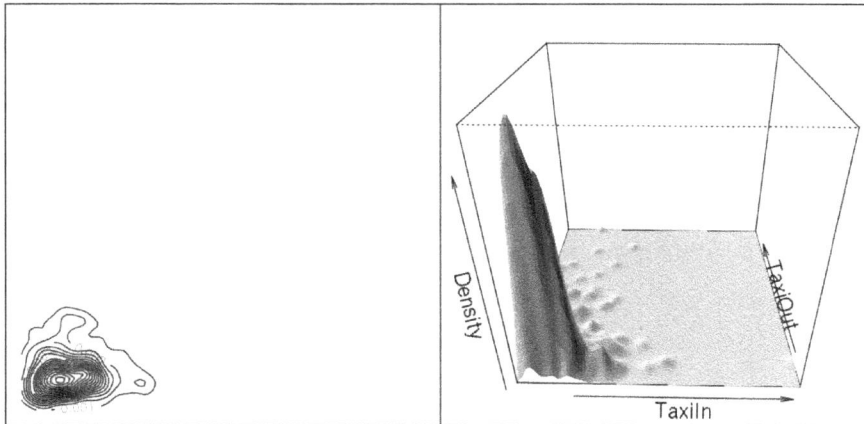

On the right plot, you can see the empirical distribution of the two variables on a perspective plot, where most cases can be found in the bottom-left corner. This means that most flights had only relatively short **TaxiIn** and **TaxiOut** times, which suggests a rather heavy-tailed distribution. The left plot shows a similar image, but from a bird's eye view: the contour lines represent a cross-section of the right-hand side 3D graph. Multivariate normal distribution looks more central, something like a 2-dimensional bell curve:

```
> set.seed(42)
> mvt <- roystonTest(MASS::mvrnorm(100, mu = c(0, 0),
+           Sigma = matrix(c(10, 3, 3, 2), 2)))
> mvnPlot(mvt, type = "contour", default = TRUE)
> mvnPlot(mvt, type = "persp", default = TRUE)
```

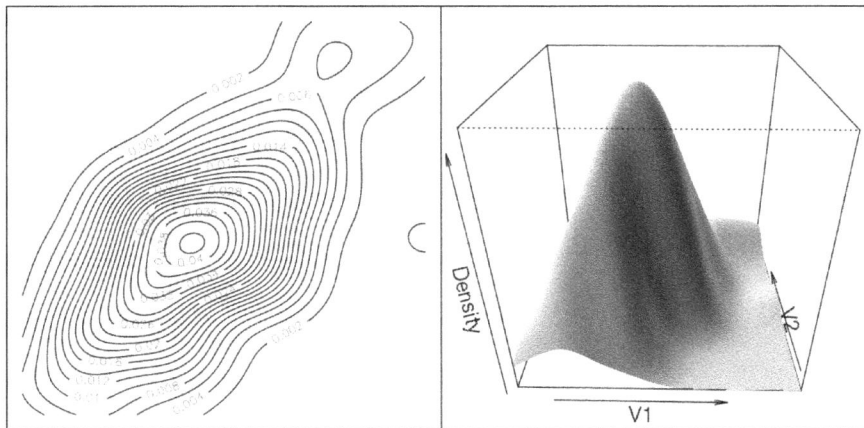

See *Chapter 13, Data Around Us* on how to create similar contour maps on spatial data.

Dependence of variables

Besides normality, relatively high correlation coefficients are desired when applying dimension reduction methods. The reason is that, if there is no statistical relationship between the variables, for example, PCA will return the exact same values without much transformation.

To this end, let's see how the numerical variables of the `hflights` dataset are correlated (the output, being a large matrix, is suppressed this time):

```
> hflights_numeric <- hflights[, which(sapply(hflights, is.numeric))]
> cor(hflights_numeric, use = "pairwise.complete.obs")
```

In the preceding example, we have created a new R object to hold only the numeric columns of the original `hflights` data frame, leaving out five character vectors. Then, we run `cor` with pair-wise deletion of missing values, which returns a matrix with 16 columns and 16 rows:

```
> str(cor(hflights_numeric, use = "pairwise.complete.obs"))
 num [1:16, 1:16] NA NA NA NA NA NA NA NA NA NA ...
 - attr(*, "dimnames")=List of 2
 ..$ : chr [1:16] "Year" "Month" "DayofMonth" "DayOfWeek" ...
 ..$ : chr [1:16] "Year" "Month" "DayofMonth" "DayOfWeek" ...
```

The number of missing values in the resulting correlation matrix seems to be very high. This is because `Year` was 2011 in all cases, thus resulting in a standard variation of zero. It's wise to exclude `Year` along with the non-numeric variables from the dataset—by not only filtering for numeric values, but also checking the variance:

```
> hflights_numeric <- hflights[,which(
+       sapply(hflights, function(x)
+           is.numeric(x) && var(x, na.rm = TRUE) != 0))]
```

Now the number of missing values is a lot lower:

```
> table(is.na(cor(hflights_numeric, use = "pairwise.complete.obs")))
FALSE   TRUE
  209    16
```

Can you guess why we still have some missing values here despite the pair-wise deletion of missing values? Well, running the preceding command results in a rather informative warning, but we will get back to this question later:

```
Warning message:
```

```
In cor(hflights_numeric, use = "pairwise.complete.obs") :
  the standard deviation is zero
```

Let's now proceed with analyzing the actual numbers in the 15x15 correlation matrix, which would be way too large to print in this book. To this end, we did not show the result of the original `cor` command shown previously, but instead, let's rather visualize those 225 numbers with the graphical capabilities of the `ellipse` package:

```
> library(ellipse)
> plotcorr(cor(hflights_numeric, use = "pairwise.complete.obs"))
```

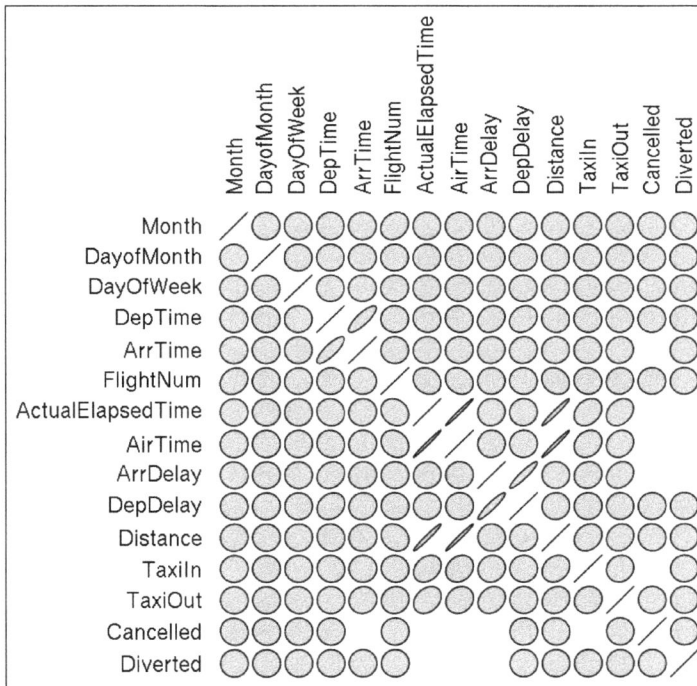

Now we see the values of the correlation matrix represented by ellipses, where:

- A perfect circle stands for the correlation coefficient of zero
- Ellipses with a smaller area reflect the relatively large distance of the correlation coefficient from zero
- The tangent represents the negative/positive sign of the coefficient

To help you with analyzing the preceding results, let's render a similar plot with a few artificially generated numbers that are easier to interpret:

```
> plotcorr(cor(data.frame(
+       1:10,
+       1:10 + runif(10),
+       1:10 + runif(10) * 5,
+       runif(10),
+       10:1,
+       check.names = FALSE)))
```

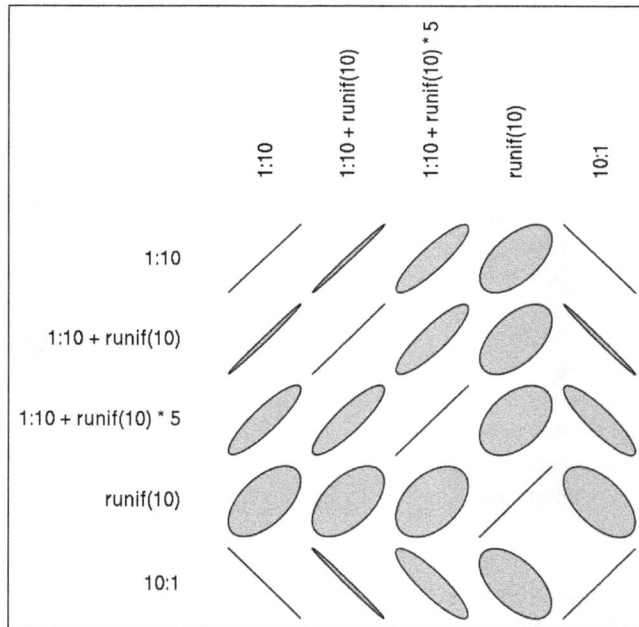

Similar plots on the correlation matrix can be created with the `corrgram` package.

But let's get back to the `hflights` dataset! On the previous diagram, some narrow ellipses are rendered for the time-related variables, which show a relatively high correlation coefficient, and even the `Month` variable seems to be slightly associated with the `FlightNum` function:

```
> cor(hflights$FlightNum, hflights$Month)
[1] 0.2057641
```

On the other hand, the plot shows perfect circles in most cases, which stand for a correlation coefficient around zero. This suggests that most variables are not correlated at all, so computing the principal components of the original dataset would not be very helpful due to the low proportion of common variance.

KMO and Barlett's test

We can verify this assumption on low communalities by a number of statistical tests; for example, the SAS and SPSS folks tend to use KMO or Bartlett's test to see if the data is suitable for PCA. Both algorithms are available in R as well via, for example, via the `psych` package:

```
> library(psych)
> KMO(cor(hflights_numeric, use = "pairwise.complete.obs"))
Error in solve.default(r) :
  system is computationally singular: reciprocal condition number = 0
In addition: Warning message:
In cor(hflights_numeric, use = "pairwise.complete.obs") :
  the standard deviation is zero
matrix is not invertible, image not found
Kaiser-Meyer-Olkin factor adequacy
Call: KMO(r = cor(hflights_numeric, use = "pairwise.complete.obs"))
Overall MSA = NA
MSA for each item =
         Month      DayofMonth      DayOfWeek
          0.5            0.5            0.5
       DepTime         ArrTime        FlightNum
          0.5             NA            0.5
ActualElapsedTime      AirTime         ArrDelay
         NA             NA             NA
       DepDelay        Distance         TaxiIn
          0.5            0.5             NA
       TaxiOut        Cancelled        Diverted
          0.5             NA             NA
```

Unfortunately, the `Overall` MSA (*Measure of Sampling Adequacy*, representing the average correlations between the variables) is not available in the preceding output due to the previously identified missing values of the correlation matrix. Let's pick a pair of variables where the correlation coefficient was NA for further analysis! Such a pair can be easily identified from the previous plot; no circle or ellipse was drawn for missing values, for example, for `Cancelled` and `AirTime`:

```
> cor(hflights_numeric[, c('Cancelled', 'AirTime')])
          Cancelled AirTime
Cancelled         1      NA
AirTime          NA       1
```

This can be explained by the fact, that if a flight is cancelled, then the time spent in the air does not vary much; furthermore, this data is not available:

```
> cancelled <- which(hflights_numeric$Cancelled == 1)
> table(hflights_numeric$AirTime[cancelled], exclude = NULL)
<NA>
2973
```

So we get missing values when calling `cor` due to these NA; similarly, we also get NA when calling `cor` with pair-wise deletion, as only the non-cancelled flights remain in the dataset, resulting in zero variance for the `Cancelled` variable:

```
> table(hflights_numeric$Cancelled)
     0      1
224523   2973
```

This suggests removing the `Cancelled` variable from the dataset before we run the previously discussed assumption tests, as the information stored in that variable is redundantly available in other columns of the dataset as well. Or, in other words, the `Cancelled` column can be computed by a linear transformation of the other columns, which can be left out from further analysis:

```
> hflights_numeric <- subset(hflights_numeric, select = -Cancelled)
```

And let's see if we still have any missing values in the correlation matrix:

```
> which(is.na(cor(hflights_numeric, use = "pairwise.complete.obs")),
+    arr.ind = TRUE)
         row col
Diverted  14   7
Diverted  14   8
Diverted  14   9
```

```
ActualElapsedTime    7   14
AirTime              8   14
ArrDelay             9   14
```

It seems that the `Diverted` column is responsible for a similar situation, and the other three variables were not available when the flight was diverted. After another subset, we are now ready to call KMO on a full correlation matrix:

```
> hflights_numeric <- subset(hflights_numeric, select = -Diverted)
> KMO(cor(hflights_numeric[, -c(14)], use = "pairwise.complete.obs"))
Kaiser-Meyer-Olkin factor adequacy
Call: KMO(r = cor(hflights_numeric[, -c(14)], use = "pairwise.complete.
obs"))
Overall MSA =  0.36
MSA for each item =
              Month       DayofMonth       DayOfWeek
               0.42             0.37            0.35
            DepTime          ArrTime       FlightNum
               0.51             0.49            0.74
  ActualElapsedTime          AirTime        ArrDelay
               0.40             0.40            0.39
           DepDelay         Distance          TaxiIn
               0.38             0.67            0.06
            TaxiOut
               0.06
```

The `Overall MSA`, or the so called **Kaiser-Meyer-Olkin (KMO)** index, is a number between 0 and 1; this value suggests whether the partial correlations of the variables are small enough to continue with data reduction methods. A general rating system or rule of a thumb for KMO can be found in the following table, as suggested by Kaiser:

Value	Description
KMO < 0.5	Unacceptable
0.5 < KMO < 0.6	Miserable
0.6 < KMO < 0.7	Mediocre
0.7 < KMO < 0.8	Middling
0.8 < KMO < 0.9	Meritorious
KMO > 0.9	Marvelous

The KMO index being below 0.5 is considered unacceptable, which basically means that the partial correlation computed from the correlation matrix suggests that the variables are not correlated enough for a meaningful dimension reduction or latent variable model.

Although leaving out some variables with the lowest MSA would improve the `Overall MSA`, and we could build some appropriate models in the following pages, for instructional purposes we won't spend any more time on data transformation for the time being, and we will use the `mtcars` dataset, which was introduced in *Chapter 3, Filtering and Summarizing Data*:

```
> KMO(mtcars)
Kaiser-Meyer-Olkin factor adequacy
Call: KMO(r = mtcars)
Overall MSA =  0.83
MSA for each item =
 mpg  cyl disp   hp drat   wt qsec   vs   am gear carb
0.93 0.90 0.76 0.84 0.95 0.74 0.74 0.91 0.88 0.85 0.62
```

It seems that the `mtcars` database is a great choice for multivariate statistical analysis. This can be also verified by the so-called Bartlett test, which suggests whether the correlation matrix is similar to an identity matrix. Or, in other words, if there is a statistical relationship between the variables. On the other hand, if the correlation matrix has only zeros except for the diagonal, then the variables are independent from each other; thus it would not make much sense to think of multivariate methods. The `psych` package provides an easy-to-use function to compute Bartlett's test as well:

```
> cortest.bartlett(cor(mtcars))
$chisq
[1] 1454.985

$p.value
[1] 3.884209e-268

$df
[1] 55
```

The very low `p-value` suggests that we reject the null-hypothesis of the Bartlett test. This means that the correlation matrix differs from the identity matrix, so the correlation coeffiecients between the variables seem to be closer to 1 than 0. This is in sync with the high KMO value.

Before focusing on the actual statistical methods, please be advised that, although the preceding assumptions make sense in most cases and should be followed as a rule of a thumb, KMO and Bartlett's tests are not always required. High communality is important for factor analysis and other latent models, while for example PCA is a mathematical transformation that will work with even low KMO values.

Principal Component Analysis

Finding the really important fields in databases with a huge number of variables may prove to be a challenging task for the data scientist. This is where **Principal Component Analysis (PCA)** comes into the picture: to find the core components of data. It was invented more than 100 years ago by Karl Pearson, and it has been widely used in diverse fields since then.

The objective of PCA is to interpret the data in a more meaningful structure with the help of orthogonal transformations. This linear transformation is intended to reveal the internal structure of the dataset with an arbitrarily designed new basis in the vector space, which best explains the variance of the data. In plain English, this simply means that we compute new variables from the original data, where these new variables include the variance of the original variables in decreasing order.

This can be either done by eigendecomposition of the covariance, correlation matrix (the so-called R-mode PCA), or singular value decomposition (the so-called Q-mode PCA) of the dataset. Each method has great advantages, such as computation performance, memory requirements, or simply avoiding the prior standardization of the data before passing it to PCA when using a correlation matrix in eigendecomposition.

Either way, PCA can successfully ship a lower-dimensional image of the data, where the uncorrelated principal components are the linear combinations of the original variables. And this informative overview can be a great help to the analyst when identifying the underlying structure of the variables; thus the technique is very often used for exploratory data analysis.

PCA results in the exact same number of extracted components as the original variables. The first component includes most of the common variance, so it has the highest importance in describing the original dataset, while the last component often only includes some unique information from only one original variable. Based on this, we would usually only keep the first few components of PCA for further analysis, but we will also see some use cases where we will concentrate on the extracted unique variance.

PCA algorithms

R provides a variety of functions to run PCA. Although it's possible to compute the components manually by `eigen` or `svd` as R-mode or Q-mode PCA, we will focus on the higher level functions for the sake of simplicity. Relying on my stats-teacher background, I think that sometimes it's more efficient to concentrate on how to run an analysis and interpreting the results rather than spending way too much time with the linear algebra background—especially with given time/page limits.

R-mode PCA can be conducted by `princomp` or `principal` from the `psych` package, while the more preferred Q-mode PCA can be called by `prcomp`. Now let's focus on the latter and see what the components of `mtcars` look like:

```
> prcomp(mtcars, scale = TRUE)
Standard deviations:
 [1] 2.57068 1.62803 0.79196 0.51923 0.47271 0.46000 0.36778 0.35057
 [9] 0.27757 0.22811 0.14847


Rotation:
            PC1        PC2        PC3         PC4        PC5        PC6
mpg   -0.36253   0.016124  -0.225744  -0.0225403   0.102845  -0.108797
cyl    0.37392   0.043744  -0.175311  -0.0025918   0.058484   0.168554
disp   0.36819  -0.049324  -0.061484   0.2566079   0.393995  -0.336165
hp     0.33006   0.248784   0.140015  -0.0676762   0.540047   0.071436
drat  -0.29415   0.274694   0.161189   0.8548287   0.077327   0.244497
wt     0.34610  -0.143038   0.341819   0.2458993  -0.075029  -0.464940
qsec  -0.20046  -0.463375   0.403169   0.0680765  -0.164666  -0.330480
vs    -0.30651  -0.231647   0.428815  -0.2148486   0.599540   0.194017
am    -0.23494   0.429418  -0.205767  -0.0304629   0.089781  -0.570817
gear  -0.20692   0.462349   0.289780  -0.2646905   0.048330  -0.243563
carb   0.21402   0.413571   0.528545  -0.1267892  -0.361319   0.183522
            PC7        PC8        PC9        PC10       PC11
mpg    0.367724  -0.7540914   0.235702   0.139285  -0.1248956
```

```
cyl    0.057278 -0.2308249  0.054035 -0.846419 -0.1406954
disp   0.214303  0.0011421  0.198428  0.049380  0.6606065
hp    -0.001496 -0.2223584 -0.575830  0.247824 -0.2564921
drat   0.021120  0.0321935 -0.046901 -0.101494 -0.0395302
wt    -0.020668 -0.0085719  0.359498  0.094394 -0.5674487
qsec   0.050011 -0.2318400 -0.528377 -0.270673  0.1813618
vs    -0.265781  0.0259351  0.358583 -0.159039  0.0084146
am    -0.587305 -0.0597470 -0.047404 -0.177785  0.0298235
gear   0.605098  0.3361502 -0.001735 -0.213825 -0.0535071
carb  -0.174603 -0.3956291  0.170641  0.072260  0.3195947
```

> Please note that we have called `prcomp` with `scale` set to TRUE, which is FALSE by default due to being backward-compatible with the S language. But in general, scaling is highly recommended. Using the scaling option is equivalent to running PCA on a dataset after scaling it previously, such as: `prcomp(scale(mtcars))`, which results in data with unit variance.

First, `prcomp` returned the standard deviations of the principal components, which shows how much information was preserved by the 11 components. The standard deviation of the first component is a lot larger than any other subsequent value, which explains more than 60 percent of the variance:

```
> summary(prcomp(mtcars, scale = TRUE))
Importance of components:
                          PC1    PC2    PC3    PC4     PC5     PC6     PC7
Standard deviation     2.571  1.628  0.792 0.5192  0.4727  0.4600  0.3678
Proportion of Variance 0.601  0.241  0.057 0.0245  0.0203  0.0192  0.0123
Cumulative Proportion  0.601  0.842  0.899 0.9232  0.9436  0.9628  0.9751
                          PC8    PC9   PC10   PC11
Standard deviation     0.3506  0.278 0.22811  0.148
Proportion of Variance 0.0112  0.007 0.00473  0.002
Cumulative Proportion  0.9863  0.993 0.99800  1.000
```

Besides the first component, only the second one has a higher standard deviation than 1, which means that only the first two components include at least as much information as the original variables did. Or, in other words: only the first two variables have a higher eigenvalue than one. The eigenvalue can be computed by the square of the standard deviation of the principal components, summing up to the number of original variables as expected:

```
> sum(prcomp(scale(mtcars))$sdev^2)
[1] 11
```

Determining the number of components

PCA algorithms always compute the same number of principal components as the number of variables in the original dataset. The importance of the component decreases from the first one to the last one.

As a rule of a thumb, we can simply keep all those components with higher standard deviation than 1. This means that we keep those components, which explains at least as much variance as the original variables do:

```
> prcomp(scale(mtcars))$sdev^2
 [1] 6.608400 2.650468 0.627197 0.269597 0.223451 0.211596 0.135262
 [8] 0.122901 0.077047 0.052035 0.022044
```

So the preceding summary suggests keeping only two components out of the 11, which explains almost 85 percent of the variance:

```
> (6.6 + 2.65) / 11
[1] 0.8409091
```

An alternative and great visualization tool to help us determine the optimal number of component is scree plot. Fortunately, there are at least two great functions in the psych package we can use here: the scree and the VSS.scree functions:

```
> VSS.scree(cor(mtcars))
```

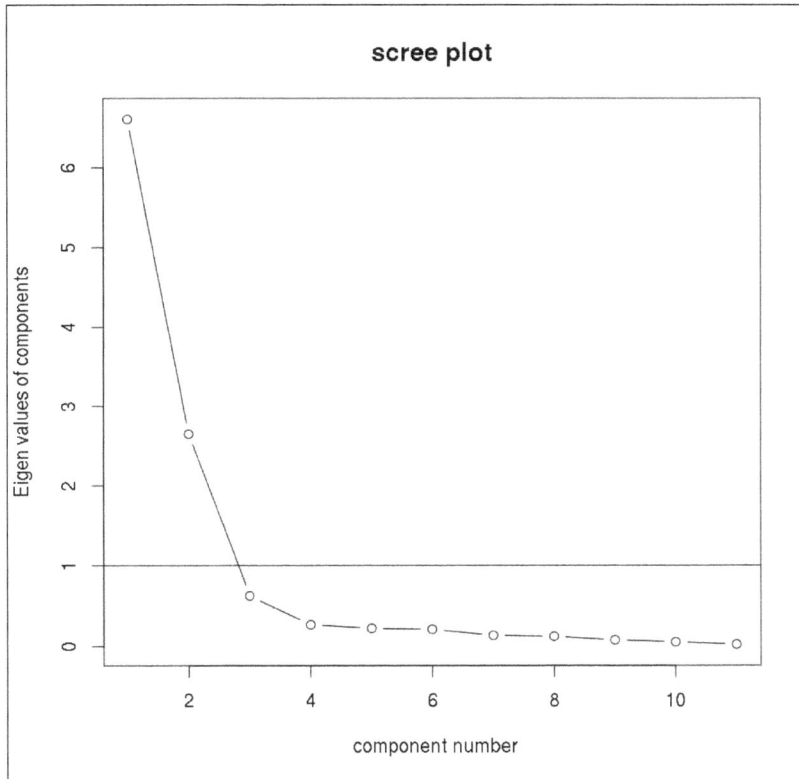

scree plot

```
> scree(cor(mtcars))
```

Parallel Analysis Scree Plots

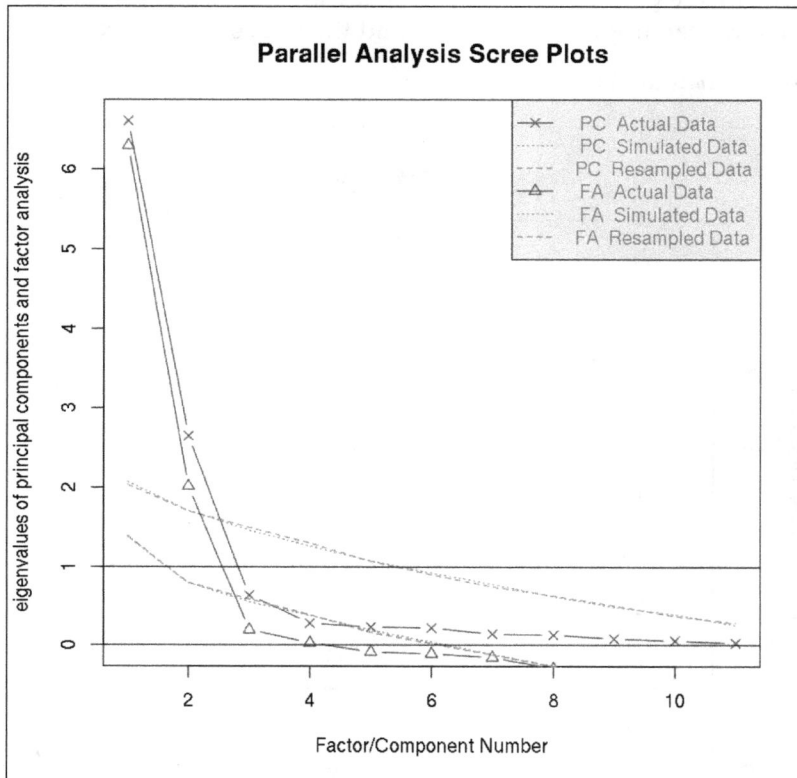

The only difference between the preceding two plots is that `scree` also shows the eigenvalues of a factor analysis besides PCA. Read more about this in the next section of this chapter.

As can be seen, `vss.scree` provides a visual overview on the eigenvalues of the principal components, and it also highlights the critical value at 1 by a horizontal line. This is usually referred to as the Kaiser criterion.

Besides this rule of a thumb, as discussed previously one can also rely on the so-called Elbow-rule, which simply suggests that the line-plot represents an arm and the optimal number of components is the point where this arm's elbow can be found. So we have to look for the point from where the curve becomes less steep. This sharp break is probably at 3 in this case instead of 2, as we have found with the Kaiser criterion.

And besides Cattell's original scree test, we can also compare the previously described `scree` of the components with a bit of a randomized data to identify the optimal number of components to keep:

```
> fa.parallel(mtcars)
```

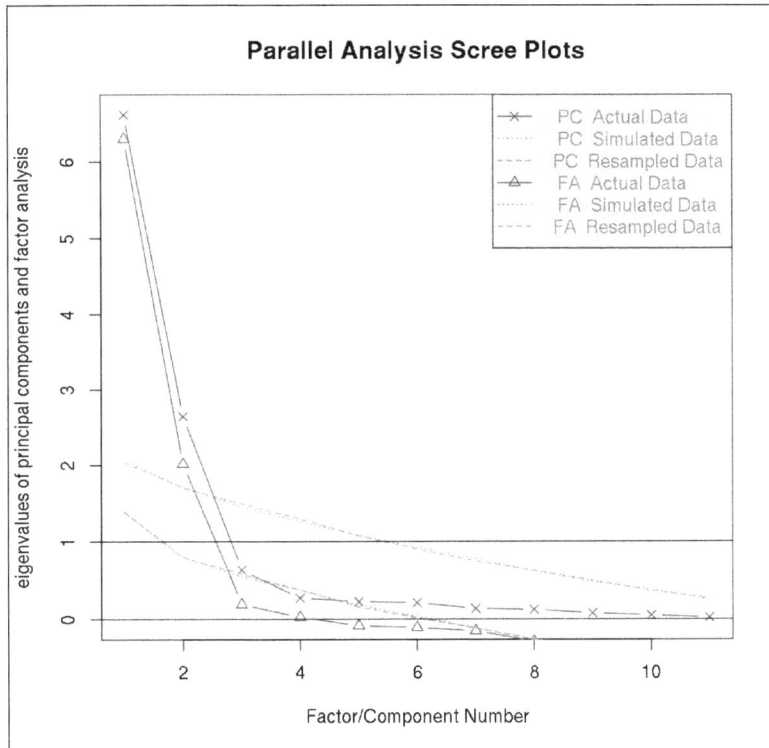

Parallel Analysis Scree Plots

Parallel analysis suggests that the number of factors = 2
and the number of components =   2

Now we have verified the optimal number of principal components to keep for further analysis with a variety of statistical tools, and we can work with only two variables instead of 11 after all, which is great! But what do these artificially created variables actually mean?

Interpreting components

The only problem with reducing the dimension of our data is that it can be very frustrating to find out what our newly created, highly compressed, and transformed data actually is. Now we have PC1 and PC2 for our 32 cars:

```
> pc <- prcomp(mtcars, scale = TRUE)
> head(pc$x[, 1:2])
                       PC1       PC2
Mazda RX4          -0.646863  1.70811
Mazda RX4 Wag      -0.619483  1.52562
Datsun 710         -2.735624 -0.14415
Hornet 4 Drive     -0.306861 -2.32580
Hornet Sportabout   1.943393 -0.74252
Valiant            -0.055253 -2.74212
```

These values were computed by multiplying the original dataset with the identified weights, so-called loadings (rotation) or the component matrix. This is a standard linear transformation:

```
> head(scale(mtcars) %*% pc$rotation[, 1:2])
                       PC1       PC2
Mazda RX4          -0.646863  1.70811
Mazda RX4 Wag      -0.619483  1.52562
Datsun 710         -2.735624 -0.14415
Hornet 4 Drive     -0.306861 -2.32580
Hornet Sportabout   1.943393 -0.74252
Valiant            -0.055253 -2.74212
```

Both variables are scaled with the mean being zero and the standard deviation as described previously:

```
> summary(pc$x[, 1:2])
      PC1               PC2
 Min.   :-4.187    Min.   :-2.742
 1st Qu.:-2.284    1st Qu.:-0.826
 Median :-0.181    Median :-0.305
 Mean   : 0.000    Mean   : 0.000
 3rd Qu.: 2.166    3rd Qu.: 0.672
 Max.   : 3.892    Max.   : 4.311
```

```
> apply(pc$x[, 1:2], 2, sd)
   PC1    PC2
2.5707 1.6280
> pc$sdev[1:2]
[1] 2.5707 1.6280
```

All scores computed by PCA are scaled, because it always returns the values transformed to a new coordinate system with an orthogonal basis, which means that the components are not correlated and scaled:

```
> round(cor(pc$x))
```

	PC1	PC2	PC3	PC4	PC5	PC6	PC7	PC8	PC9	PC10	PC11
PC1	1	0	0	0	0	0	0	0	0	0	0
PC2	0	1	0	0	0	0	0	0	0	0	0
PC3	0	0	1	0	0	0	0	0	0	0	0
PC4	0	0	0	1	0	0	0	0	0	0	0
PC5	0	0	0	0	1	0	0	0	0	0	0
PC6	0	0	0	0	0	1	0	0	0	0	0
PC7	0	0	0	0	0	0	1	0	0	0	0
PC8	0	0	0	0	0	0	0	1	0	0	0
PC9	0	0	0	0	0	0	0	0	1	0	0
PC10	0	0	0	0	0	0	0	0	0	1	0
PC11	0	0	0	0	0	0	0	0	0	0	1

To see what the principal components actually mean, it's really helpful to check the loadings matrix, as we have seen before:

```
> pc$rotation[, 1:2]
          PC1       PC2
mpg  -0.36253  0.016124
cyl   0.37392  0.043744
disp  0.36819 -0.049324
hp    0.33006  0.248784
drat -0.29415  0.274694
wt    0.34610 -0.143038
qsec -0.20046 -0.463375
vs   -0.30651 -0.231647
am   -0.23494  0.429418
gear -0.20692  0.462349
carb  0.21402  0.413571
```

Probably this analytical table might be more meaningful in some visual way, for example as a `biplot`, which shows not only the original variables but also the observations (black labels) on the same plot with the new coordinate system based on the principal components (red labels):

```
> biplot(pc, cex = c(0.8, 1.2))
> abline(h = 0, v = 0, lty = 'dashed')
```

We can conclude that PC1 includes information mostly from the number of cylinders (cyl), displacement (disp), weight (wt), and gas consumption (mpg), although the latter looks likely to decrease the value of PC1. This was found by checking the highest and lowest values on the PC1 axis. Similarly, we find that PC2 is constructed by speed-up (qsec), number of gears (gear), carburetors (carb), and the transmission type (am).

To verify this, we can easily compute the correlation coefficient between the original variables and the principal components:

```
> cor(mtcars, pc$x[, 1:2])
          PC1       PC2
mpg  -0.93195  0.026251
cyl   0.96122  0.071216
disp  0.94649 -0.080301
hp    0.84847  0.405027
drat -0.75617  0.447209
wt    0.88972 -0.232870
qsec -0.51531 -0.754386
vs   -0.78794 -0.377127
am   -0.60396  0.699103
gear -0.53192  0.752715
carb  0.55017  0.673304
```

Does this make sense? How would you name PC1 and PC2? The number of cylinders and displacement seem like engine parameters, while the weight is probably rather influenced by the body of the car. Gas consumption should be affected by both specs. The other component's variables deal with suspension, but we also have speed there, not to mention the bunch of mediocre correlation coefficients in the preceding matrix. Now what?

Rotation methods

Based on the fact that rotation methods are done in a subspace, rotation is always suboptimal compared to the previously discussed PCA. This means that the new axes after rotation will explain less variance than the original components.

On the other hand, rotation simplifies the structure of the components and thus makes it a lot easier to understand and interpret the results; thus, these methods are often used in practice.

> Rotation methods can be (and are) usually applied to both PCA and FA (more on this later). Orthogonal methods are preferred.

There are two main types of rotation:

- Orthogonal, where the new axes are orthogonal to each other. There is no correlation between the components/factors.

- Oblique, where the new axes are not necessarily orthogonal to each other; thus there might be some correlation between the variables.

Varimax rotation is one of the most popular rotation methods. It was developed by Kaiser in 1958 and has been popular ever since. It is often used because the method maximizes the variance of the loadings matrix, resulting in more interpretable scores:

```
> varimax(pc$rotation[, 1:2])
$loadings
      PC1     PC2
mpg  -0.286 -0.223
cyl   0.256  0.276
disp  0.312  0.201
hp           0.403
drat -0.402
wt    0.356  0.116
qsec  0.148 -0.483
vs          -0.375
am   -0.457  0.174
gear -0.458  0.217
carb -0.106  0.454


                  PC1   PC2
SS loadings       1.000 1.000
Proportion Var    0.091 0.091
Cumulative Var    0.091 0.182


$rotmat
          [,1]     [,2]
[1,]   0.76067 0.64914
[2,]  -0.64914 0.76067
```

Now the first component seems to be mostly affected (negatively dominated) by the transmission type, number of gears, and rear axle ratio, while the second one is affected by speed-up, horsepower, and the number of carburetors. This suggests naming PC2 as power, while PC1 instead refers to transmission. Let's see those 32 automobiles in this new coordinate system:

```
> pcv <- varimax(pc$rotation[, 1:2])$loadings
```

```
> plot(scale(mtcars) %*% pcv, type = 'n',
+     xlab = 'Transmission', ylab = 'Power')
> text(scale(mtcars) %*% pcv, labels = rownames(mtcars))
```

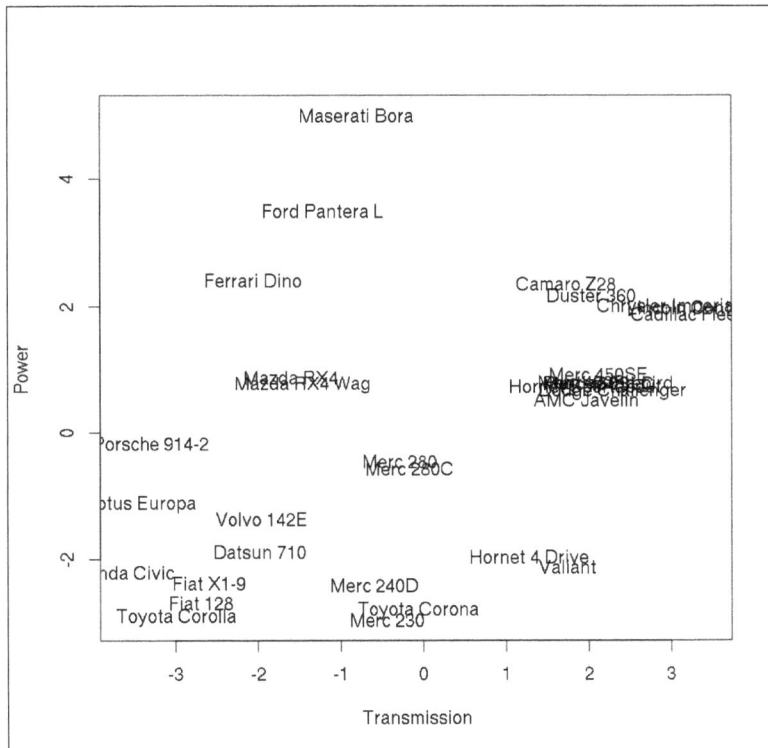

Based on the preceding plot, every data scientist should pick a car from the upper left quarter to go with the top rated models, right? Those cars have great power based on the y axis and good transmission systems, as shown on the x axis — do not forget about the transmission being negatively correlated with the original variables. But let's see some other rotation methods and the advantages of those as well!

Quartimax rotation is an orthogonal method, as well, and minimizes the number of components needed to explain each variable. This often results in a general component and additional smaller components. When a compromise between Varimax and Quartimax rotation methods is needed, you might opt for Equimax rotation.

Oblique rotation methods include Oblimin and Promax, which are not available in the base stats or even the highly used `psych` package. Instead, we can load the `GPArotation` package, which provides a wide range of rotation methods for PCA and FA as well. For demonstration purposes, let's see how Promax rotation works, which is a lot faster compared to, for example, Oblimin:

```
> library(GPArotation)
> promax(pc$rotation[, 1:2])
$loadings

Loadings:
       PC1    PC2
mpg  -0.252 -0.199
cyl   0.211  0.258
disp  0.282  0.174
hp           0.408
drat -0.416
wt    0.344
qsec  0.243 -0.517
vs          -0.380
am   -0.502  0.232
gear -0.510  0.276
carb -0.194  0.482


                 PC1   PC2
SS loadings     1.088 1.088
Proportion Var  0.099 0.099
Cumulative Var  0.099 0.198


$rotmat
          [,1]     [,2]
[1,]   0.65862 0.58828
[2,]  -0.80871 0.86123


> cor(promax(pc$rotation[, 1:2])$loadings)
         PC1       PC2
PC1  1.00000  -0.23999
PC2 -0.23999   1.00000
```

The result of the last command supports the view that oblique rotation methods generate scores that might be correlated, unlike when running an orthogonal rotation.

Outlier-detection with PCA

PCA can be used for a variety of goals besides exploratory data analysis. For example, we can use PCA to generate eigenfaces, compress images, classify observations, or to detect outliers in a multidimensional space via image filtering. Now, we will construct a simplified model discussed in a related research post published on R-bloggers in 2012: `http://www.r-bloggers.com/finding-a-pin-in-a-haystack-pca-image-filtering`.

The challenge described in the post was to detect a foreign metal object in the sand photographed by the Curiosity Rover on the Mars. The image can be found at the official NASA website at `http://www.nasa.gov/images/content/694811main_pia16225-43_full.jpg`, for which I've created a shortened URL for future use: `http://bit.ly/nasa-img`.

In the following image, you can see a strange metal object highlighted in the sand in a black circle, just to make sure you know what we are looking for. The image found at the preceding URL does not have this highlight:

And now let's use some statistical methods to identify that object without (much) human intervention! First, we need to download the image from the Internet and load it into R. The `jpeg` package will be really helpful here:

```
> library(jpeg)
> t <- tempfile()
> download.file('http://bit.ly/nasa-img', t)
trying URL 'http://bit.ly/nasa-img'
Content type 'image/jpeg' length 853981 bytes (833 Kb)
opened URL
=====================================================
downloaded 833 Kb

> img <- readJPEG(t)
> str(img)
 num [1:1009, 1:1345, 1:3] 0.431 0.42 0.463 0.486 0.49 ...
```

The `readJPEG` function returns the RGB values of every pixel in the picture, resulting in a three dimensional array where the first dimension is the row, the second is the column, and the third dimension includes the three color values.

> RGB is an additive color model that can reproduce a wide variety of colors by mixing red, green, and blue by given intensities and optional transparency. This color model is highly used in computer science.

As PCA requires a matrix as an input, we have to convert this 3-dimensional array to a 2-dimensional dataset. To this end, let's not bother with the order of pixels for the time being, as we can reconstruct that later, but let's simply list the RGB values of all pixels, one after the other:

```
> h <- dim(img)[1]
> w <- dim(img)[2]
> m <- matrix(img, h*w)
> str(m)
 num [1:1357105, 1:3] 0.431 0.42 0.463 0.486 0.49 ...
```

In a nutshell, we saved the original height of the image (in pixels) in variable h, saved the width in w, and then converted the 3D array to a matrix with 1,357,105 rows. And, after four lines of data loading and three lines of data transformation, we can call the actual, rather simplified statistical method at last:

```
> pca <- prcomp(m)
```

As we've seen before, data scientists do indeed deal with data preparation most of the time, while the actual data analysis can be done easily, right?

The extracted components seems to perform pretty well; the first component explains more than 96 percent of the variance:

```
> summary(pca)
Importance of components:
                        PC1    PC2     PC3
Standard deviation     0.277 0.0518 0.00765
Proportion of Variance 0.965 0.0338 0.00074
Cumulative Proportion  0.965 0.9993 1.00000
```

Previously, interpreting RGB values was pretty straightforward, but what do these components mean?

```
> pca$rotation
         PC1      PC2      PC3
[1,] -0.62188  0.71514  0.31911
[2,] -0.57409 -0.13919 -0.80687
[3,] -0.53261 -0.68498  0.49712
```

It seems that the first component is rather mixed with all three colors, the second component misses the green color, while the third component includes almost only green. Why not visualize that instead of trying to imagine how these artificial values look? To this end, let's extract the color intensities from the preceding component/ loading matrix by the following quick helper function:

```
> extractColors <- function(x)
+       rgb(x[1], x[2], x[3])
```

Calling this on the absolute values of the component matrix results in the hex-color codes that describe the principal components:

```
> (colors <- apply(abs(pca$rotation), 2, extractColors))
      PC1       PC2       PC3
"#9F9288" "#B623AF" "#51CE7F"
```

These color codes can be easily rendered — for example, on a pie chart, where the area of the pies represents the explained variance of the principal components:

```
> pie(pca$sdev, col = colors, labels = colors)
```

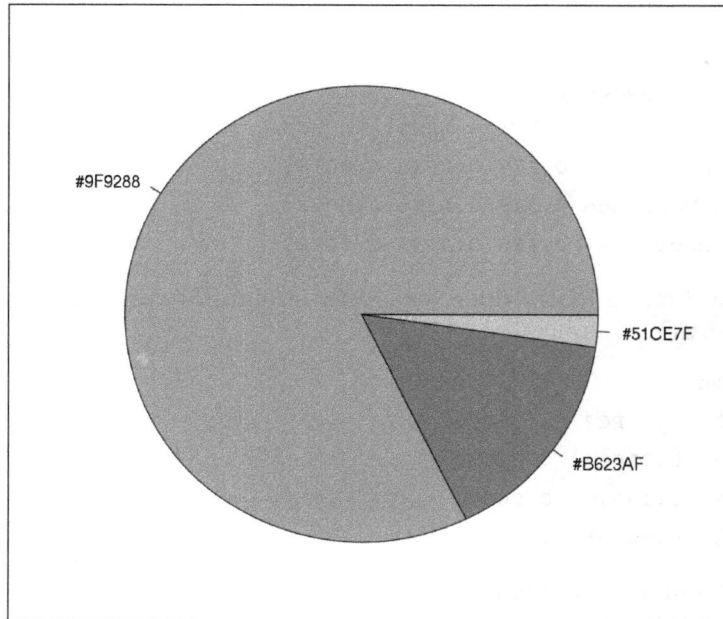

Now we no longer have red, green, or blue intensities or actual colors in the computed scores stored in pca$x; rather, the principal components describe each pixel with the visualized colors shown previously. And, as previously discussed, the third component stands for a greenish color, the second one misses green (resulting in a purple color), while the first component includes a rather high value from all RGB colors resulting in a tawny color, which is not surprising at all knowing that the photo was taken in the desert of Mars.

Now we can render the original image with monochrome colors to show the intensity of the principal components. The following few lines of code produce two modified photos of the Curiosity Rover and its environment based on PC1 and PC2:

```
> par(mfrow = c(1, 2), mar = rep(0, 4))
> image(matrix(pca$x[, 1], h), col = gray.colors(100))
> image(matrix(pca$x[, 2], h), col = gray.colors(100), yaxt = 'n')
```

Although the image was rotated by 90 degrees in some of the linear transformations, it's pretty clear that the first image was not really helpful in finding the foreign metal object in the sand. As a matter of fact, this image represents the noise in the desert area, as `PC1` included sand-like color intensities, so this component is useful for describing the variety of tawny colors.

On the other hand, the second component highlights the metal object in the sand very well! All surrounding pixels are dim, due to the low ratio of purple color in normal sand, while the anomalous object is rather dark.

I really like this piece of R code and the simplified example: although they're still basic enough to follow, they also demonstrate the power of R and how standard data analytic methods can be used to harvest information from raw data.

Factor analysis

Although the literature on confirmatory **factor analysis** (**FA**) is really impressive and is being highly used in, for example, social sciences, we will only focus on exploratory FA, where our goal is to identify some unknown, not observed variables based on other empirical data.

The latent variable model of FA was first introduced in 1904 by Spearman for one factor, and then Thurstone generalized the model for more than one factor in 1947. This statistical model assumes that the manifest variables available in the dataset are the results of latent variables that were not observed but can be tracked based on the observed data.

FA can deal with continuous (numeric) variables, and the model states that each observed variable is the sum of some unknown, latent factors.

> Please note the that normality, KMO, and Bartlett's tests are a lot more important to check before doing FA compared to PCA; the latter is a rather descriptive method while, in FA, we are actually building a model.

The most used exploratory FA method is maximum-likelihood FA, which is also available in the `factanal` function in the already installed `stats` package. Other factoring methods are made available by the `fa` functions in the `psych` package — for example, **ordinary least squares (OLS)**, **weighted least squares (WLS)**, **generalized weighted least squares (GLS)**, or principal factor solution. These functions take raw data or the covariance matrix as input.

For demonstration purposes, let's see how the default factoring method performs on a subset of `mtcars`. Let's extract all performance-related variables except for displacement, which is probably accountable for all the other relevant metrics:

```
> m <- subset(mtcars, select = c(mpg, cyl, hp, carb))
```

Now simply call and save the results of `fa` on the preceding `data.frame`:

```
> (f <- fa(m))
Factor Analysis using method =   minres
Call: fa(r = m)
Standardized loadings (pattern matrix) based upon correlation matrix
       MR1   h2   u2  com
mpg  -0.87 0.77 0.23    1
cyl   0.91 0.83 0.17    1
hp    0.92 0.85 0.15    1
carb  0.69 0.48 0.52    1

                 MR1
SS loadings     2.93
Proportion Var  0.73

Mean item complexity =   1
```

```
Test of the hypothesis that 1 factor is sufficient.


The degrees of freedom for the null model are   6
and the objective function was   3.44 with Chi Square of   99.21
The degrees of freedom for the model are 2
and the objective function was   0.42


The root mean square of the residuals (RMSR) is   0.07
The df corrected root mean square of the residuals is   0.12


The harmonic number of observations is   32
with the empirical chi square   1.92   with prob <   0.38
The total number of observations was   32
with MLE Chi Square =   11.78   with prob <   0.0028


Tucker Lewis Index of factoring reliability =   0.677
RMSEA index =   0.42
and the 90 % confidence intervals are   0.196 0.619
BIC =   4.84
Fit based upon off diagonal values = 0.99
Measures of factor score adequacy
                                                MR1
Correlation of scores with factors              0.97
Multiple R square of scores with factors        0.94
Minimum correlation of possible factor scores   0.87
```

Well, this is a rather impressive amount of information with a bunch of details! MR1 stands for the first extracted factor named after the default factoring method (Minimal Residuals or OLS). Since there is only one factor included in the model, rotation of factors is not an option. There is a test or hypothesis to check whether the numbers of factors are sufficient, and some coefficients represent a really great model fit.

The results can be summarized on the following plot:

```
> fa.diagram(f)
```

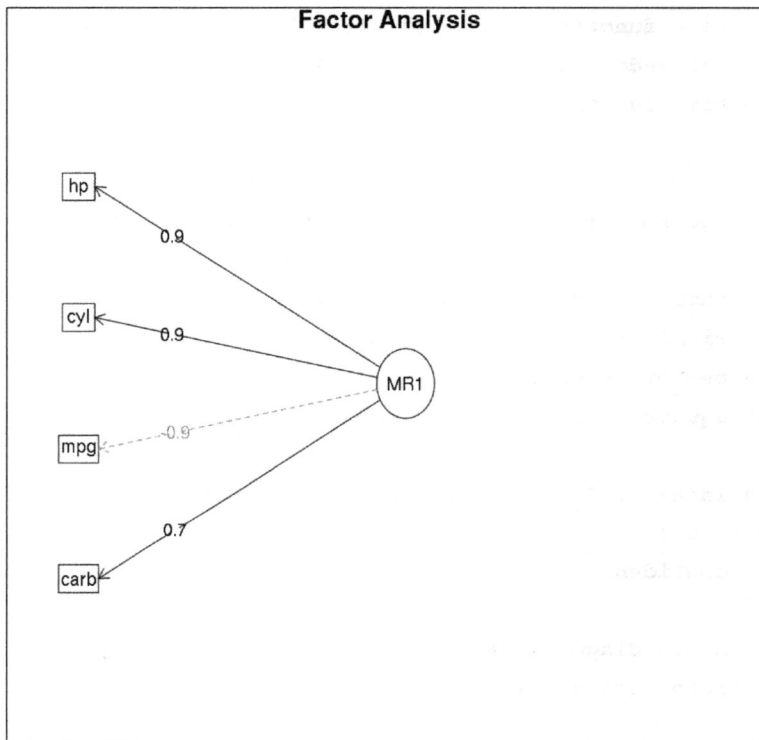

Here we see the high correlation coefficients between the latent and the observed variables, and the direction of the arrows suggests that the factor has an effect on the values found in our empirical dataset. Guess the relationship between this factor and the displacement of the car engines!

```
> cor(f$scores, mtcars$disp)
0.87595
```

Well, this seems like a good match.

Principal Component Analysis versus Factor Analysis

Unfortunately, principal components are often confused with factors, and the two terms and related methods are sometimes used as synonyms, although the mathematical background and goals of the two methods are really different.

PCA is used to reduce the number of variables by creating principal components that then can be used in further projects instead of the original variables. This means that we try to extract the essence of the dataset in the means of artificially created variables, which best describe the variance of the data:

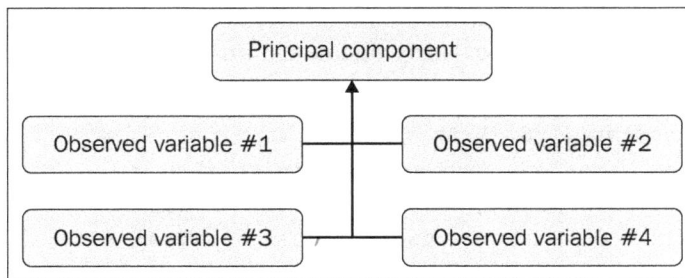

FA is the other way around, as it tries to identify unknown, latent variables to explain the original data. In plain English, we use the manifest variables from our empirical dataset to guess the internal structure of the data:

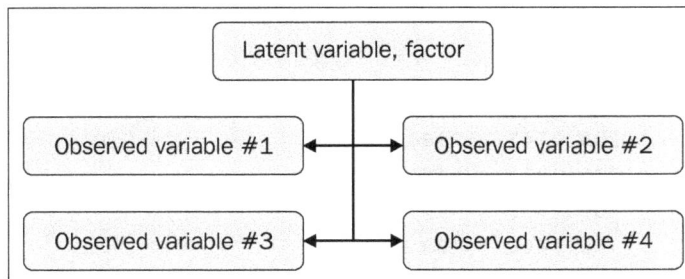

Multidimensional Scaling

Multidimensional Scaling (**MDS**) is a multivariate technique that was first used in geography. The main goal of MDS is to plot multivariate data points in two dimensions, thus revealing the structure of the dataset by visualizing the relative distance of the observations. MDA is used in diverse fields such as attitude study in psychology, sociology, and market research.

While the MASS package provides non-metric MDS via the isoMDS function, we will concentrate on the classical metric MDS, which is available in the cmdscale function offered by the stats package. Both types of MDS take a distance matrix as the main argument and can be created from any numeric tabular data by the dist function.

But before we explore more complex examples, let's see what MDS can offer us while working with an already existing distance matrix, such as the built-in eurodist dataset:

```
> as.matrix(eurodist)[1:5, 1:5]
          Athens Barcelona Brussels Calais Cherbourg
Athens         0      3313     2963   3175      3339
Barcelona   3313         0     1318   1326      1294
Brussels    2963      1318        0    204       583
Calais      3175      1326      204      0       460
Cherbourg   3339      1294      583    460         0
```

The preceding values represents the travel distance between 21 European cities in kilometers, although only the first 5-5 values were shown. Running classical MDS is fairly easy:

```
> (mds <- cmdscale(eurodist))
                      [,1]       [,2]
Athens           2290.2747   1798.803
Barcelona        -825.3828    546.811
Brussels           59.1833   -367.081
Calais            -82.8460   -429.915
Cherbourg        -352.4994   -290.908
Cologne           293.6896   -405.312
Copenhagen        681.9315  -1108.645
Geneva             -9.4234    240.406
Gibraltar       -2048.4491    642.459
Hamburg           561.1090   -773.369
Hook of Holland   164.9218   -549.367
Lisbon          -1935.0408     49.125
```

```
Lyons         -226.4232    187.088
Madrid       -1423.3537    305.875
Marseilles    -299.4987    388.807
Milan          260.8780    416.674
Munich         587.6757     81.182
Paris         -156.8363   -211.139
Rome           709.4133   1109.367
Stockholm      839.4459  -1836.791
Vienna         911.2305    205.930
```

These scores are very similar to two principal components, such as running `prcomp(eurodist)$x[, 1:2]`. As a matter of fact, PCA can be considered as the most basic MDS solution.

Anyway, we have just transformed the 21-dimensional space into 2 dimensions, which can be plotted very easily (unlike the previous matrix with 21 rows and 21 columns):

```
> plot(mds)
```

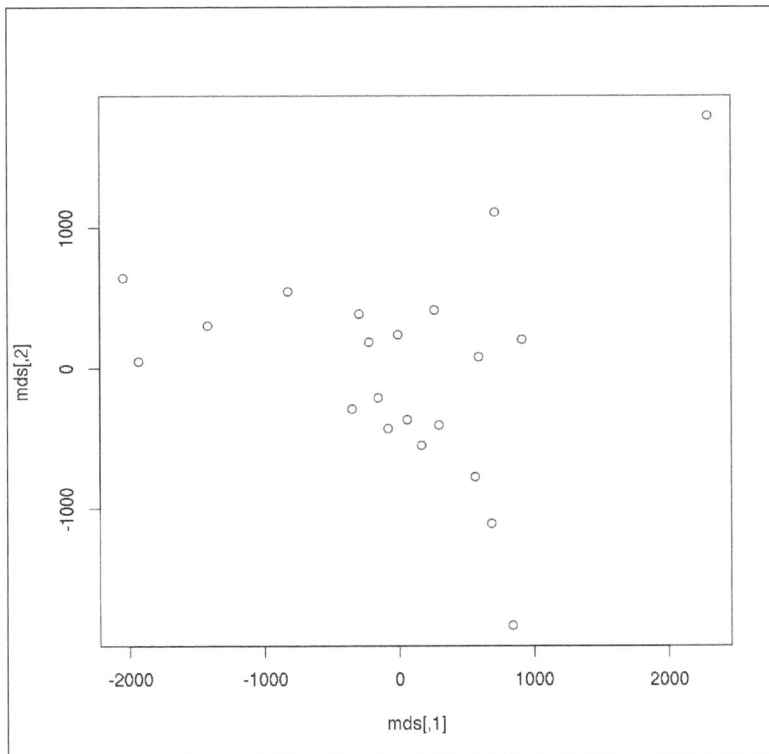

Does this ring a bell? If not, please feel free to see the following image, where the following two lines of code also show the city names instead of the anonymous points:

```
> plot(mds, type = 'n')
> text(mds[, 1], mds[, 2], labels(eurodist))
```

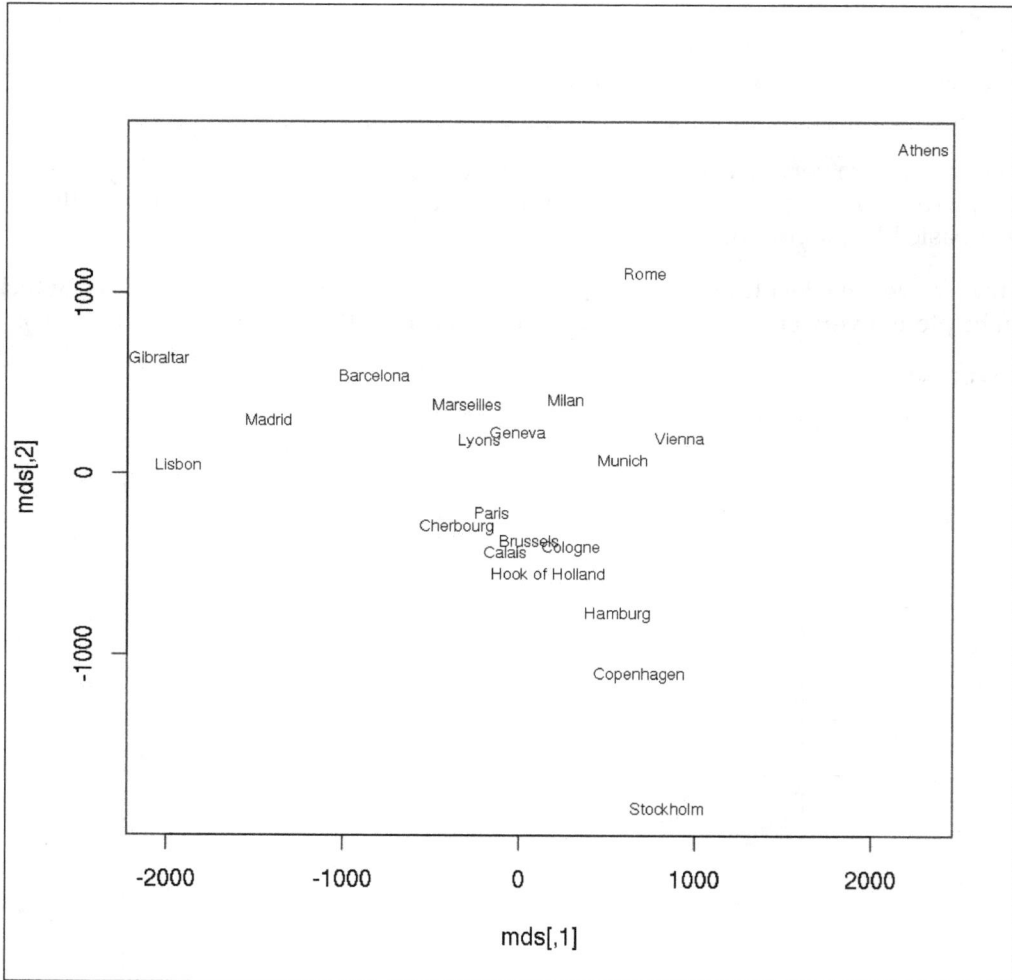

Although the y axis is flipped, which you can fix by multiplying the second argument of text by -1, we have just rendered a European map of cities from the distance matrix—without any further geographical data. I find this rather impressive.

Please find more data visualization tricks and methods in *Chapter 13, Data Around Us.*

Now let's see how to apply MDS on non-geographic data that was not prepared with a view to its being a distance matrix. Let's get back to the `mtcars` dataset:

```
> mds <- cmdscale(dist(mtcars))
> plot(mds, type = 'n')
> text(mds[, 1], mds[, 2], rownames(mds))
```

The plot shows the 32 cars of the original dataset scattered in a two-dimensional space. The distance between the elements was computed by MDS, which took into account all the 11 original variables, and it's very easy to identify the similar and very different car types. We will cover these topics in more details in the next chapter, *Chapter 10, Classification and Clustering*.

Summary

In this chapter, we covered a number of ways to deal with multivariate data to reduce the number of available dimensions in the means of artificially computed continuous variables and to identify underlying, latent, and similarly numeric variables. On the other hand, sometimes it's rather difficult to describe reality with numbers and we should rather think in categories.

The next chapter will introduce new methods to define data types (clusters) and will also demonstrate how to classify elements with the help of available training data.

10
Classification and Clustering

In the previous chapter, we concentrated on how to compress information found in a number of continuous variables into a smaller set of numbers, but these statistical methods are somewhat limited when we are dealing with categorized data, for example when analyzing surveys.

Although some methods try to convert discrete variables into numeric ones, such as by using a number of dummy or indicator variables, in most cases it's simply better to think about our research design goals instead of trying to forcibly use previously learned methods in the analysis.

> We can replace a categorical variable with a number of dummy variables by creating a new variable for each label of the original discrete variable, and then assign *1* to the related column and *0* to all the others. Such values can be used as numeric variables in statistical analysis, especially with regression models.

When we analyze a sample and target population via categorical variables, usually we are not interested in individual cases, but instead in similar elements and groups. Similar elements can be defined as rows in a dataset with similar values in the columns.

In this chapter, we will discuss different *supervised* and *unsupervised* ways to identify similar cases in a dataset, such as:

- Hierarchical clustering
- K-means clustering
- Some machine learning algorithms
- Latent class model
- Discriminant analysis
- Logistic regression

Cluster analysis

Clustering is an unsupervised data analysis method that is used in diverse fields, such as pattern recognition, social sciences, and pharmacy. The aim of cluster analysis is to make homogeneous subgroups called clusters, where the objects in the same cluster are similar, and the clusters differ from each other.

Hierarchical clustering

Cluster analysis is one of the most well known and popular pattern recognition methods; thus, there are many clustering models and algorithms analyzing the distribution, density, possible center points, and so on in the dataset. In this section we are going to examine some hierarchical clustering methods.

Hierarchical clustering can be either agglomerative or divisive. In agglomerative methods every case starts out as an individual cluster, then the closest clusters are merged together in an iterative manner, until finally they merge into one single cluster, which includes all elements of the original dataset. The biggest problem with this approach is that distances between clusters have to be recalculated at each iteration, which makes it extremely slow on large data. I'd rather not suggest trying to run the following commands on the `hflights` dataset.

Divisive methods on the other hand take a top-down approach. They start from a single cluster, which is then iteratively divided into smaller groups until they are all singletons.

The `stats` package contains the `hclust` function for hierarchical clustering that takes a distance matrix as an input. To see how it works, let's use the `mtcars` dataset that we already analyzed in *Chapter 3, Filtering and Summarizing Data* and *Chapter 9, From Big to Smaller Data*. The `dist` function is also familiar from the latter chapter:

```
> d <- dist(mtcars)
> h <- hclust(d)
> h

Call:
hclust(d = d)

Cluster method   : complete
Distance         : euclidean
Number of objects: 32
```

Well, this is a way too brief output and only shows that our distance matrix included 32 elements and the clustering method. A visual representation of the results will be a lot more useful for such a small dataset:

```
> plot(h)
```

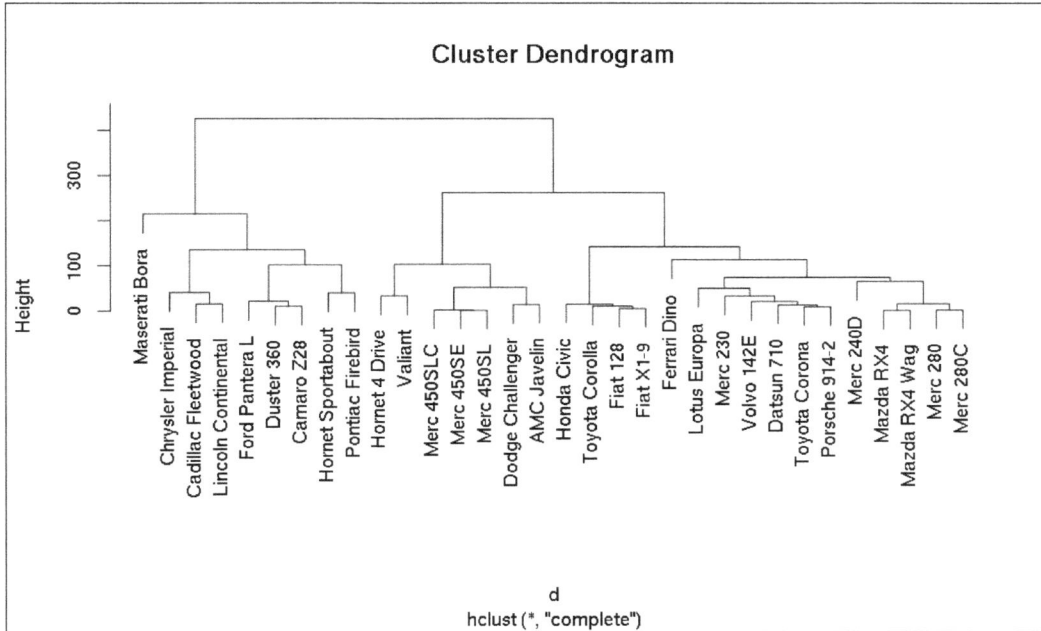

Cluster Dendrogram

d
hclust (*, "complete")

By plotting this `hclust` object, we obtained a *dendrogram*, which shows how the clusters are formed. It can be useful for determining the number of clusters, although in datasets with numerous cases it becomes difficult to interpret. A horizontal line can be drawn to any given height on the *y* axis so that the *n* number of intersections with the line provides a n-cluster solution.

R can provide very convenient ways of visualizing the clusters on the *dendrogram*. In the following plot, the red boxes show the cluster membership of a three-cluster solution on top of the previous plot:

```
> plot(h)
> rect.hclust(h, k=3, border = "red")
```

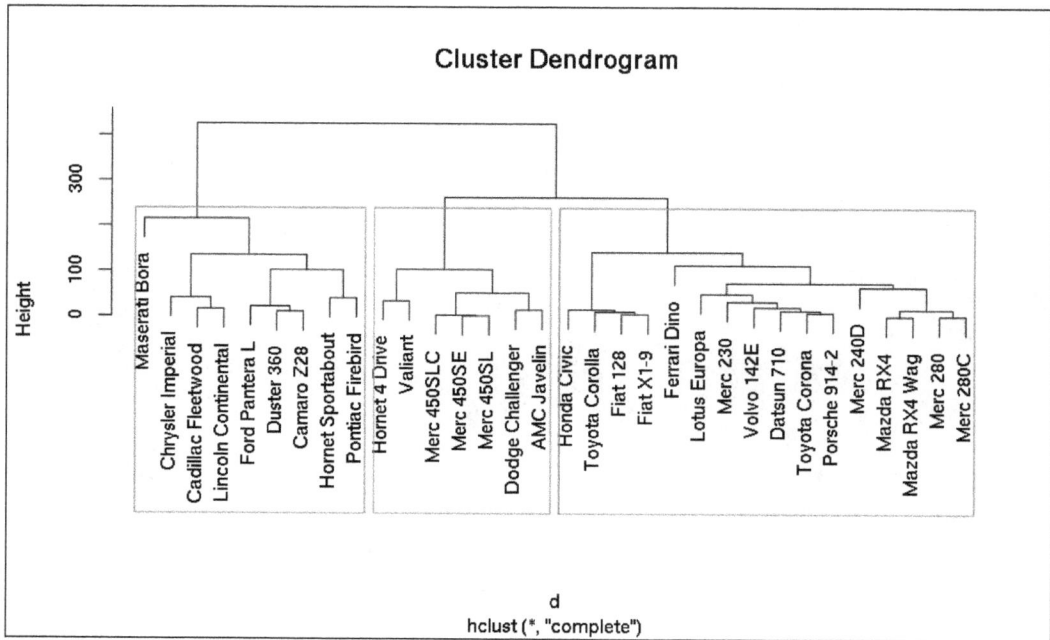

Although this graph looks nice and it is extremely useful to have similar elements grouped together, for bigger datasets, it becomes hard to see through. Instead, we might be rather interested in the actual cluster membership represented in a vector:

```
> (cn <- cutree(h, k = 3))
         Mazda RX4      Mazda RX4 Wag         Datsun 710
                 1                  1                  1
    Hornet 4 Drive   Hornet Sportabout            Valiant
                 2                  3                  2
        Duster 360           Merc 240D           Merc 230
                 3                  1                  1
          Merc 280           Merc 280C          Merc 450SE
                 1                  1                  2
        Merc 450SL         Merc 450SLC  Cadillac Fleetwood
```

2	2	3
Lincoln Continental	Chrysler Imperial	Fiat 128
3	3	1
Honda Civic	Toyota Corolla	Toyota Corona
1	1	1
Dodge Challenger	AMC Javelin	Camaro Z28
2	2	3
Pontiac Firebird	Fiat X1-9	Porsche 914-2
3	1	1
Lotus Europa	Ford Pantera L	Ferrari Dino
1	3	1
Maserati Bora	Volvo 142E	
3	1	

And the number of elements in the resulting clusters as a frequency table:

```
> table(cn)

 1  2  3
16  7  9
```

It seems that *Cluster 1*, the third cluster on the preceding plot, has the most elements. Can you guess how this group differs from the other two clusters? Well, those readers who are familiar with car names might be able to guess the answer, but let's see what the numbers actually show:

> Please note that we use the round function in the following examples to suppress the number of decimal places to 1 or 4 in the code output to fit the page width.

```
> round(aggregate(mtcars, FUN = mean, by = list(cn)), 1)
  Group.1  mpg cyl  disp    hp drat  wt qsec  vs  am gear carb
1       1 24.5 4.6 122.3  96.9  4.0 2.5 18.5 0.8 0.7  4.1  2.4
2       2 17.0 7.4 276.1 150.7  3.0 3.6 18.1 0.3 0.0  3.0  2.1
3       3 14.6 8.0 388.2 232.1  3.3 4.2 16.4 0.0 0.2  3.4  4.0
```

There's a really spectacular difference in the average performance and gas consumption between the clusters! What about the standard deviation inside the groups?

```
> round(aggregate(mtcars, FUN = sd, by = list(cn)), 1)
  Group.1 mpg cyl disp   hp drat  wt qsec  vs  am gear carb
1       1 5.0   1 34.6 31.0  0.3 0.6  1.8 0.4 0.5  0.5  1.5
```

```
2        2 2.2   1 30.2 32.5   0.2 0.3   1.2 0.5 0.0   0.0   0.9
3        3 3.1   0 58.1 49.4   0.4 0.9   1.3 0.0 0.4   0.9   1.7
```

These values are pretty low compared to the standard deviations in the original dataset:

```
> round(sapply(mtcars, sd), 1)
   mpg   cyl  disp    hp  drat    wt  qsec    vs    am  gear  carb
   6.0   1.8 123.9  68.6   0.5   1.0   1.8   0.5   0.5   0.7   1.6
```

And the same applies when compared to the standard deviation between the groups as well:

```
> round(apply(
+    aggregate(mtcars, FUN = mean, by = list(cn)),
+    2, sd), 1)
Group.1     mpg    cyl   disp     hp   drat     wt   qsec
   1.0     5.1    1.8  133.5   68.1    0.5    0.8    1.1
    vs      am   gear   carb
   0.4     0.4    0.6    1.0
```

This means that we achieved our original goal to identify similar elements of our data and organize those in groups that differ from each other. But why did we split the original data into exactly three artificially defined groups? Why not two, four, or even more?

Determining the ideal number of clusters

The NbClust package offers a very convenient way to do some exploratory data analysis on our data before running the actual cluster analysis. The main function of the package can compute 30 different indices, all designed to determine the ideal number of groups. These include:

- Single link
- Average
- Complete link
- McQuitty
- Centroid (cluster center)
- Median
- K-means
- Ward

After loading the package, let's start with a visual method representing the possible number of clusters in our data—on a knee plot, which might be familiar from *Chapter 9, From Big to Smaller Data*, where you can also find some more information about the following elbow-rule:

```
> library(NbClust)
> NbClust(mtcars, method = 'complete', index = 'dindex')
```

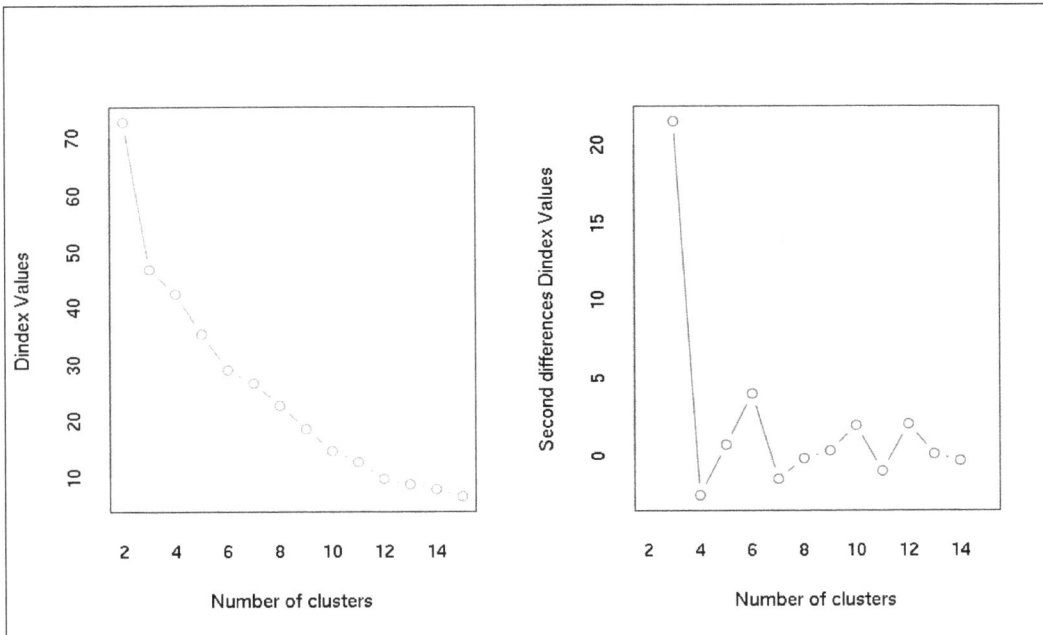

In the preceding plots, we traditionally look for the *elbow*, but the second differences plot on the right might be more straightforward for most readers. There we are interested in where the most significant peak can be found, which suggests that choosing three groups would be ideal when clustering the mtcars dataset.

Unfortunately, running all NbClust methods fails on such a small dataset. Thus, for demonstrational purposes, we are now running only a few standard methods and filtering the results for the suggested number of clusters via the related list element:

```
> NbClust(mtcars, method = 'complete', index = 'hartigan')$Best.nc
All 32 observations were used.
```

```
Number_clusters    Value_Index
        3.0000        34.1696
```

```
> NbClust(mtcars, method = 'complete', index = 'kl')$Best.nc
All 32 observations were used.
```

```
Number_clusters      Value_Index
        3.0000           6.8235
```

Both the Hartigan and Krzanowski-Lai indexes suggest sticking to three clusters. Let's view the `iris` dataset as well, which includes a lot more cases with fewer numeric columns, so we can run all available methods:

```
> NbClust(iris[, -5], method = 'complete', index = 'all')$Best.nc[1,]
All 150 observations were used.
```

```
*******************************************************************
* Among all indices:
* 2 proposed 2 as the best number of clusters
* 13 proposed 3 as the best number of clusters
* 5 proposed 4 as the best number of clusters
* 1 proposed 6 as the best number of clusters
* 2 proposed 15 as the best number of clusters


                ***** Conclusion *****


* According to the majority rule, the best number of clusters is  3

*******************************************************************
```

KL	CH	Hartigan	CCC	Scott	Marriot
4	4	3	3	3	3
TrCovW	TraceW	Friedman	Rubin	Cindex	DB
3	3	4	6	3	3
Silhouette	Duda	PseudoT2	Beale	Ratkowsky	Ball
2	4	4	3	3	3
PtBiserial	Frey	McClain	Dunn	Hubert	SDindex
3	1	2	15	0	3
Dindex	SDbw				
0	15				

The output summarizes that the ideal number of clusters is three based on the 13 methods returning that number, five further methods suggest four clusters, and a few other cluster numbers were also computed by a much smaller number of methods.

These methods are not only useful with the previously discussed hierarchical clustering, but generally used with k-means clustering as well, where the number of clusters is to be defined before running the analysis — unlike the hierarchical method, where we cut the dendogram after the heavy computations have already been run.

K-means clustering

K-means clustering is a non-hierarchical method first described by MacQueen in 1967. Its big advantage over hierarchical clustering is its great performance.

> Unlike hierarchical cluster analysis, k-means clustering requires you to determine the number of clusters before running the actual analysis.

The algorithm runs the following steps in a nutshell:

1. Initialize a predefined (k) number of randomly chosen centroids in space.
2. Assign each object to the cluster with the closest centroid.
3. Recalculate centroids.
4. Repeat the second and third steps until convergence.

We are going to use the kmeans function from the stats package. As k-means clustering requires a prior decision on the number of clusters, we can either use the NbClust function described previously, or we can come up with an arbitrary number that fits the goals of the analysis.

According to the previously defined optimal cluster number in the previous section, we are going to stick to three groups, where the within-cluster sum of squares ceases to drop significantly:

```
> (k <- kmeans(mtcars, 3))
K-means clustering with 3 clusters of sizes 16, 7, 9

Cluster means:
      mpg      cyl     disp       hp     drat       wt     qsec
1 24.50000 4.625000 122.2937  96.8750 4.002500 2.518000 18.54312
2 17.01429 7.428571 276.0571 150.7143 2.994286 3.601429 18.11857
```

```
3 14.64444 8.000000 388.2222 232.1111 3.343333 4.161556 16.40444
        vs        am      gear      carb
1 0.7500000 0.6875000 4.125000 2.437500
2 0.2857143 0.0000000 3.000000 2.142857
3 0.0000000 0.2222222 3.444444 4.000000
```

Clustering vector:

Mazda RX4	Mazda RX4 Wag	Datsun 710
1	1	1
Hornet 4 Drive	Hornet Sportabout	Valiant
2	3	2
Duster 360	Merc 240D	Merc 230
3	1	1
Merc 280	Merc 280C	Merc 450SE
1	1	2
Merc 450SL	Merc 450SLC	Cadillac Fleetwood
2	2	3
Lincoln Continental	Chrysler Imperial	Fiat 128
3	3	1
Honda Civic	Toyota Corolla	Toyota Corona
1	1	1
Dodge Challenger	AMC Javelin	Camaro Z28
2	2	3
Pontiac Firebird	Fiat X1-9	Porsche 914-2
3	1	1
Lotus Europa	Ford Pantera L	Ferrari Dino
1	3	1
Maserati Bora	Volvo 142E	
3	1	

```
Within cluster sum of squares by cluster:
[1] 32838.00 11846.09 46659.32
 (between_SS / total_SS =  85.3 %)


Available components:


[1] "cluster"       "centers"      "totss"      "withinss"
[5] "tot.withinss" "betweenss"    "size"       "iter"
[9] "ifault"
```

The cluster means show some really important characteristics for each cluster, which we generated manually for the hierarchical clusters in the previous section. We can see that, in the first cluster, the cars have high mpg (low gas consumption), on average four cylinders (in contrast to six or eight), rather low performance and so on. The output also automatically reveals the actual cluster numbers.

Let's compare these to the clusters defined by the hierarchical method:

```
> all(cn == k$cluster)
[1] TRUE
```

The results seem to be pretty stable, right?

The cluster numbers have no meaning and their order is arbitrary. In other words, the cluster membership is a nominal variable. Based on this, the preceding R command might return FALSE instead of TRUE when the cluster numbers were allocated in a different order, but comparing the actual cluster membership will verify that we have found the very same groups. See for example cbind(cn, k$cluster) to generate a table including both cluster memberships.

Visualizing clusters

Plotting these clusters is also a great way to understand groupings. To this end, we will use the `clusplot` function from the `cluster` package. For easier understanding, this function reduces the number of dimensions to two, in a similar way to when we are conducting a PCA or MDS (described in *Chapter 9, From Big to Smaller Data*):

```
> library(cluster)
> clusplot(mtcars, k$cluster, color = TRUE, shade = TRUE, labels = 2)
```

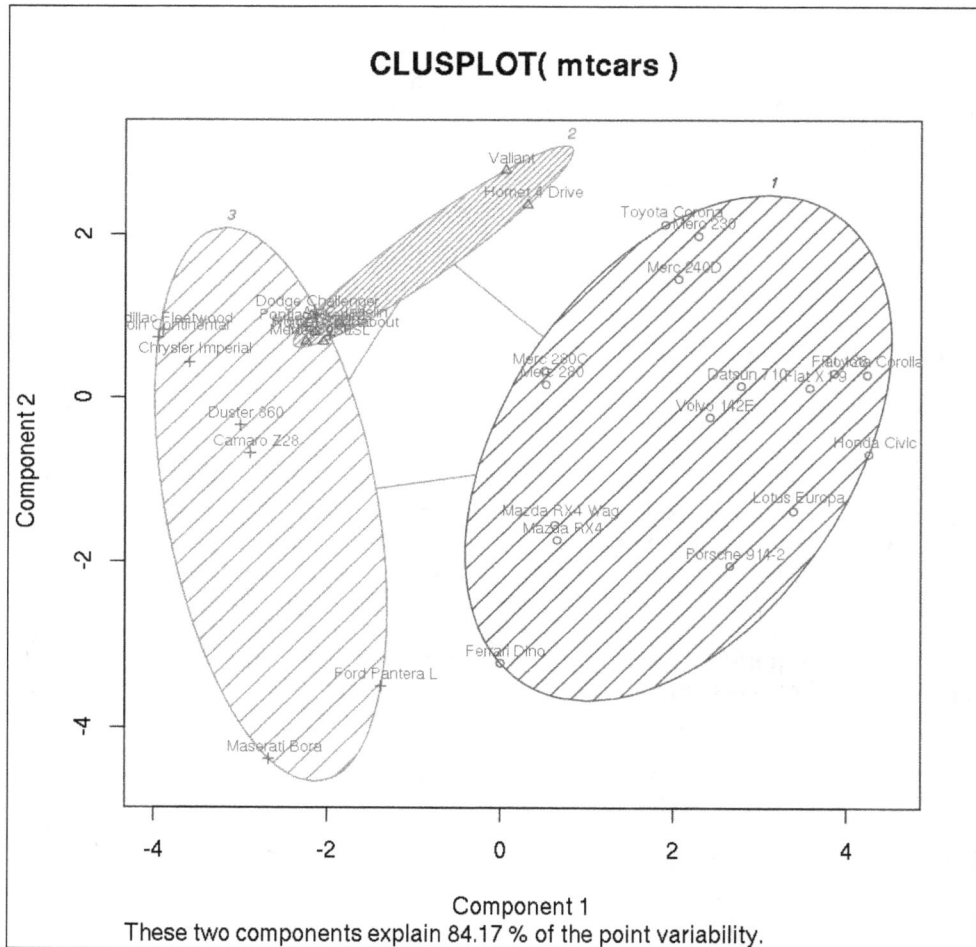

As you can see, after the dimension reduction, the two components explain 84.17 percent of variance, so this small information loss is a great trade-off in favor of an easier understanding of the clusters.

Visualizing the relative density of the ellipses with the `shade` parameter can also help us realize how similar the elements of the same groups are. And we used the labels argument to show both the points and cluster labels as well. Be sure to stick to the default of *0* (no labels) or *4* (only ellipse labels) when visualizing large number of elements.

Latent class models

Latent Class Analysis (LCA) is a method for identifying latent variables among polychromous outcome variables. It is similar to factor analysis, but can be used with discrete/categorical data. To this end, LCA is mostly used when analyzing surveys.

In this section, we are going to use the `poLCA` function from the `poLCA` package. It uses expectation-maximization and Newton-Raphson algorithms for finding the maximum likelihood for the parameters.

The `poLCA` function requires the data to be coded as integers starting from one or as a factor, otherwise it will produce an error message. To this end, let's transform some of the variables in the `mtcars` dataset to factors:

```
> factors <- c('cyl', 'vs', 'am', 'carb', 'gear')
> mtcars[, factors] <- lapply(mtcars[, factors], factor)
```

> The preceding command will overwrite the `mtcars` dataset in your current R session. To revert to the original dataset for other examples, please delete this updated dataset from the session by `rm(mtcars)` if needed.

Latent Class Analysis

Now that the data is in an appropriate format, we can conduct the LCA. The related function comes with a number of important arguments:

- First, we have to define a formula that describes the model. Depending on the formula, we can define LCA (similar to clustering but with discrete variables) or **Latent Class Regression (LCR)** model.

- The `nclass` argument specifies the number of latent classes assumed in the model, which is 2 by default. Based on the previous examples in this chapter, we will override this to 3.

- We can use the `maxiter`, `tol`, `probs.start`, and `nrep` parameters to fine-tune the model.

- The `graphs` argument can display or suppress the parameter estimates.

Let's start with basic LCA of three latent classes defined by all the available discrete variables:

```
> library(poLCA)
> p <- poLCA(cbind(cyl, vs, am, carb, gear) ~ 1,
+    data = mtcars, graphs = TRUE, nclass = 3)
```

The first part of the output (which can be also accessed via the `probs` element of the preceding saved `poLCA` list) summarizes the probabilities of the outcome variables by each latent class:

```
> p$probs
Conditional item response (column) probabilities,
 by outcome variable, for each class (row)


$cyl
              4      6 8
class 1:  0.3333 0.6667 0
class 2:  0.6667 0.3333 0
class 3:  0.0000 0.0000 1


$vs
              0      1
class 1:  0.0000 1.0000
class 2:  0.2667 0.7333
class 3:  1.0000 0.0000


$am
              0      1
class 1:  1.0000 0.0000
class 2:  0.2667 0.7333
class 3:  0.8571 0.1429


$carb
              1      2      3      4      6      8
class 1:  1.0000 0.0000 0.0000 0.0000 0.0000 0.0000
class 2:  0.2667 0.4000 0.0000 0.2667 0.0667 0.0000
class 3:  0.0000 0.2857 0.2143 0.4286 0.0000 0.0714
```

```
$gear
                 3     4      5
class 1:   1.0000 0.0 0.0000
class 2:   0.0000 0.8 0.2000
class 3:   0.8571 0.0 0.1429
```

From these probabilities, we can see that all 8-cylinder cars belong to the third class, the first one only includes cars with automatic transmission, one carburetor, three gears, and so on. The exact same values can be plotted as well by setting the graph parameter to TRUE in the function call, or by calling the plot function directly afterwards:

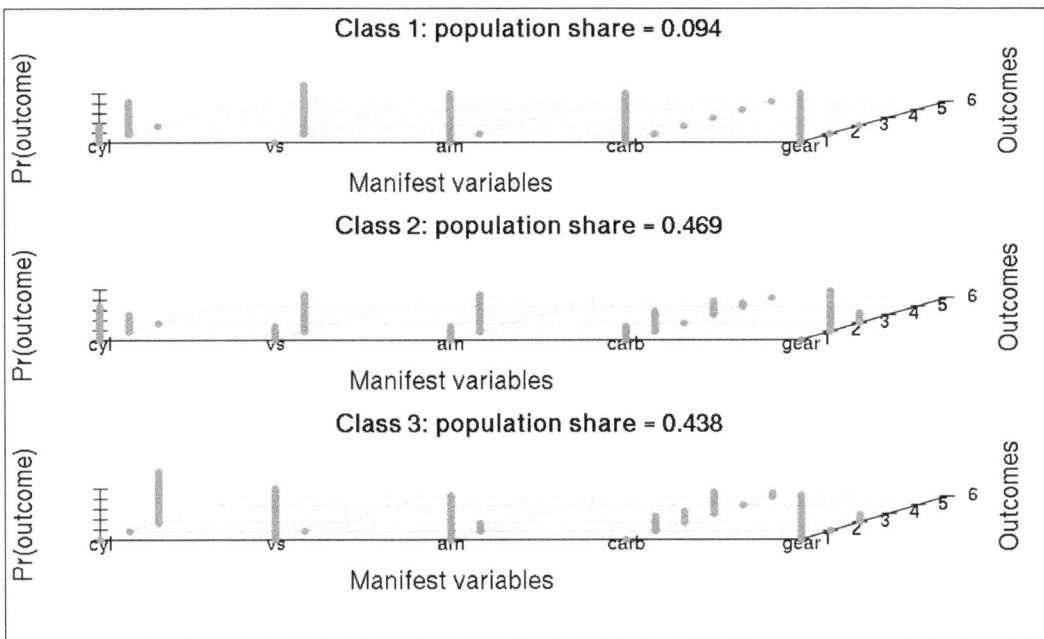

The plot is also useful in highlighting that the first latent class includes only a few elements compared to the other classes (also known as "Estimated class population shares"):

```
> p$P
[1] 0.09375 0.46875 0.43750
```

The `poLCA` object can also reveal a bunch of other important information about the results. Just to name a few, let's see the named list parts of the object, which can be extracted via the standard `$` operator:

- The `predclass` returns the most likely class memberships
- On the other hand, the posterior element is a matrix containing the class membership probabilities of each case
- The **Akaike Information Criterion** (`aic`), **Bayesian Information Criterion** (`bic`), **deviance** (`Gsq`), and `Chisq` values represent different measures of goodness of fit

LCR models

On the other hand, the LCR model is a supervised method, where we are not mainly interested in the latent variables explaining our observations at the exploratory data analysis scale, but instead we are using training data from which one or more covariates predict the probability of the latent class membership.

Discriminant analysis

Discriminant Function Analysis (DA) refers to the process of determining which continuous independent (predictor) variables discriminate between a discrete dependent (response) variable's categories, which can be considered as a reversed **Multivariate Analysis of Variance (MANOVA)**.

This suggests that DA is very similar to logistic regression (see *Chapter 6, Beyond the Linear Trend Line (authored by Renata Nemeth and Gergely Toth)* and the following section), which is more generally used because of its flexibility. While logistic regression can handle both categorical and continuous data, DA requires numeric independent variables and has a few further requirements that logistic regression does not have:

- Normal distribution is assumed
- Outliers should be eliminated
- No two variables should be highly correlated (multi-collinearity)
- The sample size of the smallest category should be higher than the number of predictor values
- The number of independent variables should not exceed the sample size

There are two different types of DA, and we will use `lda` from the `MASS` package for the linear discriminant function, and `qda` for the quadratic discriminant function.

Let us start with the dependent variable being the number of gears, and we will use all the other numeric values as independent variables. To make sure that we start with a standard `mtcars` dataset not overwritten in the preceding examples, let's clear the namespace and update the gear column to include categories instead of the actual numeric values:

```
> rm(mtcars)
> mtcars$gear <- factor(mtcars$gear)
```

Due to the low number of observations (and as we have already discussed the related options in *Chapter 9, From Big to Smaller Data*), we can now set aside conducting the normality and other tests. Let's proceed with the actual analysis.

We call the `lda` function, setting **cross validation** (**CV**) to `TRUE`, so that we can test the accuracy of the prediction. The dot in the formula refers to all variables except the explicitly mentioned gear:

```
> library(MASS)
> d <- lda(gear ~ ., data = mtcars, CV =TRUE)
```

So now we can check the accuracy of the predictions by comparing them to the original values via the confusion matrix:

```
> (tab <- table(mtcars$gear, d$class))

     3  4  5
  3 14  1  0
  4  2 10  0
  5  1  1  3
```

To present relative percentages instead of the raw numbers, we can do some quick transformations:

```
> tab / rowSums(tab)
              3          4          5
  3 0.93333333 0.06666667 0.00000000
  4 0.16666667 0.83333333 0.00000000
  5 0.20000000 0.20000000 0.60000000
```

And we can also compute the percentage of missed predictions:

```
> sum(diag(tab)) / sum(tab)
[1] 0.84375
```

After all, around 84 percent of the cases got classified into their most likely respective classes, which were made up from the actual probabilities that can be extracted by the `posterior` element of the list:

```
> round(d$posterior, 4)
```

	3	4	5
Mazda RX4	0.0000	0.8220	0.1780
Mazda RX4 Wag	0.0000	0.9905	0.0095
Datsun 710	0.0018	0.6960	0.3022
Hornet 4 Drive	0.9999	0.0001	0.0000
Hornet Sportabout	1.0000	0.0000	0.0000
Valiant	0.9999	0.0001	0.0000
Duster 360	0.9993	0.0000	0.0007
Merc 240D	0.6954	0.2990	0.0056
Merc 230	1.0000	0.0000	0.0000
Merc 280	0.0000	1.0000	0.0000
Merc 280C	0.0000	1.0000	0.0000
Merc 450SE	1.0000	0.0000	0.0000
Merc 450SL	1.0000	0.0000	0.0000
Merc 450SLC	1.0000	0.0000	0.0000
Cadillac Fleetwood	1.0000	0.0000	0.0000
Lincoln Continental	1.0000	0.0000	0.0000
Chrysler Imperial	1.0000	0.0000	0.0000
Fiat 128	0.0000	0.9993	0.0007
Honda Civic	0.0000	1.0000	0.0000
Toyota Corolla	0.0000	0.9995	0.0005
Toyota Corona	0.0112	0.8302	0.1586
Dodge Challenger	1.0000	0.0000	0.0000
AMC Javelin	1.0000	0.0000	0.0000
Camaro Z28	0.9955	0.0000	0.0044
Pontiac Firebird	1.0000	0.0000	0.0000
Fiat X1-9	0.0000	0.9991	0.0009
Porsche 914-2	0.0000	1.0000	0.0000
Lotus Europa	0.0000	0.0234	0.9766
Ford Pantera L	0.9965	0.0035	0.0000
Ferrari Dino	0.0000	0.0670	0.9330
Maserati Bora	0.0000	0.0000	1.0000
Volvo 142E	0.0000	0.9898	0.0102

Now we can run `lda` again without cross validation to see the actual discriminants and how the different categories of `gear` are structured:

```
> d <- lda(gear ~ ., data = mtcars)
> plot(d)
```

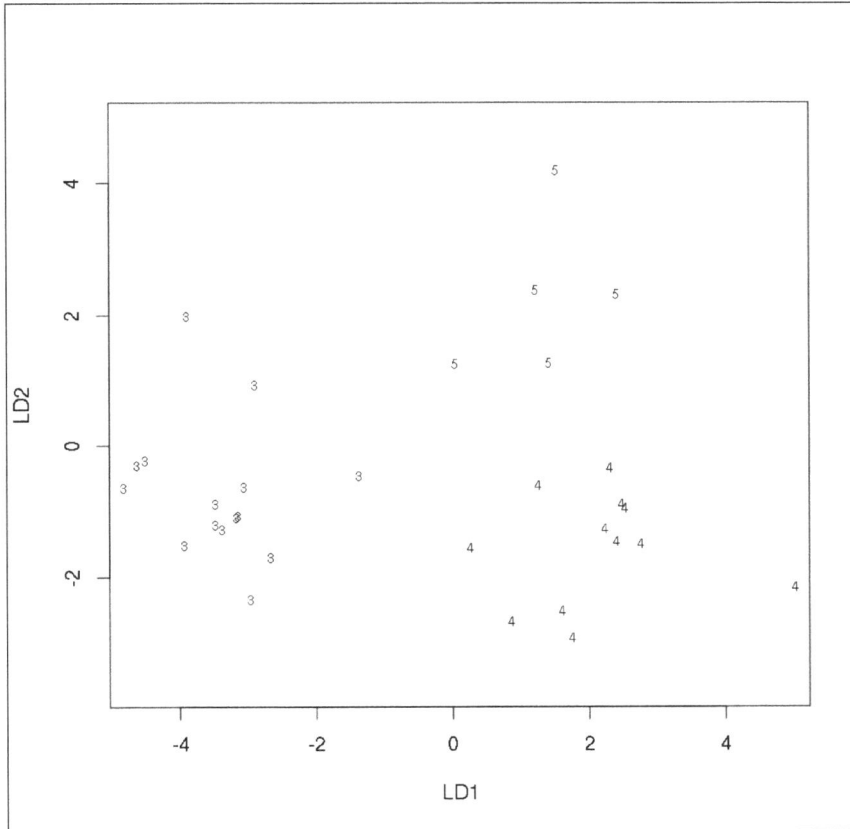

The numbers in the preceding plot stand for the cars in the `mtcars` dataset presented by the actual number of gears. It is really straightforward that the elements rendered by the two discriminants highlight the similarity of cars with the same number of gears and the difference between those with unequal values in the `gear` column.

These discriminants can be also extracted from the d object by calling `predict`, or can directly be rendered on a histogram to see the distribution of this continuous variable by the categories of the independent variable:

```
> plot(d, dimen = 1, type = "both" )
```

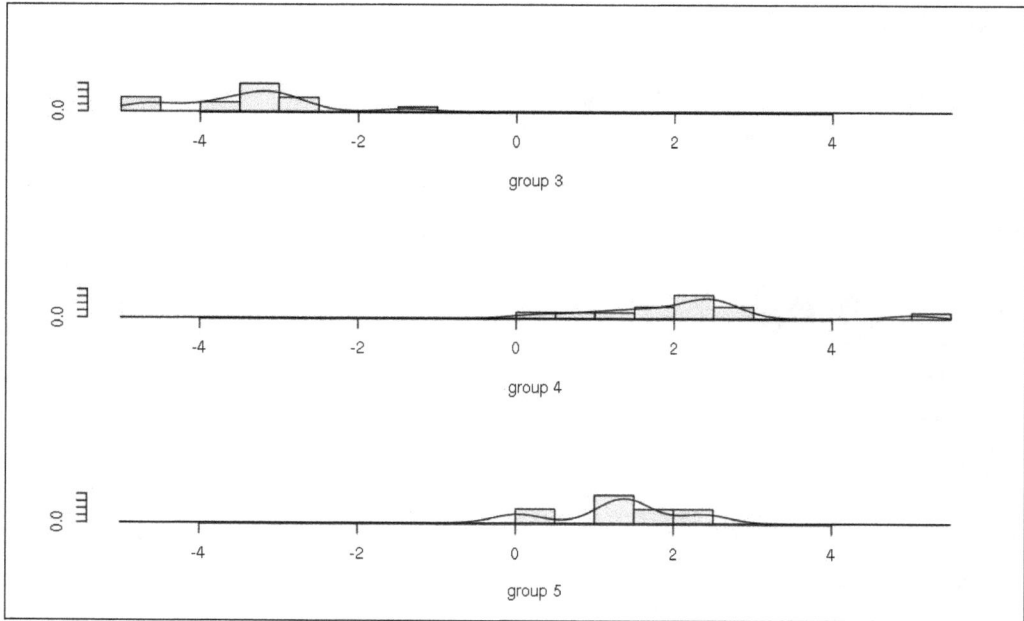

Logistic regression

Although logistic regression was partly covered in *Chapter 6, Beyond the Linear Trend Line (authored by Renata Nemeth and Gergely Toth)*, as it's often used to solve classification problems we will revisit this topic again with some related examples and some notes on — for example — the multinomial version of logistic regression, which was not introduced in the previous chapters.

Our data often does not meet the requirements of the *discriminant analysis*. In such cases, using logistic, logit, or probit regression can be a reasonable choice, as these methods are not sensitive to non-normal distribution and unequal variances within each group; on the other hand, they require much larger sample sizes. For small sample sizes, discriminant analysis is much more reliable.

As a rule of thumb, you should have at least 50 observations for each independent variable, which means that, if we want to build a logistic regression model for the `mtcars` dataset as earlier, we will need at least 500 observations — but we have only 32.

To this end, we will restrict this section to one or two quick examples on how to conduct a logit regression—for example, to estimate whether a car has automatic or manual transmission based on the performance and weight of the automobile:

```
> lr <- glm(am ~ hp + wt, data = mtcars, family = binomial)
> summary(lr)


Call:
glm(formula = am ~ hp + wt, family = binomial, data = mtcars)


Deviance Residuals:
    Min       1Q    Median       3Q       Max
-2.2537   -0.1568   -0.0168   0.1543    1.3449


Coefficients:
            Estimate Std. Error z value Pr(>|z|)
(Intercept) 18.86630    7.44356   2.535  0.01126 *
hp           0.03626    0.01773   2.044  0.04091 *
wt          -8.08348    3.06868  -2.634  0.00843 **
---
Signif. codes:  0 '***' 0.001 '**' 0.01 '*' 0.05 '.' 0.1 ' ' 1


(Dispersion parameter for binomial family taken to be 1)


    Null deviance: 43.230  on 31  degrees of freedom
Residual deviance: 10.059  on 29  degrees of freedom
AIC: 16.059


Number of Fisher Scoring iterations: 8
```

The most important table from the preceding output is the coefficients table, which describes whether the model and the independent variables significantly contribute to the value of the independent variable. We can conclude that:

- A 1-unit increase of horsepower increases the log odds of having a manual transmission (at least back in 1974, when the data was collected)

- A 1-unit increase of weight (in pounds), on the other hand, decreases the same log odds by 8

It seems that, despite (or rather due to) the low sample size, the model fits the data very well, and the horsepower and weight of the cars can explain whether a car has an automatic transmission or manual shift:

```
> table(mtcars$am, round(predict(lr, type = 'response')))

     0  1
  0 18  1
  1  1 12
```

But running the preceding command on the number of gears instead of transmission would fail, as logit regression by default expects a dichotomous variable. We can overcome this by fitting multiple models on the data, such as verifying whether a car has 3/4/5 gears or not with dummy variables, or by fitting a multinomial logistic regression. The nnet package has a very convenient function to do so:

```
> library(nnet)
> (mlr <- multinom(factor(gear) ~ ., data = mtcars))
# weights:  36 (22 variable)
initial   value 35.155593
iter  10 value 5.461542
iter  20 value 0.035178
iter  30 value 0.000631
final   value 0.000000
converged
Call:
multinom(formula = factor(gear) ~ ., data = mtcars)

Coefficients:
   (Intercept)       mpg        cyl       disp         hp      drat
4   -12.282953 -1.332149  -10.29517  0.2115914 -1.7284924  15.30648
5     7.344934  4.934189  -38.21153  0.3972777 -0.3730133  45.33284
          wt        qsec         vs        am       carb
4  21.670472   0.1851711   26.46396  67.39928  45.79318
5  -4.126207 -11.3692290  -38.43033  32.15899  44.28841

Residual Deviance: 4.300374e-08
AIC: 44
```

As expected, it returns a highly fitted model to our small dataset:

```
> table(mtcars$gear, predict(mlr))

     3  4  5
  3 15  0  0
  4  0 12  0
  5  0  0  5
```

However, due to the small sample size, this model is extremely limited. Before proceeding to the next examples, please remove the updated mtcars dataset from the current R session to avoid unexpected errors:

```
> rm(mtcars)
```

Machine learning algorithms

Machine learning (ML) is a collection of data-driven algorithms that work without being explicitly programmed for a specific task. Unlike non-ML algorithms, they require (and learn by) the training data. ML algorithms are classified into supervised and unsupervised types.

Supervised learning means that the training data consists of input vectors and their corresponding output value as well. This means that the task is to establish relationships between inputs and outputs in a historical database, called the training set, and thus make it possible to predict outputs for future input values.

For example, banks have vast databases on previous loan transaction details. The input vector is comprised of personal information—such as age, salary, marital status and so on—while the output (target) variable shows whether the payment deadlines were kept or not. In this case, a supervised algorithm may detect different groups of people who may be prone to not being able to keep the deadlines, which may serve as a screening of applicants.

Unsupervised learning has different goals. As the output values are not available in the historical dataset, the aim is to identify underlying correlations between the inputs, and define arbitrary groups of cases.

The K-Nearest Neighbors algorithm

K-Nearest Neighbors (k-NN), unlike the hierarchical or k-means clustering, is a supervised classification algorithm. Although it is often confused with k-means clustering, k-NN classification is a completely different method. It is mostly used in pattern recognition and business analytics. A big advantage of k-NN is that it is not sensitive to outliers, and the usage is extremely straightforward — just like with most machine learning algorithms.

The main idea of k-NN is that it identifies the *k* number of nearest neighbors of the observation in the historical dataset, then it defines the class of the observation to match the majority of the neighbors mentioned earlier.

As a sample analysis, we are going to use the `knn` function from the `class` package. The `knn` function takes 4 parameters, where `train` and `test` are the training and test datasets respectively, `cl` is the class membership of the training data, and `k` is the number of neighbors to take into account when classifying the elements of the test dataset.

The default value of `k` is `1`, which always works without a problem — although usually with a rather low accuracy. When defining a higher number of neighbors to be used in the analysis for improved accuracy, it's wise to select an integer that is not a multiple of the number of classes.

Let's split the `mtcars` dataset into two parts: training and test data. For the sake of simplicity, half of the cars will belong to the training set, and the other half to the test set:

```
> set.seed(42)
> n       <- nrow(mtcars)
> train <- mtcars[sample(n, n/2), ]
```

> We used `set.seed` to configure the random generator's state to a (well) known number for the sake of reproducibility: so that the exact same *random* numbers will be generated on all machines.

So we sampled 16 integers between 1 and 32 to select 50 percent of the rows from the `mtcars` dataset. Some might consider the following `dplyr` (discussed in *Chapter 3, Filtering and Summarizing Data* and in *Chapter 4, Restructuring Data*) code snippet more appealing for the task:

```
> library(dplyr)
> train <- sample_n(mtcars, n / 2)
```

Then let's select the rest of the rows with the difference of the newly created `data.frame` compared to the original data:

```
> test <- mtcars[setdiff(row.names(mtcars), row.names(train)), ]
```

Now we have to define the class memberships of the observations in the training data, what we would like to predict in the test dataset in the means of classification. To this end, we might use what we have learned in the previous section and, instead of an already known characteristic of the cars, we could run a clustering method to define the class membership of each element in the training data—but that's not something we should do for instructional purposes. You could also run the clustering algorithm on your test data as well, right? The major difference between the supervised and unsupervised methods is that we have empirical data with the former methods to feed the classification models.

So, instead, let's use the number of gears in the cars as the class membership and, based on the information found in the training data, let's predict the number of gears in the test dataset:

```
> library(class)
> (cm <- knn(
+     train = subset(train, select = -gear),
+     test  = subset(test, select = -gear),
+     cl    = train$gear,
+     k     = 5))
[1] 4 4 4 4 3 4 4 3 3 3 3 3 4 4 4 3
Levels: 3 4 5
```

The test cases have just got classified into the preceding classes. We can check the accuracy of the classification, for example, by calculating the correlation coefficient between the real and predicted number of gears:

```
> cor(test$gear, as.numeric(as.character(cm)))

[1] 0.5459487
```

Well, this might have been a lot better, especially if the training data had been a lot larger. Machine learning algorithms typically use millions of rows from historical databases, as opposed to our meager dataset with only 16 cases. But let's see where the model failed to provide accurate predictions by computing the confusion matrix:

```
> table(test$gear, as.numeric(as.character(cm)))

    3 4
  3 6 1
  4 0 6
  5 1 2
```

So it seems that the k-NN classification algorithm could predict the number of gears very accurately (one miss out of 13) for all those cars with three or four gears, but it ultimately failed with the ones with five gears. This can be explained by the number of related cars in the original dataset:

```
> table(train$gear)

3 4 5

8 6 2
```

Well, the training data had only two cars with 5 gears, which is indeed really tight when it comes to building a model providing accurate predictions.

Classification trees

An alternative ML method for supervised classification is the use of recursive partitioning via decision trees. The great advantage of this method is that visualizing decision rules can significantly improve understanding of the underlying data, and running the algorithm can be extremely easy in most cases.

Let's load the rpart package and build a classification tree with the response variable being the gear function again:

```
> library(rpart)
> ct <- rpart(factor(gear) ~ ., data = train, minsplit = 3)
> summary(ct)
Call:
rpart(formula = factor(gear) ~ ., data = train, minsplit = 3)
  n= 16


    CP nsplit rel error xerror      xstd
1 0.75      0      1.00  1.000 0.2500000
2 0.25      1      0.25  0.250 0.1653595
3 0.01      2      0.00  0.125 0.1210307


Variable importance
drat qsec  cyl disp   hp  mpg   am carb
  18   16   12   12   12   12    9    9


Node number 1: 16 observations,    complexity param=0.75
  predicted class=3  expected loss=0.5  P(node) =1
```

```
    class counts:      8     6     2
  probabilities: 0.500 0.375 0.125
  left son=2 (10 obs) right son=3 (6 obs)
  Primary splits:
      drat < 3.825 to the left,  improve=6.300000, (0 missing)
      disp < 212.8 to the right, improve=4.500000, (0 missing)
      am   < 0.5   to the left,  improve=3.633333, (0 missing)
      hp   < 149   to the right, improve=3.500000, (0 missing)
      qsec < 18.25 to the left,  improve=3.500000, (0 missing)
  Surrogate splits:
      mpg  < 22.15 to the left,  agree=0.875, adj=0.667, (0 split)
      cyl  < 5     to the right, agree=0.875, adj=0.667, (0 split)
      disp < 142.9 to the right, agree=0.875, adj=0.667, (0 split)
      hp   < 96    to the right, agree=0.875, adj=0.667, (0 split)
      qsec < 18.25 to the left,  agree=0.875, adj=0.667, (0 split)

Node number 2: 10 observations,    complexity param=0.25
  predicted class=3  expected loss=0.2  P(node) =0.625
    class counts:      8     0     2
  probabilities: 0.800 0.000 0.200
  left son=4 (8 obs) right son=5 (2 obs)
  Primary splits:
      am   < 0.5   to the left,  improve=3.200000, (0 missing)
      carb < 5     to the left,  improve=3.200000, (0 missing)
      qsec < 16.26 to the right, improve=1.866667, (0 missing)
      hp   < 290   to the left,  improve=1.422222, (0 missing)
      disp < 325.5 to the right, improve=1.200000, (0 missing)
  Surrogate splits:
      carb < 5     to the left,  agree=1.0, adj=1.0, (0 split)
      qsec < 16.26 to the right, agree=0.9, adj=0.5, (0 split)

Node number 3: 6 observations
  predicted class=4  expected loss=0  P(node) =0.375
    class counts:      0     6     0
  probabilities: 0.000 1.000 0.000
```

```
Node number 4: 8 observations
  predicted class=3  expected loss=0  P(node) =0.5
    class counts:    8    0    0
  probabilities: 1.000 0.000 0.000


Node number 5: 2 observations
  predicted class=5  expected loss=0  P(node) =0.125
    class counts:    0    0    2
  probabilities: 0.000 0.000 1.000
```

The resulting object is a rather simple *decision tree*—despite the fact that we have specified an extremely low `minsplit` parameter, to be able to generate more than one node. Running the preceding call without this argument would not even result in a decision tree, as the 16 cases of our train data would fit in a single node due to the default minimum value of 20 elements per node.

But we have built a decision tree where the most important rule to determine the number of gears is the rear axle ratio and whether the car has automatic or manual transmission:

```
> plot(ct); text(ct)
```

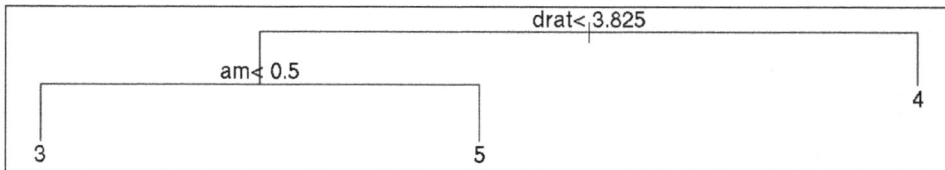

To translate this into plain and simple English:

- A car with a high rear axle ratio has four gears
- All other cars with automatic transmission have three gears
- Cars with manual shift have five gears

Well, this rule is indeed very basic due to the low number of cases and the confusion matrix also reveals the serious limitation of the model, namely that it cannot successfully identify cars with 5 gears:

```
> table(test$gear, predict(ct, newdata = test, type = 'class'))
    3 4 5
  3 7 0 0
  4 1 5 0
  5 0 2 1
```

But 13 out of 16 cars were classified perfectly, which is quite impressive and a bit better than the previous k-NN example!

Let's improve the preceding code, rather minimalist graph a bit by either calling the `main` function from the `rpart.plot` package on the preceding object, or loading the `party` package, which provides a very neat plotting function for `party` objects. One option might be to call `as.party` on the previously computed `ct` object via the `partykit` package; alternatively, we can recreate the classification tree with its `ctree` function. Based on the previous experiences, let's pass only the preceding highlighted variables to the model:

```
> library(party)
> ct <- ctree(factor(gear) ~ drat, data = train,
+     controls = ctree_control(minsplit = 3))
> plot(ct, main = "Conditional Inference Tree")
```

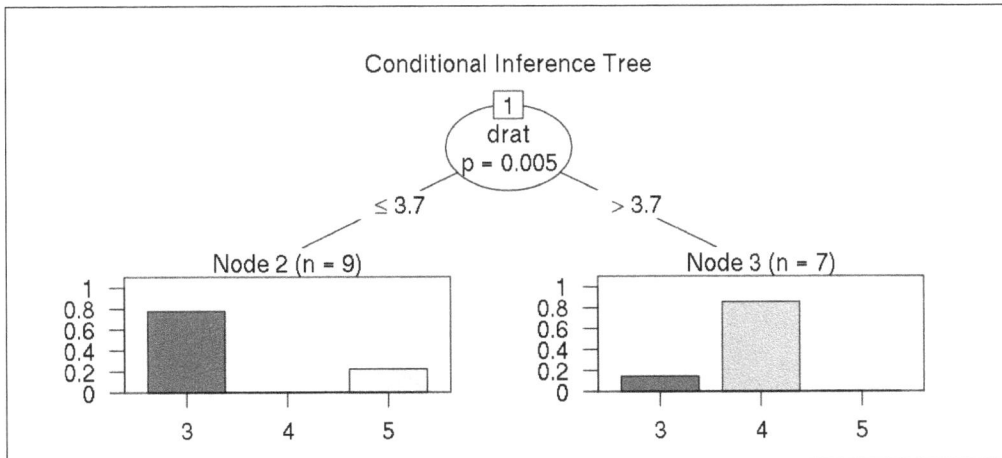

It seems that this model decides on the number of gears solely based on the rear axle ratio with a lot lower accuracy:

```
> table(test$gear, predict(ct, newdata = test, type = 'node'))

    2 3

  3 7 0

  4 1 5

  5 0 3
```

Now let's see which additional ML algorithms can provide more accurate and/or reliable models!

Random forest

The main idea behind **random forest** is that, instead of building a deep decision tree with an ever-growing number of nodes that might risk overfitting the data, we instead generate multiple trees to minimize the variance instead of maximizing the accuracy. This way the results are expected to be noisier compared to a well-trained decision tree, but on average these results are more reliable.

This can be achieved in a similar way to the preceding examples in R, via for example the randomForest package, which provides very user-friendly access to the classical random forest algorithm:

```
> library(randomForest)
> (rf <- randomForest(factor(gear) ~ ., data = train, ntree = 250))
Call:
 randomForest(formula = factor(gear) ~ ., data = train, ntree = 250)
               Type of random forest: classification
                     Number of trees: 250
No. of variables tried at each split: 3

        OOB estimate of  error rate: 25%
Confusion matrix:
   3 4 5 class.error
3  7 1 0   0.1250000
4  1 5 0   0.1666667
5  2 0 0   1.0000000
```

This function is very convenient to use: it automatically returns the confusion matrix and also computes the estimated error rate—although we can of course, generate our own based on the other subset of mtcars:

```
> table(test$gear, predict(rf, test))

    3 4 5
  3 7 0 0
  4 1 5 0
  5 1 2 0
```

But this time, the plotting function returns something new:

```
> plot(rf)
> legend('topright',
```

```
+    legend = colnames(rf$err.rate),
+    col    = 1:4,
+    fill   = 1:4,
+    bty    = 'n')
```

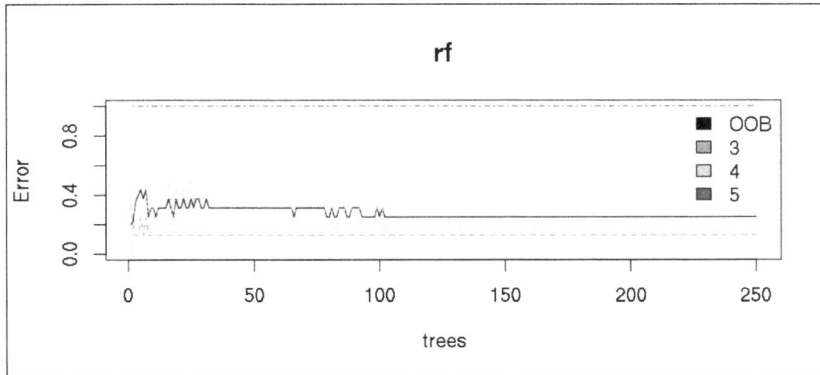

We see how the mean squared error of the model changes over time as we generate more and more decision trees on random subsamples of the training data, where the error rate does not seem to change after a while, and there's not much sense in generating more than a given number of random samples.

Well, this is really straightforward for such small example, as the combination of the possible subsamples is limited. It's also worth mentioning that the error rate of cars with five gears (blue line) did not change at all over time, which highlights again the main limitation of our training dataset.

Other algorithms

Although it would be great to continue discussing the wide variety of related ML algorithms (for example, the ID3 and Gradient Boosting algorithms from the gbm or xgboost packages) and how to call, say, Weka from the R console to use C4.5, in this chapter I can focus on only one last practical example on how to use a general interface for all these algorithms via the caret package:

```
> library(caret)
```

This package bundles some really useful functions and methods, which can be used as general, algorithm-independent tools for predictive models. This means that all the previous models could be run without actually calling the rpart, ctree, or randomForest functions, and we can simply rely on the train function of caret, which takes the algorithm definition as an argument.

For a quick example, let's see how the improved version and open-source implementation of C4.5 performs with our training data:

```
> library(C50)
> C50 <- train(factor(gear) ~ ., data = train, method = 'C5.0')
> summary(C50)


C5.0 [Release 2.07 GPL Edition]     Fri Mar 20 23:22:10 2015
-------------------------------


Class specified by attribute `outcome'

Read 16 cases (11 attributes) from undefined.data

-----  Trial 0:  -----

Rules:

Rule 0/1: (8, lift 1.8)
  drat <= 3.73
  am <= 0
  ->  class 3   [0.900]

Rule 0/2: (6, lift 2.3)
  drat > 3.73
  ->  class 4   [0.875]

Rule 0/3: (2, lift 6.0)
  drat <= 3.73
  am > 0
  ->  class 5   [0.750]

Default class: 3

*** boosting reduced to 1 trial since last classifier is very accurate
```

```
*** boosting abandoned (too few classifiers)


Evaluation on training data (16 cases):


        Rules
    ----------------
     No      Errors

      3     0( 0.0%)    <<


   (a)    (b)    (c)     <-classified as
   ----   ----   ----
     8                  (a): class 3
            6           (b): class 4
                   2    (c): class 5


   Attribute usage:

  100.00%  drat
   62.50%  am
```

This output seems extremely compelling as the error rate is exactly zero, which means that we have just created a model that perfectly fits out training data with three simple rules:

- Cars with a large rear axle ratio have four gears
- The others have either three (manual shift) or five (automatic transmission)

Well, a second look at the results reveals that we have not found the Holy Grail yet:

```
> table(test$gear, predict(C50, test))

    3 4 5
  3 7 0 0
  4 1 5 0
  5 0 3 0
```

So the overall performance of this algorithm with our test dataset resulted in 12 hits out of the 16 cars, which is a good example of how a single decision tree might over-fit the training data.

Summary

This chapter introduced a wide variety of ways to cluster and classify data, discussed which analysis procedures and models are very important, and generally used elements of a data scientist's toolbox. In the next chapter, we will focus on a less general, but still important, field— how to analyze graphs and network data.

11
Social Network Analysis of the R Ecosystem

Although the concept of social networks has a pretty long history, starting at the beginning of the last century, **social network analysis (SNA)** became extremely popular only in the last decade, probably due to the success of huge social media sites and the availability of related data. In this chapter, we are going to take a look on how to retrieve and load such data, then analyze and visualize such networks by heavily using the `igraph` package.

`Igraph` is an open source network analysis tool made by Gábor Csárdi. The software ships with a wide variety of network analysis methods, and it can be used in R, C, C++, and Python as well.

In this chapter, we will cover the following topics with some examples on the R ecosystem:

- Loading and handling network data
- Network centrality metrics
- Visualizing network graphs

Loading network data

Probably the easiest way to retrieve network-flavored information on the R ecosystem is to analyze how R packages depend on each other. Based on *Chapter 2, Getting the Data*, we could try to load this data via HTTP parsing of the CRAN mirrors but, luckily, R has a built-in function to return all available R packages from CRAN with some useful meta-information as well:

[💡 The number of packages hosted on CRAN is growing from day to day. As we are working with live data, the actual results you see might be slightly different.]

```
> library(tools)
> pkgs <- available.packages()
> str(pkgs)
 chr [1:6548, 1:17] "A3" "abc" "ABCanalysis" "abcdeFBA" ...
 - attr(*, "dimnames")=List of 2
  ..$ : chr [1:6548] "A3" "abc" "ABCanalysis" "abcdeFBA" ...
  ..$ : chr [1:17] "Package" "Version" "Priority" "Depends" ...
```

So we have a matrix with more than 6,500 rows, and the fourth column includes the dependencies in a comma-separated list. Instead of parsing those strings and cleaning the data from the package versions and other relatively unimportant characters, let's use another handy function from the tools package to do the dirty work:

```
> head(package.dependencies(pkgs), 2)
$A3
        [,1]      [,2] [,3]
[1,] "R"         ">=" "2.15.0"
[2,] "xtable"   NA   NA
[3,] "pbapply"  NA   NA


$abc
        [,1]      [,2] [,3]
[1,] "R"         ">=" "2.10"
[2,] "nnet"     NA   NA
[3,] "quantreg" NA   NA
[4,] "MASS"     NA   NA
[5,] "locfit"   NA   NA
```

So the `package.dependencies` function returns a long named list of matrixes: one for each R package, which includes the required package name and version to install and load the referred package. Besides the very same function can retrieve the list of packages that are imported or suggested by others via the `depLevel` argument. We will use this information to build a richer dataset with different types of connections between the R packages.

The following script creates the `data.frame`, in which each line represents a connection between two R packages. The `src` column shows which R package refers to the `dep` package, and the label describes the type of connection:

```
> library(plyr)
> edges <- ldply(
+    c('Depends', 'Imports', 'Suggests'), function(depLevel) {
+      deps <- package.dependencies(pkgs, depLevel = depLevel)
+      ldply(names(deps), function(pkg)
+         if (!identical(deps[[pkg]], NA))
+            data.frame(
+                src    = pkg,
+                dep    = deps[[pkg]][, 1],
+                label = depLevel,
+                stringsAsFactors = FALSE))
+ })
```

Although this code snippet might seem complex at first sight, we simply look up the dependencies of each package (like in a loop), return a row of `data.frame`, and nest it in another loop, which iterates through all previously mentioned R package connection types. The resulting R object is really straightforward to understand:

```
> str(edges)
'data.frame':   26960 obs. of   3 variables:
 $ src   : chr   "A3" "A3" "A3" "abc" ...
 $ dep   : chr   "R" "xtable" "pbapply" "R" ...
 $ label: chr   "Depends" "Depends" "Depends" "Depends" ...
```

Centrality measures of networks

So we have identified almost 30,000 relations between our 6,500 packages. Is it a sparse or dense network? In other words, how many connections do we have out of all possible package dependencies? What if all the packages depend on all other packages? We do not really need any feature-rich package to calculate that:

```
> nrow(edges) / (nrow(pkgs) * (nrow(pkgs) - 1))
[1] 0.0006288816
```

This is a rather low percentage, which makes the life of R sysadmins rather easy compared to maintaining a dense network of R software. But who are the central players in this game? Which are the top-most dependent R packages?

We can also compute a rather trivial metric to answer this question without any serious SNA knowledge, as this can be defined as "Which R package is mentioned the most times in the dep column of the edges dataset"? Or, in plain English: "Which package has the most reverse dependencies?"

```
> head(sort(table(edges$dep), decreasing = TRUE))
      R   methods     MASS    stats testthat  lattice
   3702       933      915      601      513      447
```

It seems that almost 50 percent of the packages depend on a minimal version of R. So as not to distort our directed network, let's remove these edges:

```
> edges <- edges[edges$dep != 'R', ]
```

And now it's time to transform our list of connections into a real graph object to compute more advanced metrics, and also to visualize the data:

```
> library(igraph)
> g <- graph.data.frame(edges)
> summary(g)
IGRAPH DN-- 5811 23258 --
attr: name (v/c), label (e/c)
```

After loading the package, the graph.data.frame function transforms various data sources into an igraph object. This is an extremely useful class with a variety of supported methods. The summary simply prints the number of vertices and edges, which shows that around 700 R packages have no dependencies. Let's compute the previously discussed and manually computed metrics with igraph:

```
> graph.density(g)
[1] 0.0006888828
> head(sort(degree(g), decreasing = TRUE))
 methods      MASS     stats testthat  ggplot2  lattice
     933       923       601      516      459      454
```

It's not that surprising to see the methods package at the top of the list, as it's often required in packages with complex S4 methods and classes. The MASS and stats packages include most of the often used statistical methods, but what about the others? The lattice and ggplot2 packages are extremely smart and feature-full graphing engines, and testthat is one of the most popular unit-testing extensions of R; this must be mentioned in the package descriptions before submitting new packages to the central CRAN servers.

But `degree` is only one of the available centrality metrics for social networks. Unfortunately, computing closeness, which shows the distance of each node from the others, is not really meaningful when it comes to dependency, but `betweenness` is a really interesting comparison to the preceding results:

```
> head(sort(betweenness(g), decreasing = TRUE))
   Hmisc     nlme   ggplot2      MASS multcomp       rms
 943085.3 774245.2 769692.2 613696.9 453615.3 323629.8
```

This metric shows the number of times each package acts as a bridge (the only connecting node between two others) in the shortest path between the other packages. So it's not about having a lot of depending packages; rather, it shows the importance of the packages from a more global perspective. Just imagine if a package with a high `betweenness` was deprecated and removed from CRAN; not only the directly dependent packages, but also all other packages in the dependency tree would be in a rather awkward situation.

Visualizing network data

To compare these two metrics, let's draw a simple scatter plot showing each R package by `degree` and `betweenness`:

```
> plot(degree(g), betweenness(g), type = 'n',
+    main = 'Centrality of R package dependencies')
> text(degree(g), betweenness(g), labels = V(g)$name)
```

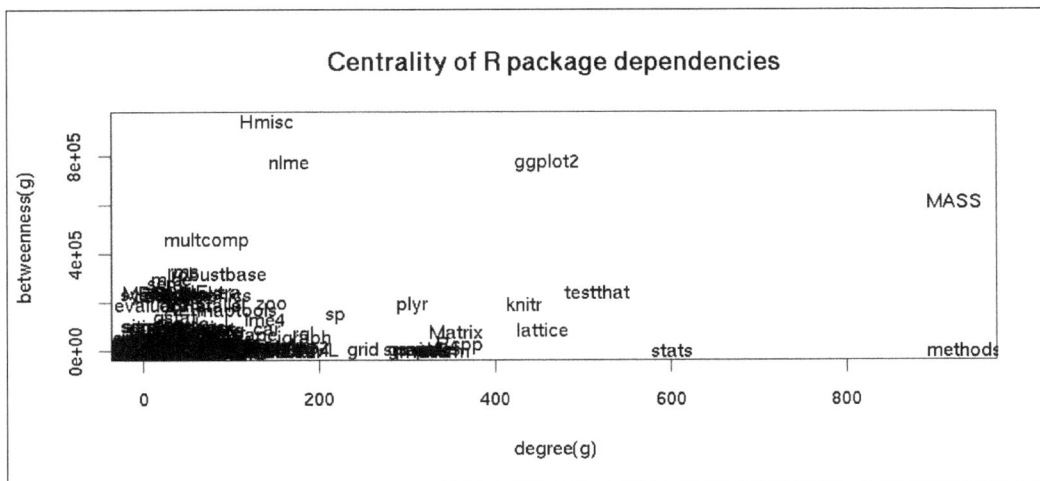

Relax; we will be soon able to generate much more spectacular and instructive plots in a few minutes! But the preceding plot shows that there are some packages with a rather low number of direct dependents that still have a great impact on the global R ecosystem.

Before we proceed, let's filter our dataset and graph to include far fewer vertices by building the dependency tree of the igraph package, including all packages it depends on or imports from:

> The following short list of igraph dependencies was generated in April 2015. Since then, a major new version of igraph has been released with a lot more dependencies due to importing from the magrittr and NMF packages, so the following examples repeated on your computer will return a much larger network and graphs. For educational purposes, we are showing the smaller network in the following outputs.

```
> edges <- edges[edges$label != 'Suggests', ]
> deptree <- edges$dep[edges$src == 'igraph']
> while (!all(edges$dep[edges$src %in% deptree] %in% deptree))
+    deptree <- union(deptree, edges$dep[edges$src %in% deptree])
> deptree
[1] "methods"    "Matrix"     "graphics"   "grid"       "stats"
[6] "utils"      "lattice"    "grDevices"
```

So we need the previously mentioned eight packages to be able to use the igraph package. Please note that not all of these are direct dependencies; some are dependencies from other packages. To draw a visual representation of this dependency tree, let's create the related graph object and plot it:

```
> g <- graph.data.frame(edges[edges$src %in% c('igraph', deptree), ])
> plot(g)
```

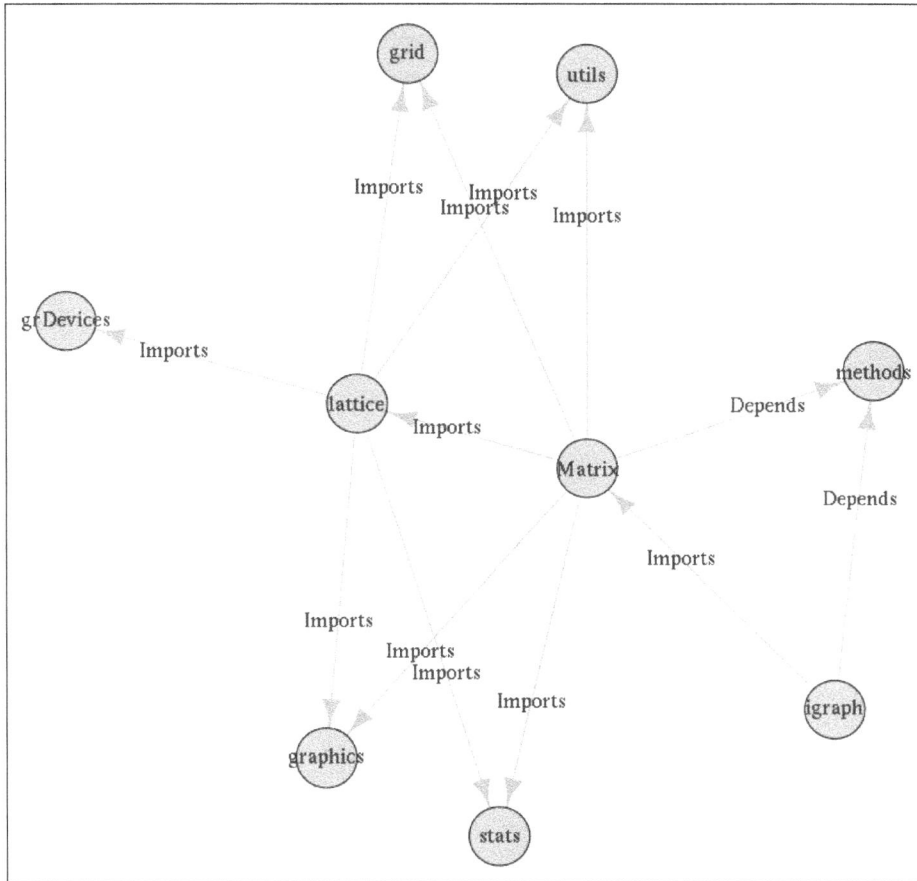

Well, the `igraph` package literally depends on only one package, although it also imports some functions from the `Matrix` package. All the other previously mentioned packages are dependencies of the latter.

To draw a more intuitive version of the preceding plot to suggest this statement, we might consider removing the dependency labels and represent that aspect by colors, and we can also emphasize the direct dependencies of `igraph` by `vertex` colors. We can modify the attributes of vertices and edges via the `V` and `E` functions:

```
> V(g)$label.color <- 'orange'
> V(g)$label.color[V(g)$name == 'igraph'] <- 'darkred'
> V(g)$label.color[V(g)$name %in%
```

```
+              edges$dep[edges$src == 'igraph']] <- 'orangered'
> E(g)$color <- c('blue', 'green')[factor(df$label)]
> plot(g, vertex.shape = 'none', edge.label = NA)
```

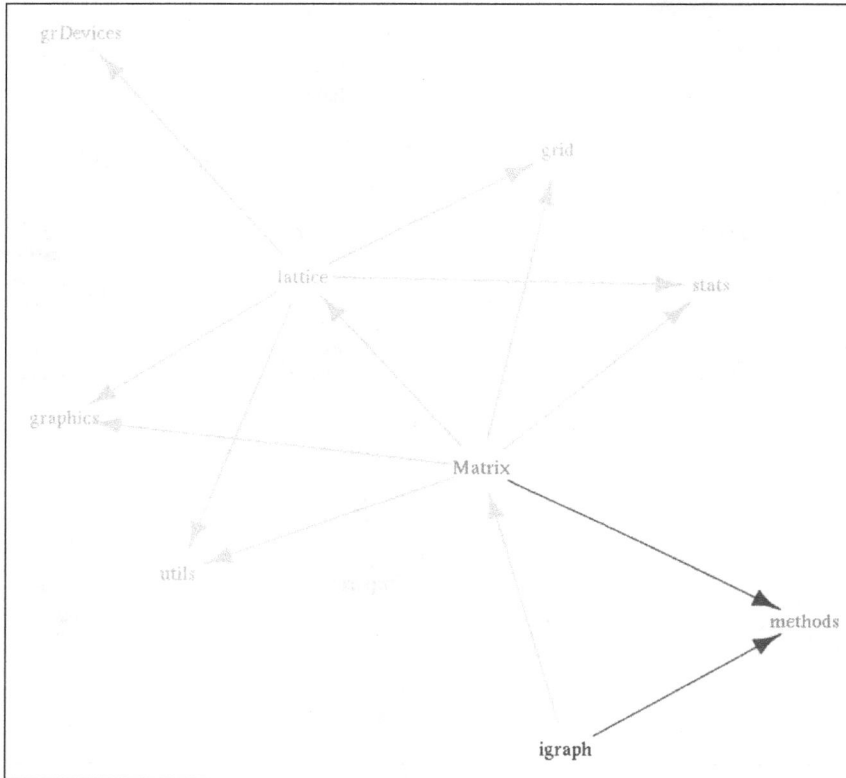

Much better! Our central topic, the `igraph` package, is highlighted in dark red, the two direct dependencies are marked in dark orange, and all the other dependencies are colored in lighter orange. Similarly, we emphasize the `Depends` relations in blue compared to the vast majority of other Imports connections.

Interactive network plots

What if you do not like the order of the vertices in the preceding plot? Feel free to rerun the last command to produce new results, or draw with `tkplot` for a dynamic plot, where you can design your custom layout by dragging-and-dropping the vertices:

```
> tkplot(g, edge.label = NA)
```

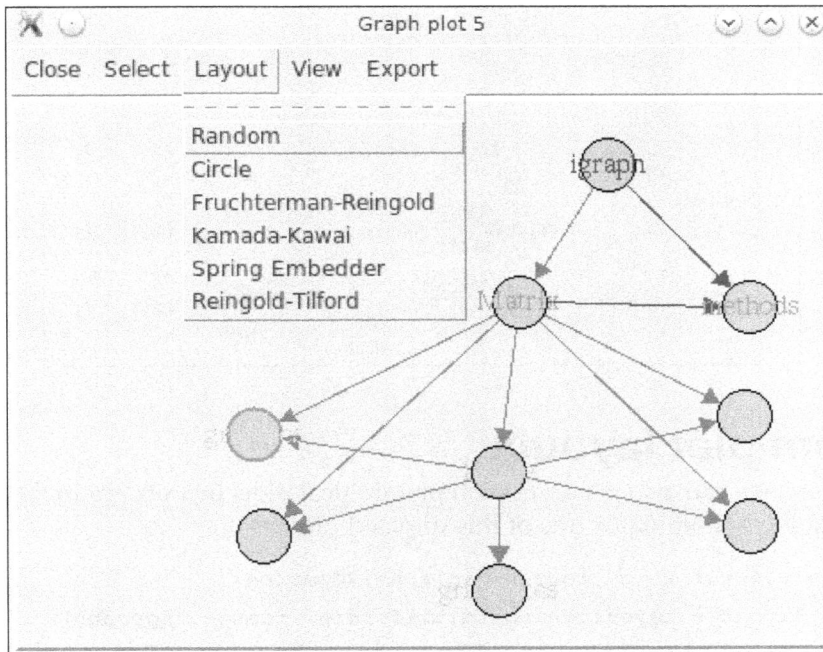

Can we do any better? Although this result is extremely useful, it lacks the immediate appeal of the currently trending, JavaScript-empowered interactive plots. So let's recreate this interactive plot with JavaScript, right from R! `htmlwidgets` and the `visNetwork` package, discussed in more detail in the *Chapter 13, Data Around Us*, can help us with this task, even without any JavaScript knowledge. Simply pass the extracted nodes and edge datasets to the `visNetwork` function:

```
> library(visNetwork)
> nodes <- get.data.frame(g, 'vertices')
> names(nodes) <- c('id', 'color')
```

```
> edges <- get.data.frame(g)
> visNetwork(nodes, edges)
```

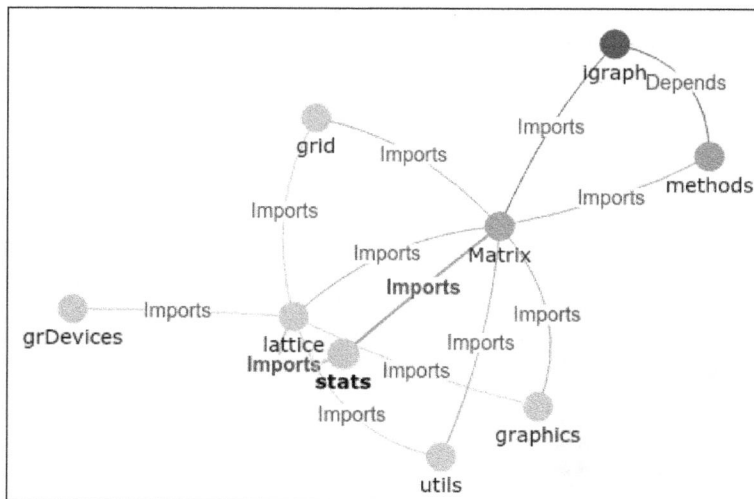

Custom plot layouts

Alternatively, we can also generate such hierarchical plots in a programmatic way, by drawing the denominator tree of this directed plot:

```
> g <- dominator.tree(g, root = "igraph")$domtree
> plot(g, layout = layout.reingold.tilford(g, root = "igraph"),
+    vertex.shape = 'none')
```

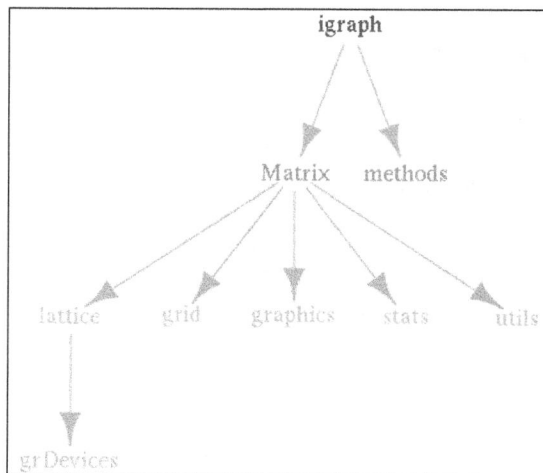

Analyzing R package dependencies with an R package

As we are using R, a statistical programming environment whose most exciting and useful feature is its community, we might prefer to look for other, already implemented solutions for this research. After a quick Google search, and having looked up a few questions on StackOverflow or posts on http://www.r-bloggers.com/, it's pretty easy to find the Revolution Analytics miniCRAN package, which has some related and useful functions:

```
> library(miniCRAN)
> pkgs <- pkgAvail()
> pkgDep('igraph', availPkgs = pkgs, suggests = FALSE,
+    includeBasePkgs = TRUE)
[1] "igraph"    "methods"    "Matrix"    "graphics"   "grid"
[6] "stats"     "utils"      "lattice"   "grDevices"
> plot(makeDepGraph('igraph', pkgs, suggests = FALSE,
+    includeBasePkgs = TRUE))
```

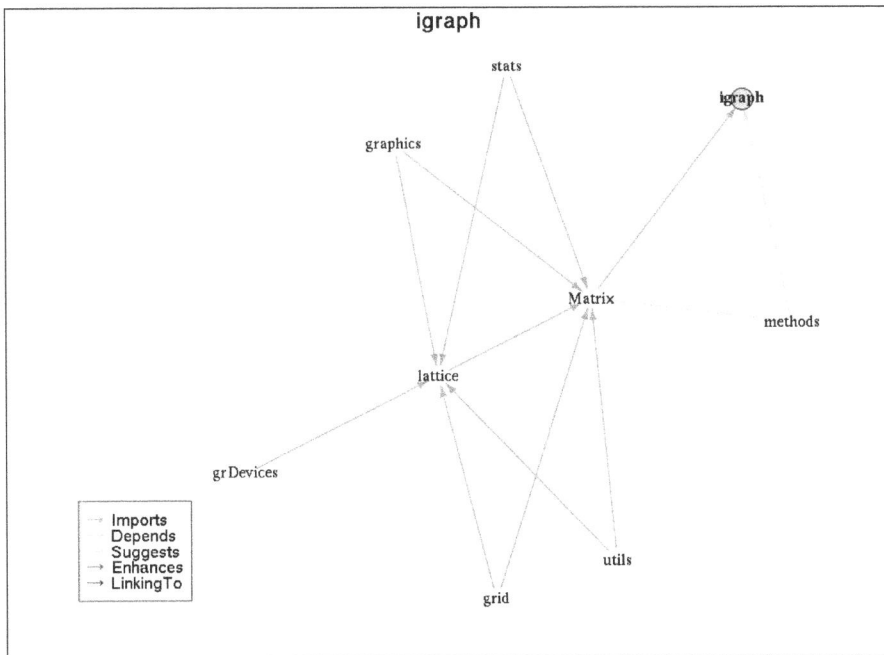

But let's get back to the original question: How do we analyze network data?

Further network analysis resources

Besides its really impressive and useful data visualization, the `igraph` package has a lot more to offer. Unfortunately, this short chapter cannot provide a decent introduction to network analysis theory, but I suggest that you skim through the package documentation as it comes with useful, self-explanatory examples and good references.

In short, network analysis provides various ways to compute centrality and density metrics, like we did at the beginning of this chapter, and also to identify bridges and simulate changes in the network; there are really powerful methods to segment the nodes in the network as well.

For example, in the *Financial Networks* chapter of the *Introduction to R for Quantitative Finance* book, which I coauthored, we developed R scripts to identify **systemically important financial institutions(SIFI)** in Hungary based on the transaction-level network data of the interbank lending market. This dataset and network theory help us to model and potentially predict future financial crises, and also to simulate the effect of central intervention.

A more detailed, freely available summary on this research was presented at the R/ Finance 2015 conference in Chicago `http://www.rinfinance.com/agenda/2015/ talk/GergelyDaroczi.pdf`, along with a Shiny application `https://bit.ly/ rfin2015-hunbanks`, and a related, simulation-based infection-model was described in the *Systemic Risk* chapter of the *Mastering R for Quantitative Finance* book as well.

The main idea behind this joint research was to identify core, peripheral, and semi-peripheral financial institutions based on the network formed by interbank lending transactions. The nodes being banks, the edges are defined as lend events between those, so we can interpret the bridges between periphery nodes as the intermediary bank between smaller banks, which usually do not lend money to each other directly.

The interesting question, after resolving some technical issues with the dataset, was to simulate what happens if an intermediary bank defaults, and if this unfortunate event might also affect other financial institutions.

Summary

This short chapter introduced a new data structure in the form of graph datasets, and we visualized small networks with various R packages, including static and interactive methods as well. In the next two chapters, we will familiarize ourselves with two other frequently used data types: first we will analyze temporal, then spatial data.

12
Analyzing Time-series

A time-series is a sequence of data points ordered in time, often used in economics or, for example, in social sciences. The great advantage of collecting data over a long period of time compared to cross-sectional observations is that we can analyze the collected values of the exact same object over time instead of comparing different observations.

This special characteristic of the data requires new methods and data structures for time-series analysis. We will cover these in this chapter:

- First, we learn how to load or transform observations into time-series objects
- Then we visualize them and try to improve the plots by smoothing and filtering the observations
- Besides seasonal decomposition, we introduce forecasting methods based on time-series models, and we also cover methods to identify outliers, extreme values, and anomalies in time-series

Creating time-series objects

Most tutorials on time-series analysis start with the `ts` function of the `stats` package, which can create time-series objects in a very straightforward way. Simply pass a vector or matrix of numeric values (time-series analysis mostly deals with continuous variables), specify the frequency of your data, and it's all set!

The frequency refers to the natural time-span of the data. Thus, for monthly data, you should set it to 12, 4 for quarterly and 365 or 7 for daily data, depending on the most characteristic seasonality of the events. For example, if your data has a significant weekly seasonality, which is pretty usual in social sciences, it should be 7, but if the calendar date is the main differentiator, such as with weather data, it should be 365.

In the forthcoming pages, let's use daily summary statistics from the `hflights` dataset. First let's load the related dataset and transform it to `data.table` for easy aggregation. We also have to create a date variable from the provided `Year`, `Month`, and `DayofMonth` columns:

```
> library(hflights)
> library(data.table)
> dt <- data.table(hflights)
> dt[, date := ISOdate(Year, Month, DayofMonth)]
```

Now let's compute the number of flights and the overall sum of arrival delays, number of cancelled flights and the average distance of the related flights for each day in 2011:

```
> daily <- dt[, list(
+      N          = .N,
+      Delays     = sum(ArrDelay, na.rm = TRUE),
+      Cancelled  = sum(Cancelled),
+      Distance   = mean(Distance)
+ ), by = date]
> str(daily)
Classes 'data.table' and 'data.frame':  365 obs. of  5 variables:
 $ date     : POSIXct, format: "2011-01-01 12:00:00" ...
 $ N        : int  552 678 702 583 590 660 661 500 602 659 ...
 $ Delays   : int  5507 7010 4221 4631 2441 3994 2571 1532 ...
 $ Cancelled: int  4 11 2 2 3 0 2 1 21 38 ...
 $ Distance : num  827 787 772 755 760 ...
 - attr(*, ".internal.selfref")=<externalptr>
```

Visualizing time-series

This is in a very familiar data structure: 365 rows for each day in 2011 and five columns to store the four metrics for the dates stored in the first variable. Let's transform that to a time-series object and plot it right away:

```
> plot(ts(daily))
```

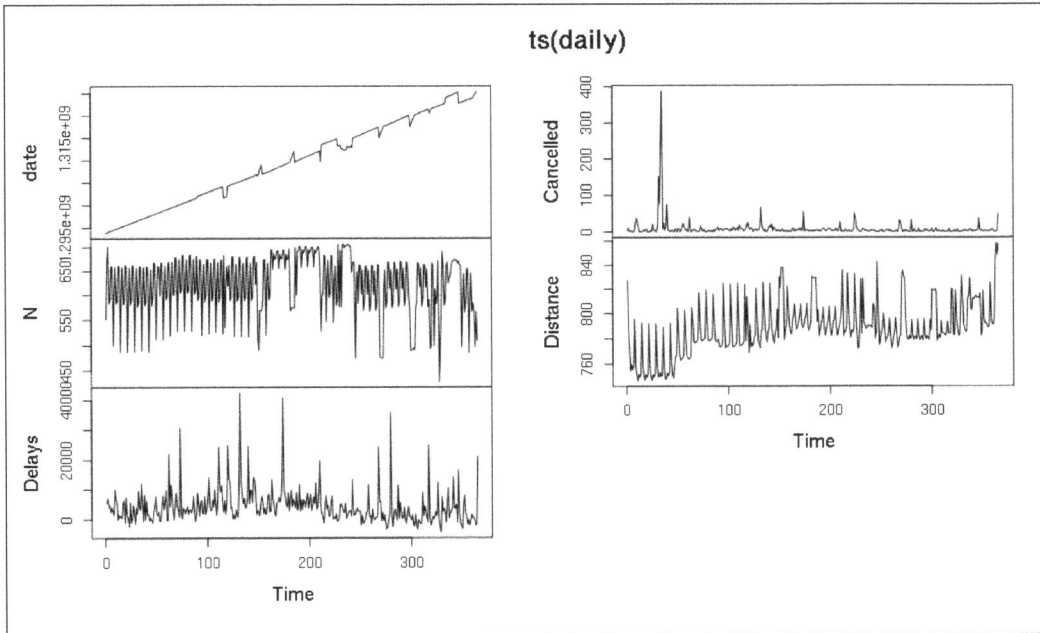

It was easy, right? We have just plotted several independent time-series on a line chart. But what's shown on the first plot? The x axis is indexed from 1 to 365 because `ts` did not automatically identify that the first column stores our dates. On the other hand, we find the date transformed to timestamps on the y axis. Shouldn't the points form a linear line?

This is one of the beauties of data visualization: a simple plot revealed a major issue with our data. It seems we have to sort the data by date:

```
> setorder(daily, date)
> plot(ts(daily))
```

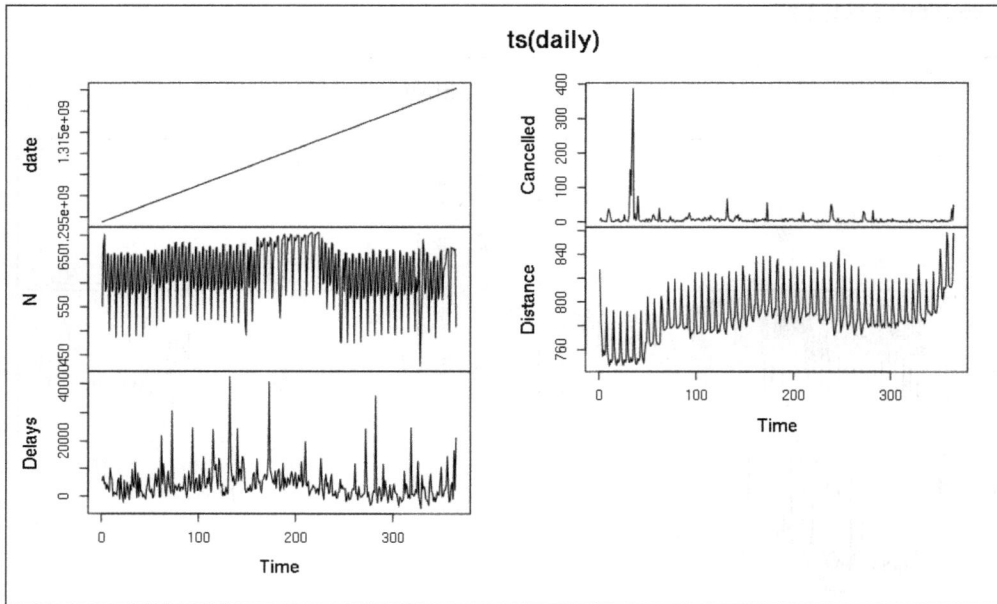

Much better! Now that the values are in the right order, we can focus on the actual time-series data one by one at a time. First let's see how the number of flights looked from the first day of 2011 with a daily frequency:

```
> plot(ts(daily$N, start = 2011, frequency = 365),
+       main = 'Number of flights from Houston in 2011')
```

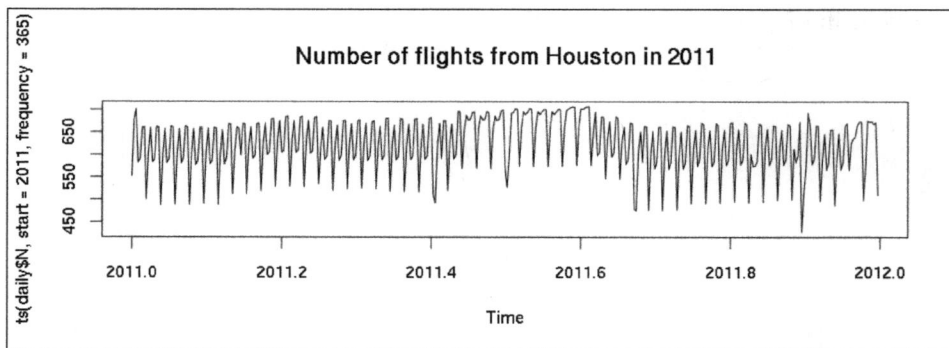

Seasonal decomposition

Well, it looks like the number of flights fluctuates a lot on weekdays, which is indeed a dominant characteristic of human-related activities. Let's verify that by identifying and removing the weekly seasonality by decomposing this time-series into the seasonal, trend, and random components with moving averages.

Although this can be done manually by utilizing the `diff` and `lag` functions, there's a much more straightforward way to do so with the `decompose` function from the `stats` package:

```
> plot(decompose(ts(daily$N, frequency = 7)))
```

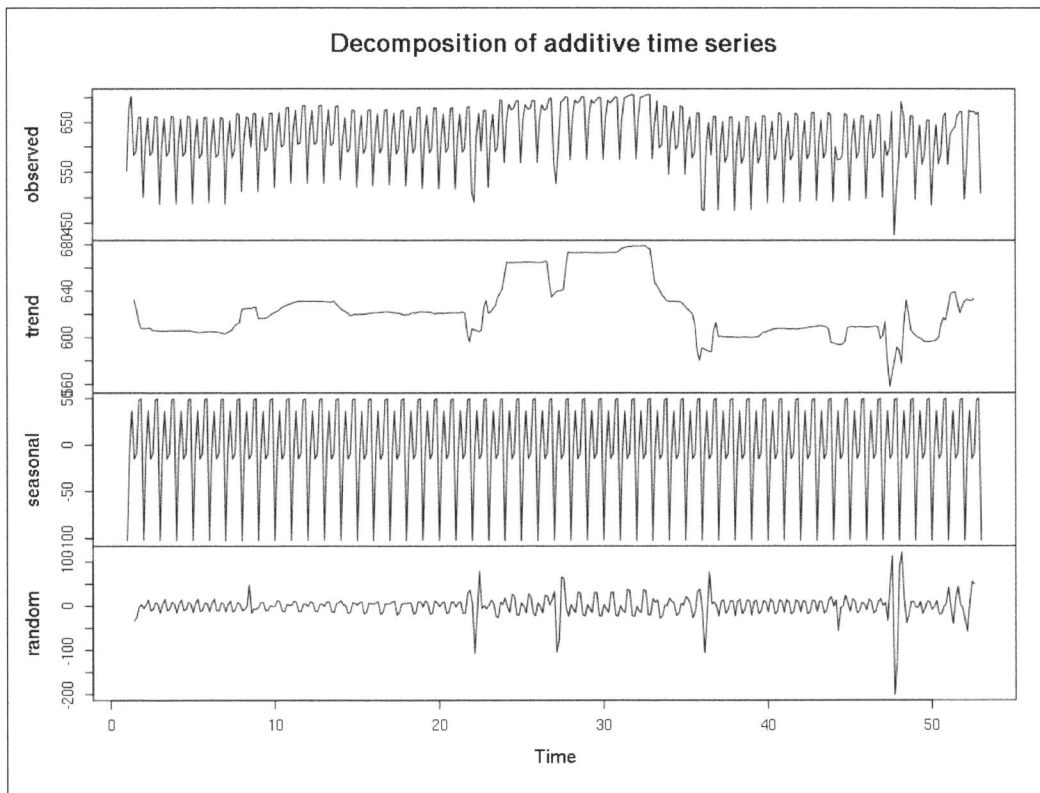

Removing the spikes in the means of weekly seasonality reveals the overall trend of the number of flights in 2011. As the *x* axis shows the number of weeks since January 1 (based on the frequency being 7), the peak interval between 25 and 35 refers to the summertime, and the lowest number of flights happened on the 46th week – probably due to Thanksgiving Day.

But the weekly seasonality is probably more interesting. Well, it's pretty hard to spot anything on the preceding plot as the very same 7-day repetition can be seen 52 times on the seasonal plot. So, instead, let's extract that data and show it in a table with the appropriate headers:

```
> setNames(decompose(ts(daily$N, frequency = 7))$figure,
+         weekdays(daily$date[1:7]))
    Saturday      Sunday      Monday     Tuesday    Wednesday
 -102.171776   -8.051328   36.595731  -14.928941    -9.483886
    Thursday      Friday
   48.335226   49.704974
```

So the seasonal effects (the preceding numbers representing the relative distance from the average) suggest that the greatest number of flights happened on Monday and the last two weekdays, while there is only a relatively small number of flights on Saturdays.

Unfortunately, we cannot decompose the yearly seasonal component of this time-series, as we have data only for one year, and we need data for at least two time periods for the given frequency:

```
> decompose(ts(daily$N, frequency = 365))
Error in decompose(ts(daily$N, frequency = 365)) :
  time series has no or less than 2 periods
```

For more advanced seasonal decomposition, see the `stl` function of the `stats` package, which uses polynomial regression models on the time-series data. The next section will cover some of this background.

Holt-Winters filtering

We can similarly remove the seasonal effects of a time-series by Holt-Winters filtering. Setting the `beta` parameter of the `HoltWinters` function to `FALSE` will result in a model with exponential smoothing practically suppressing all the outliers; setting the `gamma` argument to `FALSE` will result in a non-seasonal model. A quick example:

```
> nts <- ts(daily$N, frequency = 7)
> fit <- HoltWinters(nts, beta = FALSE, gamma = FALSE)
> plot(fit)
```

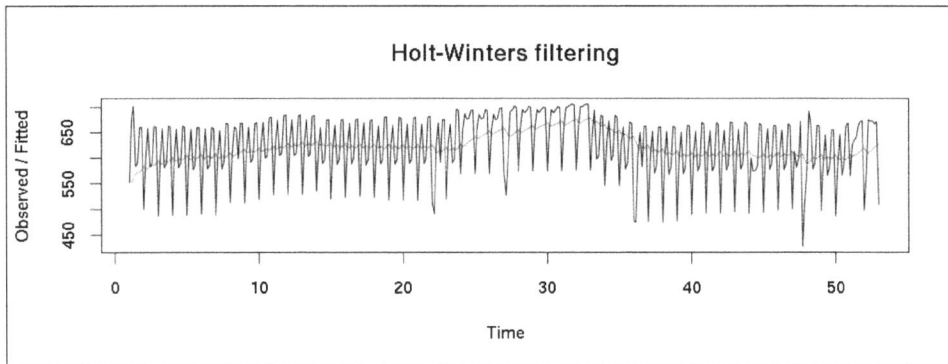

The red line represents the filtered time-series. We can also fit a double or triple exponential model on the time-series by enabling the `beta` and `gamma` parameters, resulting in a far better fit:

```
> fit <- HoltWinters(nts)
> plot(fit)
```

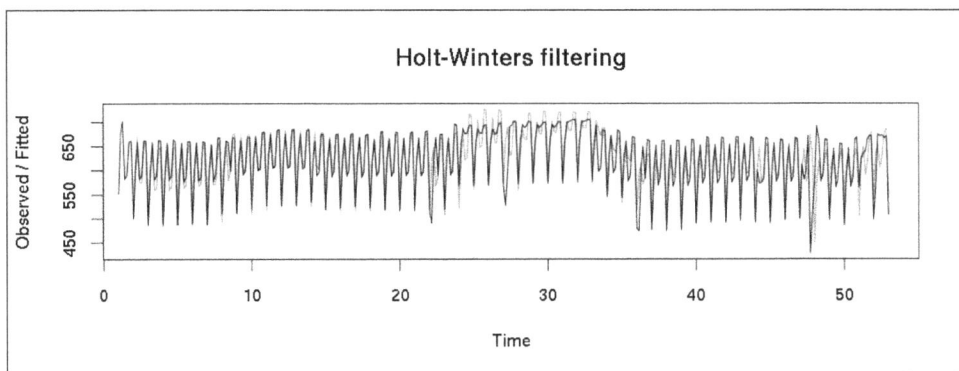

As this model provides extremely similar values compared to our original data, it can be used to predict future values as well. For this end, we will use the `forecast` package. By default, the `forecast` function returns a prediction for the forthcoming 2*frequency values:

```
> library(forecast)
> forecast(fit)
         Point Forecast     Lo 80     Hi 80     Lo 95     Hi 95
53.14286        634.0968  595.4360  672.7577  574.9702  693.2235
```

53.28571	673.6352	634.5419	712.7286	613.8471	733.4233
53.42857	628.2702	588.7000	667.8404	567.7528	688.7876
53.57143	642.5894	602.4969	682.6820	581.2732	703.9057
53.71429	678.2900	637.6288	718.9511	616.1041	740.4758
53.85714	685.8615	644.5848	727.1383	622.7342	748.9889
54.00000	541.2299	499.2901	583.1697	477.0886	605.3712
54.14286	641.8039	598.0215	685.5863	574.8445	708.7633
54.28571	681.3423	636.8206	725.8639	613.2523	749.4323
54.42857	635.9772	590.6691	681.2854	566.6844	705.2701
54.57143	650.2965	604.1547	696.4382	579.7288	720.8642
54.71429	685.9970	638.9748	733.0192	614.0827	757.9113
54.85714	693.5686	645.6194	741.5178	620.2366	766.9005
55.00000	548.9369	500.0147	597.8592	474.1169	623.7570

These are estimates for the first two weeks of 2012, where (besides the exact point predictions) we get the confidence intervals as well. Probably it's more meaningful at this time to visualize these predictions and confidence intervals:

```
> plot(forecast(HoltWinters(nts), 31))
```

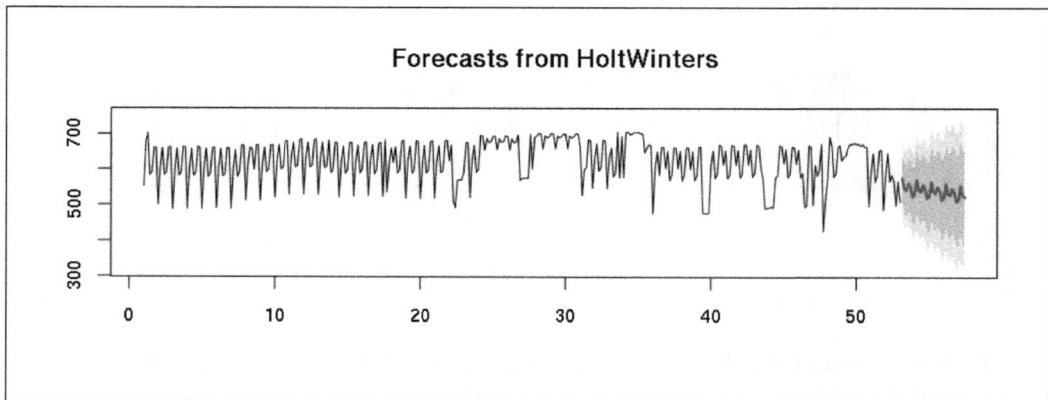

The blue points shows the estimates for the 31 future time periods and the gray area around that covers the confidence intervals returned by the forecast function.

Autoregressive Integrated Moving Average models

We can achieve similar results with **Autoregressive Integrated Moving Average (ARIMA)** models. To predict future values of a time-series, we usually have to *stationarize* it first, which means that the data has a constant mean, variance, and autocorrelation over time. In the past two sections, we used seasonal decomposition and the Holt-Winters filter to achieve this. Now let's see how the generalized version of the **Autoregressive Moving Average (ARMA)** model can help with this data transformation.

ARIMA(p, d, q) actually includes three models with three non-negative integer parameters:

- *p* refers to the autoregressive part of the model
- *d* refers to the integrated part
- *q* refers to the moving average parts

As ARIMA also includes an integrated (differencing) part over ARMA, it can deal with non-stationary time-series as well, as they naturally become stationary after differencing—in other words, when the *d* parameter is larger than zero.

Traditionally, choosing the best ARIMA model for a time-series is required to build multiple models with a variety of parameters and compare model fits. On the other hand, the `forecast` package comes with a very useful function that can select the best fitting ARIMA model for a time-series by running unit root tests and minimizing the **maximum-likelihood (ML)** and the **Akaike Information Criterion (AIC)** of the models:

```
> auto.arima(nts)
Series: ts
ARIMA(3,0,0)(2,0,0)[7] with non-zero mean

Coefficients:
         ar1      ar2     ar3    sar1    sar2  intercept
      0.3205  -0.1199  0.3098  0.2221  0.1637   621.8188
s.e.  0.0506   0.0538  0.0538  0.0543  0.0540     8.7260

sigma^2 estimated as 2626:  log likelihood=-1955.45
AIC=3924.9   AICc=3925.21   BIC=3952.2
```

It seems that the *AR(3)* model has the highest AIC with *AR(2)* seasonal effects. But checking the manual of `auto.arima` reveals that the information criteria used for the model selection were approximated due to the large number (more than 100) of observations. Re-running the algorithm and disabling approximation returns a different model:

```
> auto.arima(nts, approximation = FALSE)
Series: ts
ARIMA(0,0,4)(2,0,0)[7] with non-zero mean
```

```
Coefficients:
         ma1      ma2     ma3     ma4     sar1     sar2   intercept
      0.3257  -0.0311  0.2211  0.2364  0.2801  0.1392    621.9295
s.e.  0.0531   0.0531  0.0496  0.0617  0.0534  0.0557      7.9371
```

```
sigma^2 estimated as 2632:  log likelihood=-1955.83
AIC=3927.66    AICc=3928.07    BIC=3958.86
```

Although it seems that the preceding seasonal ARIMA model fits the data with a high AIC, we might want to build a real ARIMA model by specifying the *D* argument resulting in an integrated model via the following estimates:

```
> plot(forecast(auto.arima(nts, D = 1, approximation = FALSE), 31))
```

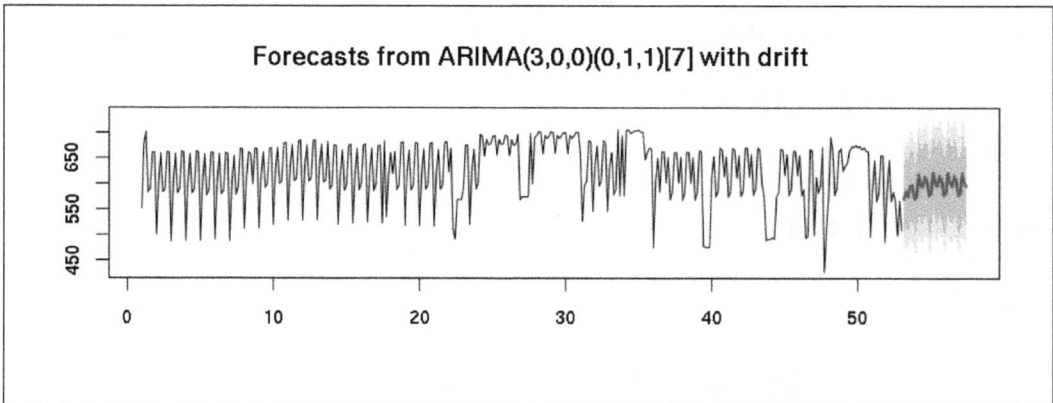

Forecasts from ARIMA(3,0,0)(0,1,1)[7] with drift

Although time-series analysis can sometimes be tricky (and finding the optimal model with the appropriate parameters requires a reasonable experience with these statistical methods), the preceding short examples proved that even a basic understanding of the time-series objects and related methods will usually provide some impressive results on the patterns of data and adequate predictions.

Outlier detection

Besides forecasting, another time-series related major task is identifying suspicious or abnormal data in a series of observations that might distort the results of our analysis. One way to do so is to build an ARIMA model and analyze the distance between the predicted and actual values. The `tsoutliers` package provides a very convenient way to do so. Let's build a model on the number of cancelled flights in 2011:

```
> cts <- ts(daily$Cancelled)
> fit <- auto.arima(cts)
> auto.arima(cts)
Series: ts
ARIMA(1,1,2)

Coefficients:
          ar1      ma1      ma2
      -0.2601  -0.1787  -0.7752
s.e.   0.0969   0.0746   0.0640

sigma^2 estimated as 539.8:  log likelihood=-1662.95
AIC=3333.9   AICc=3334.01   BIC=3349.49
```

So now we can use an *ARIMA(1,1,2)* model and the `tso` function to highlight (and optionally remove) the outliers from our dataset:

> Please note that the following `tso` call can run for several minutes with a full load on a CPU core as it may be performing heavy computations in the background.

```
> library(tsoutliers)
> outliers <- tso(cts, tsmethod = 'arima',
+    args.tsmethod  = list(order = c(1, 1, 2)))
```

```
> plot(outliers)
```

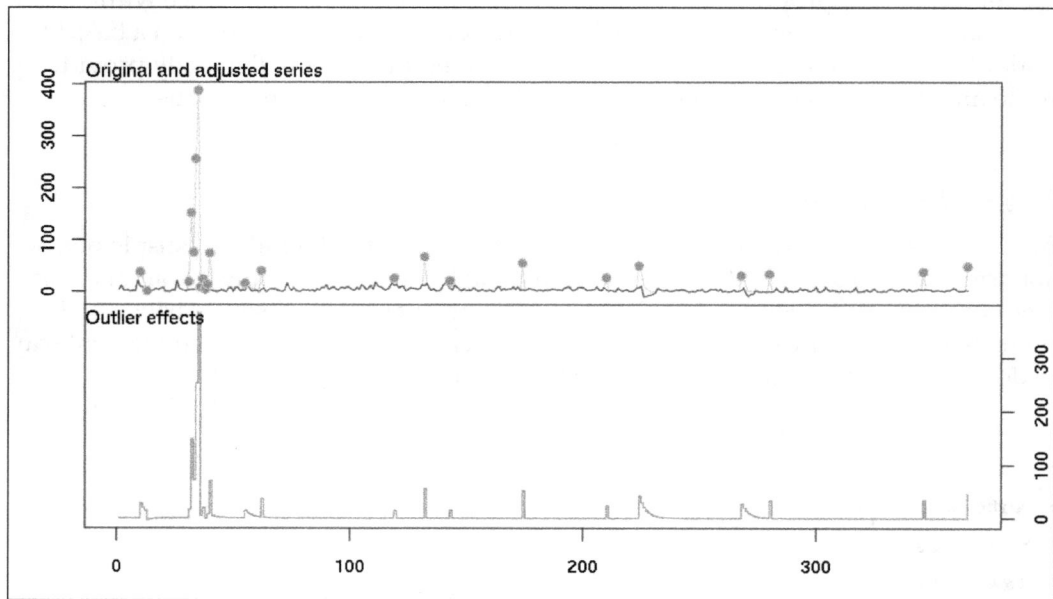

Alternatively, we can run all the preceding steps in one go by automatically calling auto.arima inside tso without specifying any extra arguments besides the time-series object:

```
> plot(tso(ts(daily$Cancelled)))
```

Anyway, the results show that all observations with a high number of cancelled flights are outliers and so should be removed from the dataset. Well, considering any day with many cancelled flights as outlier sounds really optimistic! But this is very useful information; it suggests that, for example, forecasting an outlier event is not manageable with the previously discussed methods.

Traditionally, time-series analysis deals with trends and seasonality of data, and how to *stationarize* the time-series. If we are interested in deviations from normal events, some other methods need to be used.

Twitter recently published one of its R packages to detect anomalies in time-series. Now we will use its AnomalyDetection package to identify the preceding outliers in a much faster way. As you may have noticed, the tso function was really slow to run, and it cannot really handle large amount of data – while the AnomalyDetection package performs pretty well.

We can provide the input data as a vector of a `data.frame` with the first column storing the timestamps. Unfortunately, the `AnomalyDetectionTs` function does not really work well with `data.table` objects, so let's revert to the traditional `data.frame` class:

```
> dfc <- as.data.frame(daily[, c('date', 'Cancelled'), with = FALSE])
```

Now let's load the package and plot the anomalies identified among the observations:

```
> library(AnomalyDetection)
> AnomalyDetectionTs(dfc, plot = TRUE)$plot
```

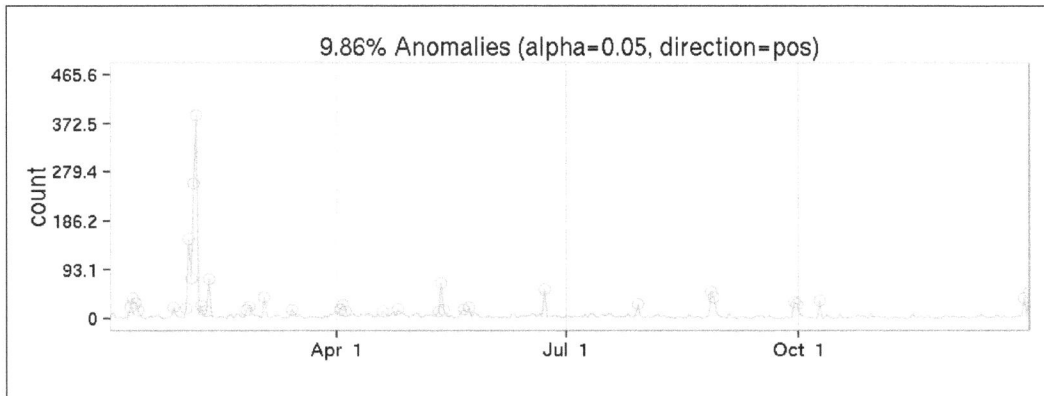

The results are very similar to the previous plots, but there are two things to note that you might have already noticed. The computation was extremely quick and, on the other hand, this plot includes human-friendly dates instead of some lame indexes on the *x* axis.

More complex time-series objects

The main limitation of the `ts` time-series R object class (besides the aforementioned *x* axis issue) is that it cannot deal with irregular time-series. To overcome this problem, we have several alternatives in R.

The `zoo` package and its reverse dependent `xts` packages are `ts`-compatible classes with tons of extremely useful methods. For a quick example, let's build a `zoo` object from our data, and see how it's represented by the default plot:

```
> library(zoo)
> zd <- zoo(daily[, -1, with = FALSE], daily[[1]])
```

```
> plot(zd)
```

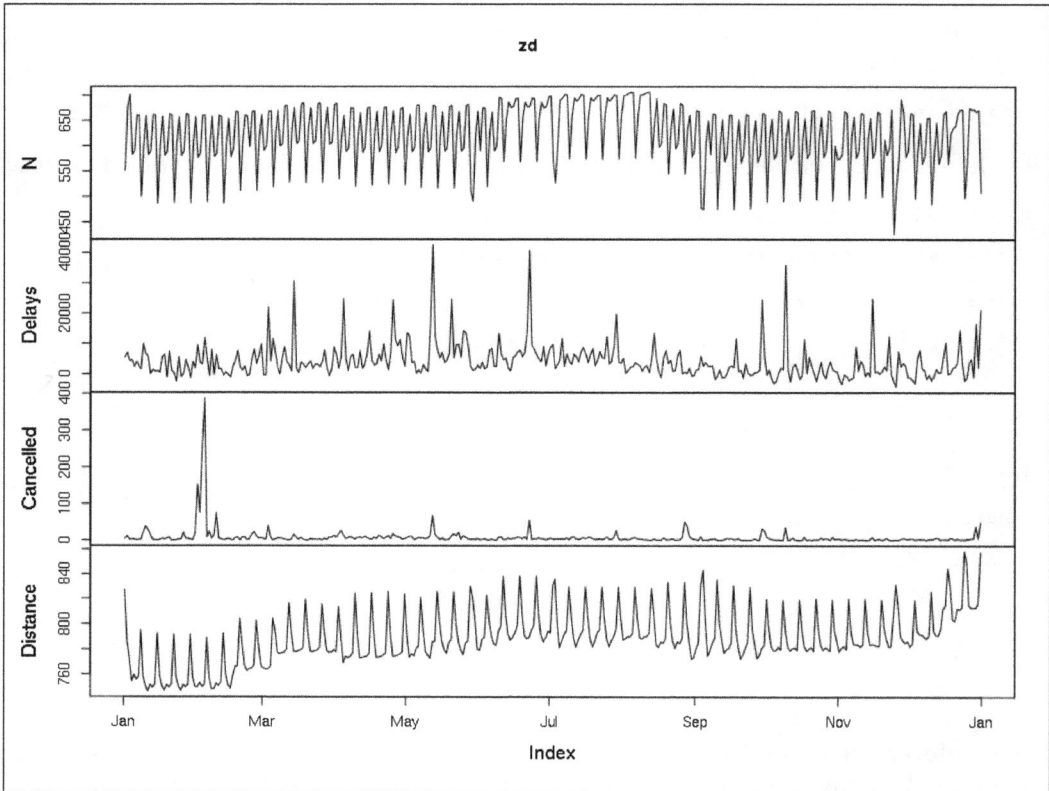

As we have defined the `date` column to act as the timestamp of the observations, it's not shown here. The *x* axis has a nice human-friendly date annotation, which is really pleasant after having checked a bunch of integer-annotated plots in the previous pages.

Of course, `zoo` supports most of the `ts` methods, such as `diff`, `lag` or cumulative sums; these can be very useful for visualizing data velocity:

```
> plot(cumsum(zd))
```

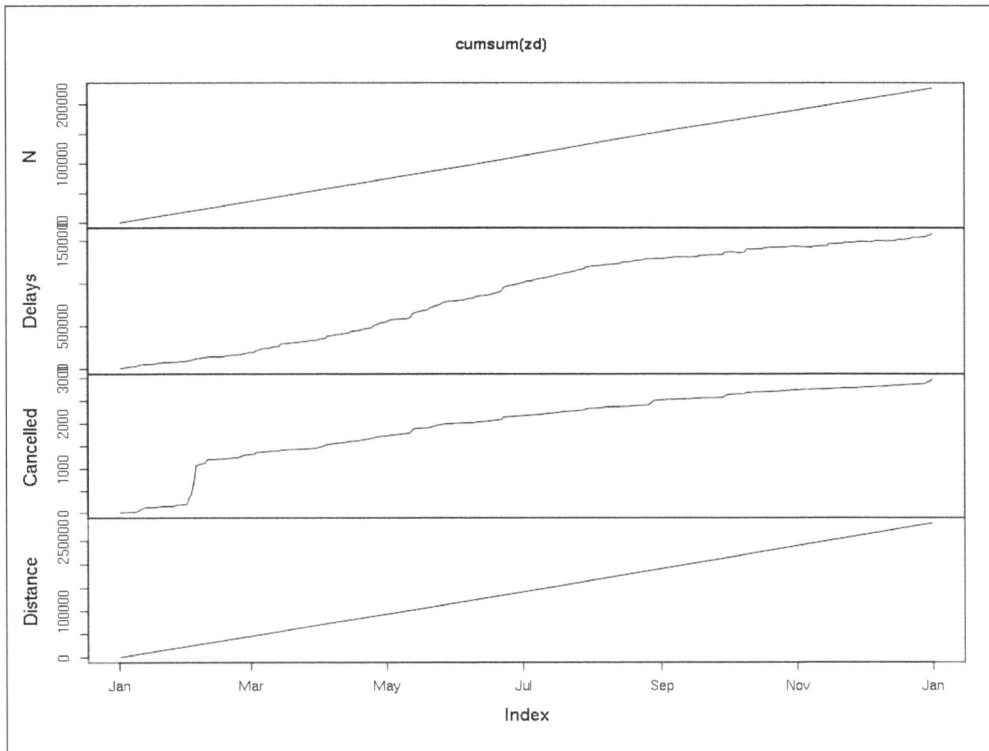

Here, the linear line for the **N** variable suggests that we do not have any missing values and our dataset includes exactly one data point per day. On the other hand, the steep elevation of the **Cancelled** line in February highlights that a single day contributed a lot to the overall number of cancelled flights in 2011.

Advanced time-series analysis

Unfortunately, this short chapter cannot provide a more detailed introduction to time-series analysis. To be honest, even two or three times the length of this chapter would not be enough for a decent tutorial, as time-series analysis, forecasting, and anomaly detection are one of the most complex topics of statistical analysis.

But the good news is that there are plenty of great books on the topics! One of the best resources — and the ultimate free online tutorial on this subject — can be found at `https://www.otexts.org/fpp`. This is a really practical and detailed online tutorial on forecasting and general time-series analysis, and I heartily recommend it to anyone who would like to build more complex and realizable time-series models in the future.

Summary

This chapter focused on how to load, visualize, and model time-related data. Although we could not cover all aspects of this challenging topic, we discussed the most widely used smoothing and filtering algorithms, seasonal decompositions, and ARIMA models; we also computed some forecasts and estimates based on these.

The next chapter is somewhat similar to this one, as we will cover another domain-independent area on another important dimension of datasets: instead of when, we will focus on *where* the observations were captured.

13
Data Around Us

Spatial data, also known as geospatial data, identifies geographic locations, such as natural or constructed features around us. Although all observations have some spatial content, such as the location of the observation, but this is out of most data analysis tools' range due to the complex nature of spatial information; alternatively, the spatiality might not be that interesting (at first sight) in the given research topic.

On the other hand, analyzing spatial data can reveal some very important underlying structures of the data, and it is well worth spending time visualizing the differences and similarities between close or far data points.

In this chapter, we are going to help with this and will use a variety of R packages to:

- Retrieve geospatial information from the Internet
- Visualize points and polygons on a map
- Compute some spatial statistics

Geocoding

As in the previous chapters, we will use the `hflights` dataset to demonstrate how one can deal with data bearing spatial information. To this end, let's aggregate our dataset, just like we did in *Chapter 12, Analyzing Time-series*, but instead of generating daily data, let's view the aggregated characteristics of the airports. For the sake of performance, we will use the `data.table` package again as introduced in *Chapter 3, Filtering and Summarizing Data* and *Chapter 4, Restructuring Data*:

```
> library(hflights)
> library(data.table)
> dt <- data.table(hflights)[, list(
+     N          = .N,
```

```
+       Cancelled = sum(Cancelled),
+       Distance  = Distance[1],
+       TimeVar   = sd(ActualElapsedTime, na.rm = TRUE),
+       ArrDelay  = mean(ArrDelay, na.rm = TRUE)) , by = Dest]
```

So we have loaded and then immediately transformed the `hfights` dataset to a `data.table` object. At the same time, we aggregated by the destination of the flights to compute:

- The number of rows
- The number of cancelled flights
- The distance
- The standard deviation of the elapsed time of the flights
- The arithmetic mean of the delays

The resulting R object looks like this:

```
> str(dt)
Classes 'data.table' and 'data.frame': 116 obs. of 6 variables:
 $ Dest     : chr  "DFW" "MIA" "SEA" "JFK" ...
 $ N        : int  6653 2463 2615 695 402 6823 4893 5022 6064 ...
 $ Cancelled: int  153 24 4 18 1 40 40 27 33 28 ...
 $ Distance : int  224 964 1874 1428 3904 305 191 140 1379 862 ...
 $ TimeVar  : num  10 12.4 16.5 19.2 15.3 ...
 $ ArrDelay : num  5.961 0.649 9.652 9.859 10.927 ...
 - attr(*, ".internal.selfref")=<externalptr>
```

So we have 116 observations all around the world and five variables describing those. Although this seems to be a spatial dataset, we have no geospatial identifiers that a computer can understand per se, so let's fetch the *geocodes* of these airports from the Google Maps API via the `ggmap` package. First, let's see how it works when we are looking for the geo-coordinates of Houston:

```
> library(ggmap)
> (h <- geocode('Houston, TX'))
Information from URL : http://maps.googleapis.com/maps/api/geocode/json?a
ddress=Houston,+TX&sensor=false
        lon       lat
1  -95.3698  29.76043
```

So the `geocode` function can return the matched latitude and longitude of the string we sent to Google. Now let's do the very same thing for all flight destinations:

```
> dt[, c('lon', 'lat') := geocode(Dest)]
```

Well, this took some time as we had to make 116 separate queries to the Google Maps API. Please note that Google limits you to 2,500 queries a day without authentication, so do not run this on a large dataset. There is a helper function in the package, called `geocodeQueryCheck`, which can be used to check the remaining number of free queries for the day.

Some of the methods and functions that we plan to use in some later sections of this chapter do not support `data.table`, so let's fall back to the traditional `data.frame` format and also print the structure of the current object:

```
> str(setDF(dt))
'data.frame':  116 obs. of  8 variables:
 $ Dest     : chr  "DFW" "MIA" "SEA" "JFK" ...
 $ N        : int  6653 2463 2615 695 402 6823 4893 5022 6064 ...
 $ Cancelled: int  153 24 4 18 1 40 40 27 33 28 ...
 $ Distance : int  224 964 1874 1428 3904 305 191 140 1379 862 ...
 $ TimeVar  : num  10 12.4 16.5 19.2 15.3 ...
 $ ArrDelay : num  5.961 0.649 9.652 9.859 10.927 ...
 $ lon      : num  -97 136.5 -122.3 -73.8 -157.9 ...
 $ lat      : num  32.9 34.7 47.5 40.6 21.3 ...
```

This was pretty quick and easy, wasn't it? Now that we have the longitude and latitude values of all the airports, we can try to show these points on a map.

Visualizing point data in space

For the first time, let's keep it simple and load some package-bundled polygons as the base map. To this end, we will use the `maps` package. After loading it, we use the `map` function to render the polygons of the United States of America, add a title, and then some points for the airports and also for Houston with a slightly modified symbol:

```
> library(maps)
> map('state')
> title('Flight destinations from Houston,TX')
```

```
> points(h$lon, h$lat, col = 'blue', pch = 13)
> points(dt$lon, dt$lat, col = 'red', pch = 19)
```

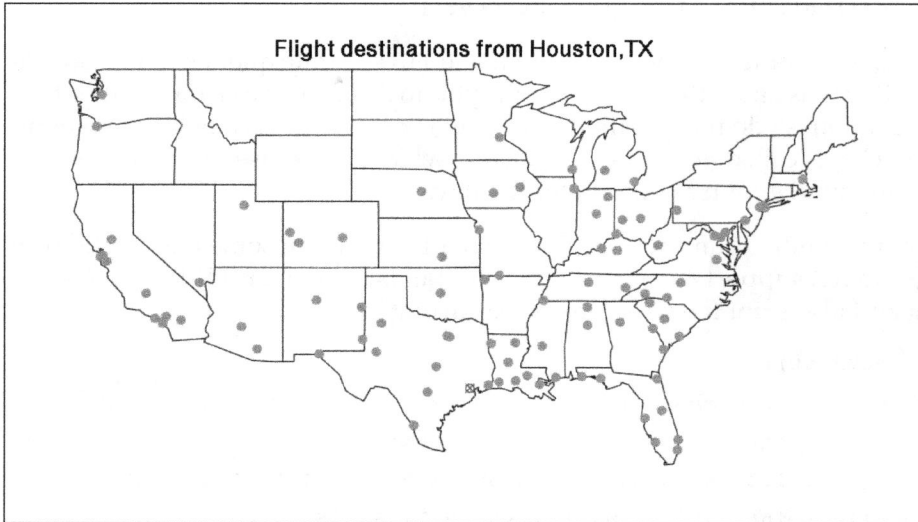

Flight destinations from Houston,TX

And showing the airport names on the plot is pretty easy as well: we can use the well-known functions from the base `graphics` package. Let's pass the three character names as labels to the text function with a slightly increased *y* value to shift the preceding text the previously rendered data points:

```
> text(dt$lon, dt$lat + 1, labels = dt$Dest, cex = 0.7)
```

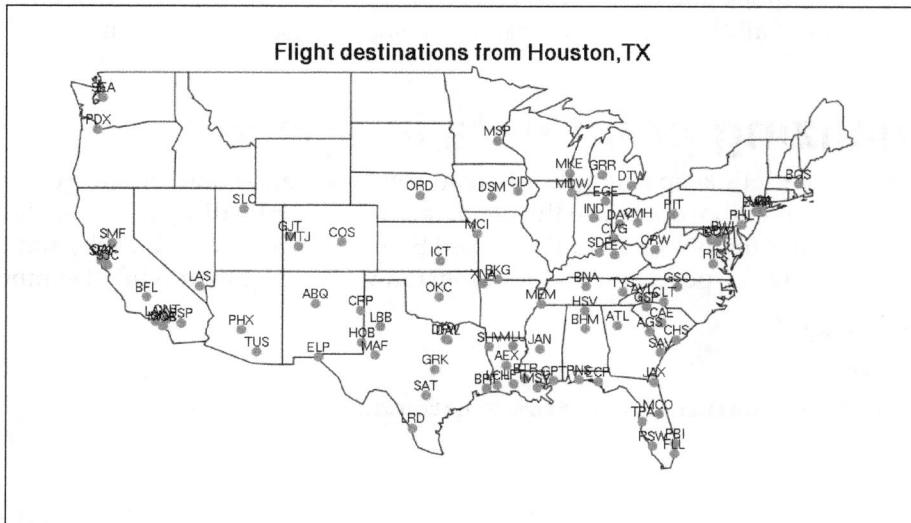

Flight destinations from Houston,TX

Now, we can also specify the color of the points to be rendered. This feature can be used to plot our first meaningful map to highlight the number of flights in 2011 to different parts of the USA:

```
> map('state')
> title('Frequent flight destinations from Houston,TX')
> points(h$lon, h$lat, col = 'blue', pch = 13)
> points(dt$lon, dt$lat, pch = 19,
+   col = rgb(1, 0, 0, dt$N / max(dt$N)))
> legend('bottomright', legend = round(quantile(dt$N)), pch = 19,
+   col = rgb(1, 0, 0, quantile(dt$N) / max(dt$N)), box.col = NA)
```

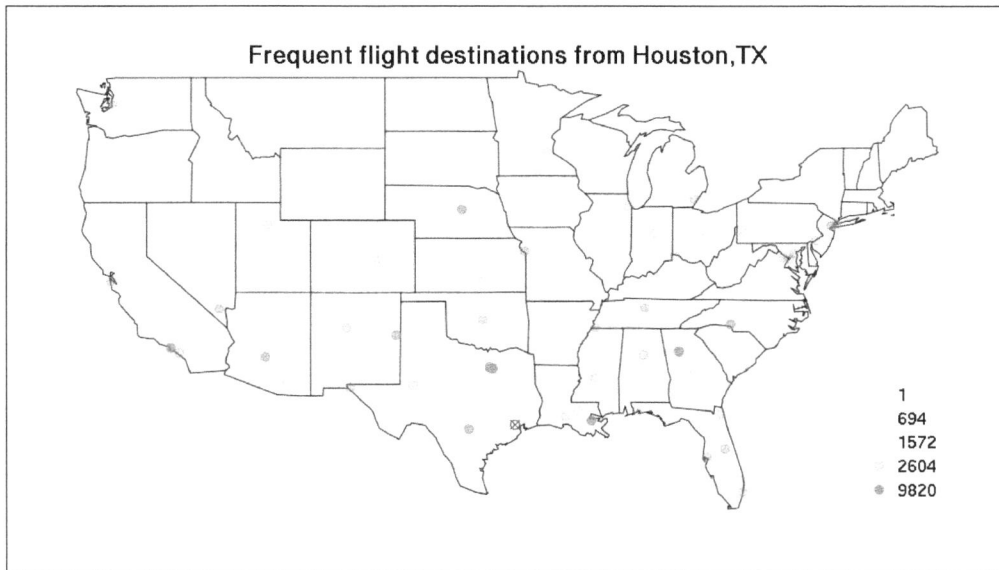

Frequent flight destinations from Houston,TX

	1
	694
	1572
	2604
●	9820

So the intensity of red shows the number of flights to the given points (airports); the values range from 1 to almost 10,000. Probably it would be more meaningful to compute these values on a state level as there are many airports, very close to each other, which might be better aggregated at a higher administrative area level. To this end, we load the polygon of the states, match the points of interest (airports) with the overlaying polygons (states), and render the polygons as a thematic map instead of points, like we did on the previous pages.

Finding polygon overlays of point data

We already have all the data we need to identify the parent state of each airport. The dt dataset includes the geo-coordinates of the locations, and we managed to render the states as polygons with the map function. Actually, this latter function can return the underlying dataset without rendering a plot:

```
> str(map_data <- map('state', plot = FALSE, fill = TRUE))
List of 4
 $ x    : num [1:15599] -87.5 -87.5 -87.5 -87.5 -87.6 ...
 $ y    : num [1:15599] 30.4 30.4 30.4 30.3 30.3 ...
 $ range: num [1:4] -124.7 -67 25.1 49.4
 $ names: chr [1:63] "alabama" "arizona" "arkansas" "california" ...
 - attr(*, "class")= chr "map"
```

So we have around 16,000 points describing the boundaries of the US states, but this map data is more detailed than we actually need (see for example the name of the polygons starting with Washington):

```
> grep('^washington', map_data$names, value = TRUE)
[1] "washington:san juan island" "washington:lopez island"
[3] "washington:orcas island"    "washington:whidbey island"
[5] "washington:main"
```

In short, the non-connecting parts of a state are defined as separate polygons. To this end, let's save a list of the state names without the string after the colon:

```
> states <- sapply(strsplit(map_data$names, ':'), '[[', 1)
```

We will use this list as the basis of aggregation from now on. Let's transform this map dataset into another class of object, so that we can use the powerful features of the sp package. We will use the maptools package to do this transformation:

```
> library(maptools)
> us <- map2SpatialPolygons(map_data, IDs = states,
+     proj4string = CRS("+proj=longlat +datum=WGS84"))
```

> An alternative way of getting the state polygons might be to directly load those instead of transforming from other data formats as described earlier. To this end, you may find the raster package especially useful to download free map **shapefiles** from gadm.org via the getData function. Although these maps are way too detailed for such a simple task, you can always simplify those — for example, with the gSimplify function of the rgeos package.

So we have just created an object called `us`, which includes the polygons of `map_data` for each state with the given **projection**. This object can be shown on a map just like we did previously, although you should use the general `plot` method instead of the `map` function:

```
> plot(us)
```

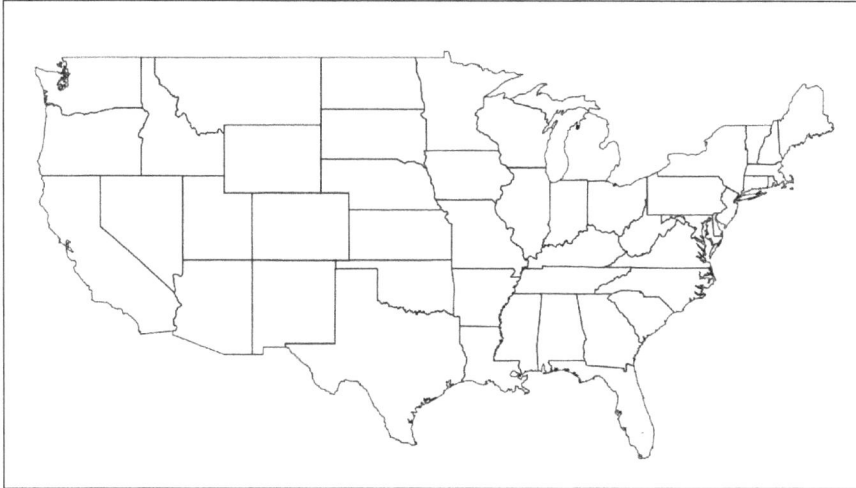

Besides this, however, the `sp` package supports so many powerful features! For example, it's very easy to identify the overlay polygons of the provided points via the `over` function. As this function name conflicts with the one found in the `grDevices` package, it's better to refer to the function along with the namespace using a double colon:

```
> library(sp)
> dtp <- SpatialPointsDataFrame(dt[, c('lon', 'lat')], dt,
+   proj4string = CRS("+proj=longlat +datum=WGS84"))
> str(sp::over(us, dtp))
'data.frame':   49 obs. of  8 variables:
 $ Dest     : chr  "BHM" "PHX" "XNA" "LAX" ...
 $ N        : int  2736 5096 1172 6064 164 NA NA 2699 3085 7886 ...
 $ Cancelled: int  39 29 34 33 1 NA NA 35 11 141 ...
 $ Distance : int  562 1009 438 1379 926 NA NA 1208 787 689 ...
 $ TimeVar  : num  10.1 13.61 9.47 15.16 13.82 ...
 $ ArrDelay : num  8.696 2.166 6.896 8.321 -0.451 ...
 $ lon      : num  -86.8 -112.1 -94.3 -118.4 -107.9 ...
 $ lat      : num  33.6 33.4 36.3 33.9 38.5 ...
```

What happened here? First, we passed the coordinates and the whole dataset to the `SpatialPointsDataFrame` function, which stored our data as spatial points with the given longitude and latitude values. Next, we called the `over` function to left-join the values of `dtp` to the US states.

> An alternative way of identifying the state of a given airport is to ask for more detailed information from the Google Maps API. By changing the default `output` argument of the `geocode` function, we can get all address components for the matched spatial object, which of course includes the state as well. Look for example at the following code snippet:
>
> `geocode('LAX','all')$results[[1]]$address_components`
>
> Based on this, you might want to get a similar output for all airports and filter the list for the short name of the state. The `rlist` package would be extremely useful in this task, as it offers some very convenient ways of manipulating lists in R.

The only problem here is that we matched only one airport to the states, which is definitely not okay. See for example the fourth column in the earlier output: it shows LAX as the matched airport for `California` (returned by `states[4]`), although there are many others there as well.

To overcome this issue, we can do at least two things. First, we can use the `returnList` argument of the `over` function to return all matched rows of `dtp`, and we will then post-process that data:

```
> str(sapply(sp::over(us, dtp, returnList = TRUE),
+     function(x) sum(x$Cancelled)))
 Named int [1:49] 51 44 34 97 23 0 0 35 66 149 ...
 - attr(*, "names")= chr [1:49] "alabama" "arizona" "arkansas" ...
```

So we created and called an anonymous function that will `sum` up the `Cancelled` values of the `data.frame` in each element of the list returned by `over`.

Another, probably cleaner, approach is to redefine `dtp` to only include the related values and pass a function to `over` to do the summary:

```
> dtp <- SpatialPointsDataFrame(dt[, c('lon', 'lat')],
+     dt[, 'Cancelled', drop = FALSE],
+     proj4string = CRS("+proj=longlat +datum=WGS84"))
> str(cancels <- sp::over(us, dtp, fn = sum))
'data.frame':   49 obs. of  1 variable:
 $ Cancelled: int  51 44 34 97 23 NA NA 35 66 149 ...
```

Either way, we have a vector to merge back to the US state names:

```
> val <- cancels$Cancelled[match(states, row.names(cancels))]
```

And to update all missing values to zero (as the number of cancelled flights in a state without any airport is not missing data, but exactly zero for sure):

```
> val[is.na(val)] <- 0
```

Plotting thematic maps

Now we have everything to create our first *thematic* map. Let's pass the `val` vector to the previously used `map` function (or `plot` it using the `us` object), specify a plot title, add a blue point for Houston, and then create a legend, which shows the quantiles of the overall number of cancelled flights as a reference:

```
> map("state", col = rgb(1, 0, 0, sqrt(val/max(val))), fill = TRUE)
> title('Number of cancelled flights from Houston to US states')
> points(h$lon, h$lat, col = 'blue', pch = 13)
> legend('bottomright', legend = round(quantile(val)),
+     fill = rgb(1, 0, 0, sqrt(quantile(val)/max(val))), box.col = NA)
```

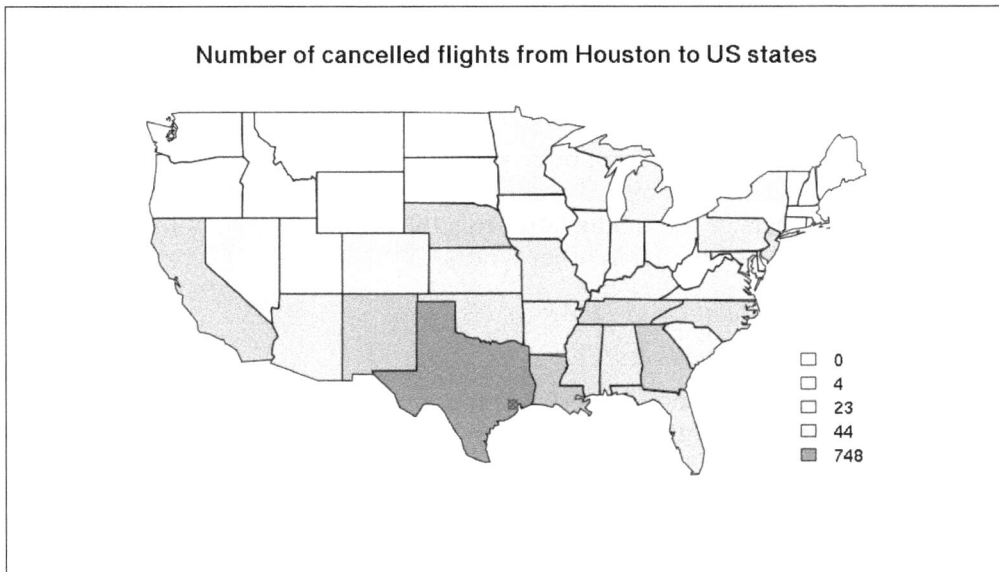

Please note that, instead of a linear scale, we have decided to compute the square root of the relative values to define the intensity of the fill color, so that we can visually highlight the differences between the states. This was necessary as most flight cancellations happened in Texas (`748`), and there were no more than 150 cancelled flights in any other state (with the average being around 45).

> You can also easily load ESRI shape files or other geospatial vector data formats into R as points or polygons with a bunch of packages already discussed and a few others as well, such as the `maptools`, `rgdal`, `dismo`, `raster`, or `shapefile` packages.

Another, probably easier, way to generate country-level thematic maps, especially choropleth maps, is to load the `rworldmap` package made by Andy South, and rely on the convenient `mapCountryData` function.

Rendering polygons around points

Besides thematic maps, another really useful way of presenting spatial data is to draw artificial polygons around the data points based on the data values. This is especially useful if there is no available polygon shape file to be used to generate a thematic map.

A level plot, contour plot, or isopleths, might be an already familiar design from tourist maps, where the altitude of the mountains is represented by a line drawn around the center of the hill at the very same levels. This is a very smart approach having maps present the height of hills—projecting this third dimension onto a 2-dimensional image.

Now let's try to replicate this design by considering our data points as mountains on the otherwise flat map. We already know the heights and exact geo-coordinates of the geometric centers of these hills (airports); the only challenge here is to draw the actual shape of these objects. In other words:

- Are these *mountains* connected?
- How steep are the *hillsides*?
- Should we consider any underlying spatial effects in the data? In other words, can we actually render these as *mountains* with a 3D shape instead of plotting independent points in space?

If the answer for the last question is positive, then we can start trying to answer the other questions by fine-tuning the plot parameters. For now, let's simply suppose that there is a spatial effect in the underlying data, and it makes sense to visualize the data in such a way. Later, we will have the chance to disprove or support this statement either by analyzing the generated plots, or by building some geo-spatial models—some of these will be discussed later, in the *Spatial Statistics* section.

Contour lines

First, let's expand our data points into a matrix with the `fields` package. The size of the resulting R object is defined arbitrarily but, for the given number of rows and columns, which should be a lot higher to generate higher resolution images, 256 is a good start:

```
> library(fields)
> out <- as.image(dt$ArrDelay, x = dt[, c('lon', 'lat')],
+    nrow = 256, ncol = 256)
```

The `as.image` function generates a special R object, which in short includes a 3-dimensional matrix-like data structure, where the x and y axes represent the longitude and latitude ranges of the original data respectively. To simplify this even more, we have a matrix with 256 rows and 256 columns, where each of those represents a discrete value evenly distributed between the lowest and highest values of the latitude and longitude. And on the z axis, we have the `ArrDelay` values—which are in most cases of course missing:

```
> table(is.na(out$z))

FALSE   TRUE
   112 65424
```

What does this matrix look like? It's better to see what we have at the moment:

```
> image(out)
```

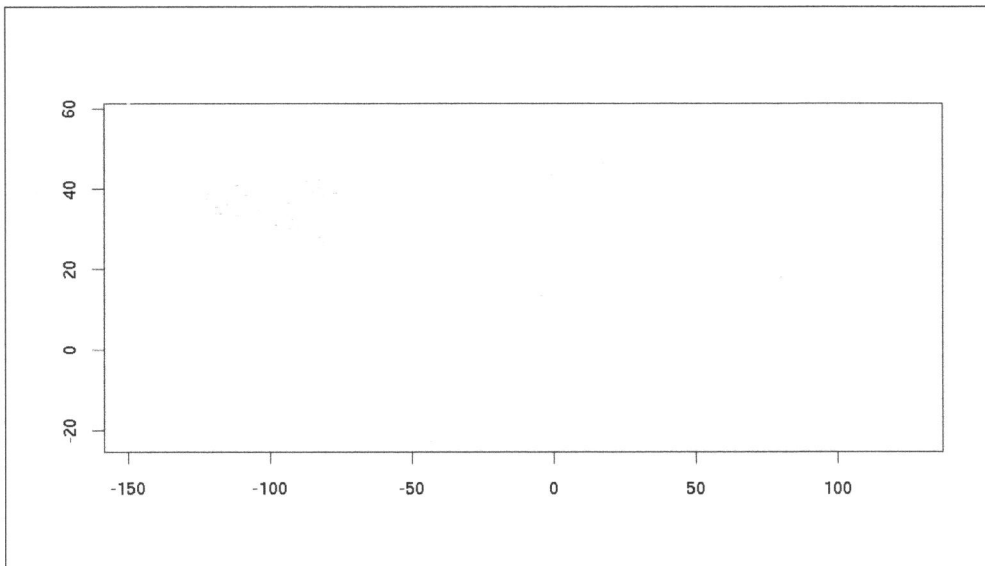

Well, this does not seem to be useful at all. What is shown there? We rendered the *x* and *y* dimensions of the matrix with *z* colors here, and most tiles of this map are empty due to the high amount of missing values in *z*. Also, it's pretty straightforward now that the dataset includes many airports outside the USA as well. How does it look if we focus only on the USA?

```
> image(out, xlim = base::range(map_data$x, na.rm = TRUE),
+            ylim = base::range(map_data$y, na.rm = TRUE))
```

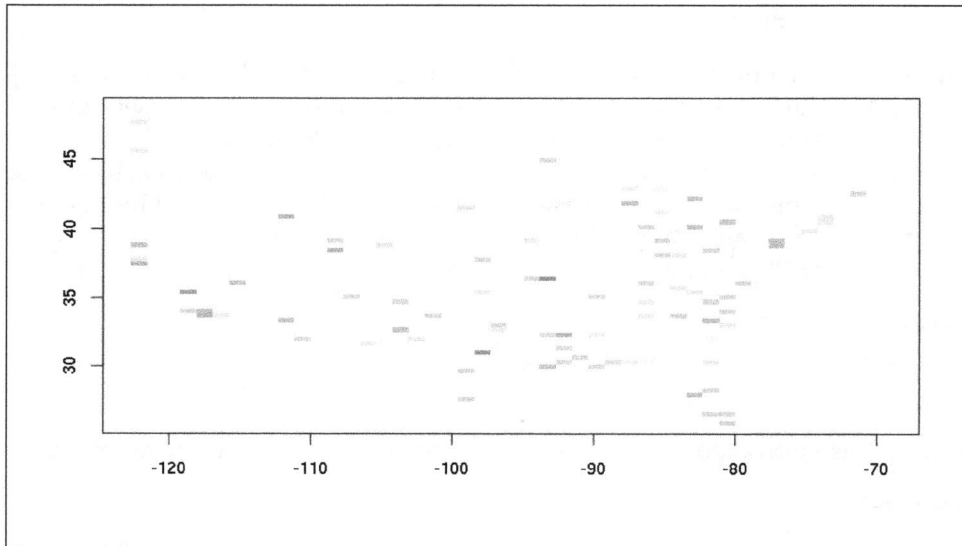

An alternative and more elegant approach to rendering only the US part of the matrix would be to drop the non-US airports from the database before actually creating the out R object. Although we will continue with this example for didactic purposes, with real data make sure that you concentrate on the target subset of your data instead of trying to smooth and model unrelated data points as well.

A lot better! So we have our data points as a tile, now let's try to identify the slope of these mountain peaks, to be able to render them on a future map. This can be done by smoothing the matrix:

```
> look <- image.smooth(out, theta = .5)
> table(is.na(look$z))
FALSE   TRUE
14470 51066
```

As can be seen in the preceding table, this algorithm successfully eliminated many missing values from the matrix. The `image.smooth` function basically reused our initial data point values in the neighboring tiles, and computed some kind of average for the conflicting overrides. This smoothing algorithm results in the following arbitrary map, which does not respect any political or geographical boundaries:

```
> image(look)
```

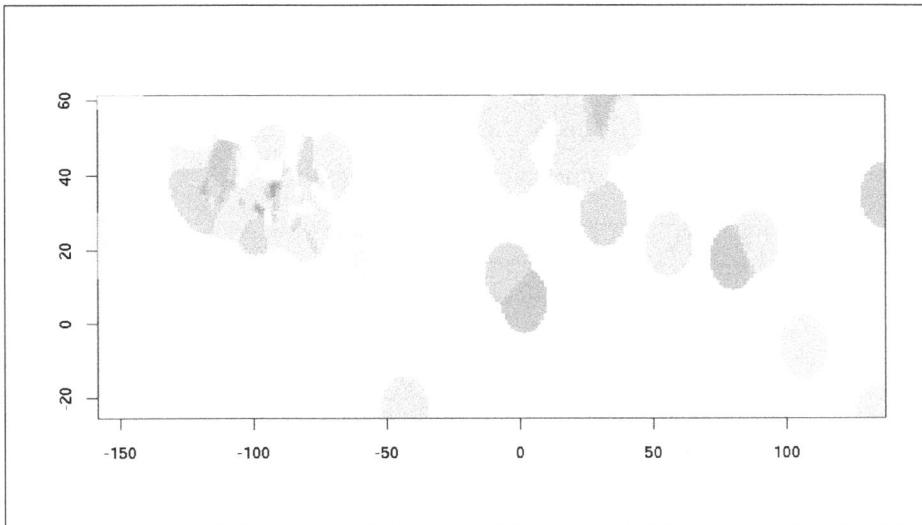

It would be really nice to plot these artificial polygons along with the administrative boundaries, so let's clear out all cells that do not belong to the territory of the USA. We will use the `point.in.polygon` function from the `sp` package to do so:

```
> usa_data <- map('usa', plot = FALSE, region = 'main')
> p <- expand.grid(look$x, look$y)
> library(sp)
> n <- which(point.in.polygon(p$Var1, p$Var2,
+   usa_data$x, usa_data$y) == 0)
> look$z[n] <- NA
```

In a nutshell, we have loaded the main polygon of the USA without any sub-administrative areas, and verified our cells in the `look` object, if those are overlapping the polygon. Then we simply reset the value of the cell, if not.

The next step is to render the boundaries of the USA, plot our smoothed contour plot, then add some eye-candy in the means of the US states and, the main point of interest, the airport:

```
> map("usa")
> image(look, add = TRUE)
> map("state", lwd = 3, add = TRUE)
> title('Arrival delays of flights from Houston')
> points(dt$lon, dt$lat, pch = 19, cex = .5)
> points(h$lon, h$lat, pch = 13)
```

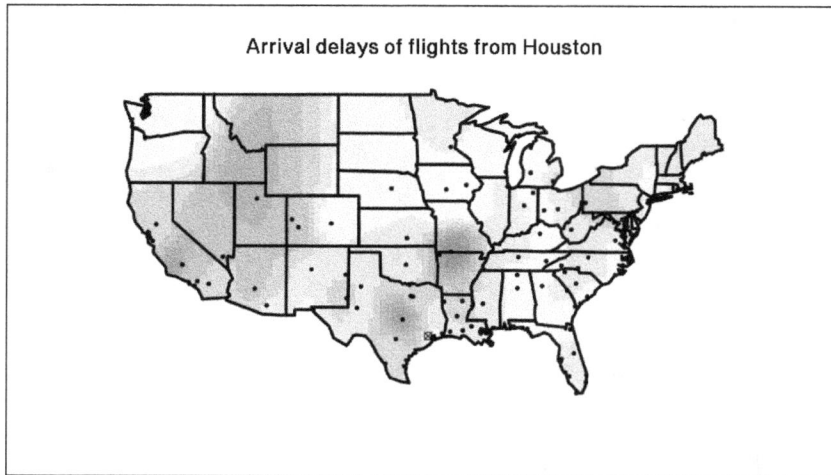

Now this is pretty neat, isn't it?

Voronoi diagrams

An alternative way of visualizing point data with polygons is to generate *Voronoi* cells between them. In short, the Voronoi map partitions the space into regions around the data points by aligning all parts of the map to one of the regions to minimize the distance from the central data points. This is extremely easy to interpret, and also to implement in R. The `deldir` package provides a function with the very same name for Delaunay triangulation:

```
> library(deldir)
> map("usa")
> plot(deldir(dt$lon, dt$lat), wlines = "tess", lwd = 2,
+     pch = 19, col = c('red', 'darkgray'), add = TRUE)
```

Here, we represented the airports with red dots, as we did before, but also added the Dirichlet tessellation (Voronoi cells) rendered as dark-gray dashed lines. For more options on how to fine-tune the results, see the `plot.deldir` method.

In the next section, let's see how to improve this plot by adding a more detailed background map to it.

Satellite maps

There are many R packages on CRAN that can fetch data from Google Maps, Stamen, Bing, or OpenStreetMap—even some of the packages that we have previously used in this chapter, such as the `ggmap` package, can do this. Similarly, the `dismo` package also comes with both geo-coding and Google Maps API integration capabilities, and there are some other packages focused on that latter, such as the `RgoogleMaps` package.

Now we will use the `OpenStreetMap` package, mainly because it supports not only the awesome OpenStreetMap database back-end, but also a bunch of other formats as well. For example, we can render really nice terrain maps via Stamen:

```
> library(OpenStreetMap)
> map <- openmap(c(max(map_data$y, na.rm = TRUE),
+                  min(map_data$x, na.rm = TRUE)),
+                c(min(map_data$y, na.rm = TRUE),
+                  max(map_data$x, na.rm = TRUE)),
+                type = 'stamen-terrain')
```

So we defined the left upper and right lower corners of the map we need, and also specified the map style to be a satellite map. As the data by default arrives from the remote servers with the Mercator projections, we first have to transform that to WGS84 (we used this previously), so that we can render the points and polygons on the top of the fetched map:

```
> map <- openproj(map,
+    projection = '+proj=longlat +ellps=WGS84 +datum=WGS84 +no_defs')
```

And showtime at last:

```
> plot(map)
> plot(deldir(dt$lon, dt$lat), wlines = "tess", lwd = 2,
+    col = c('red', 'black'), pch = 19, cex = 0.5, add = TRUE)
```

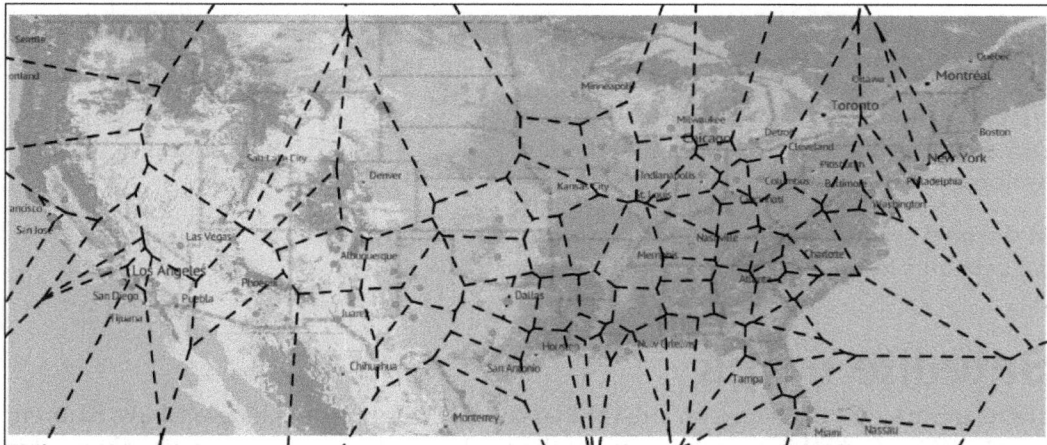

This seems to be a lot better compared to the outline map we created previously. Now you can try some other map styles as well, such as `mapquest-aerial`, or some of the really nice-looking `cloudMade` designs.

Interactive maps

Besides being able to use Web-services to download map tiles for the background of the maps created in R, we can also rely on some of those to generate truly interactive maps. One of the best known related services is the Google Visualization API, which provides a platform for hosting visualizations made by the community; you can also use it to share maps you've created with others.

Querying Google Maps

In R, you can access this API via the `googleVis` package written and maintained by Markus Gesmann and Diego de Castillo. Most functions of the package generate HTML and JavaScript code that we can directly view in a Web browser as an SVG object with the `base` plot function; alternatively, we can integrate them in a Web page, for example via the IFRAME HTML tag.

The `gvisIntensityMap` function takes a `data.frame` with country ISO or USA state codes and the actual data to create a simple intensity map. We will use the `cancels` dataset we created in the *Finding Polygon Overlays of Point Data* section, but before that, we have to do some data transformations. Let's add the state name as a new column to the `data.frame`, and replace the missing values with zero:

```
> cancels$state <- rownames(cancels)
> cancels$Cancelled[is.na(cancels$Cancelled)] <- 0
```

Now it's time to load the package and pass the data along with a few extra parameters, signifying that we want to generate a state-level US map:

```
> library(googleVis)
> plot(gvisGeoChart(cancels, 'state', 'Cancelled',
+                 options = list(
+                        region       = 'US',
+                        displayMode = 'regions',
+                        resolution  = 'provinces')))
```

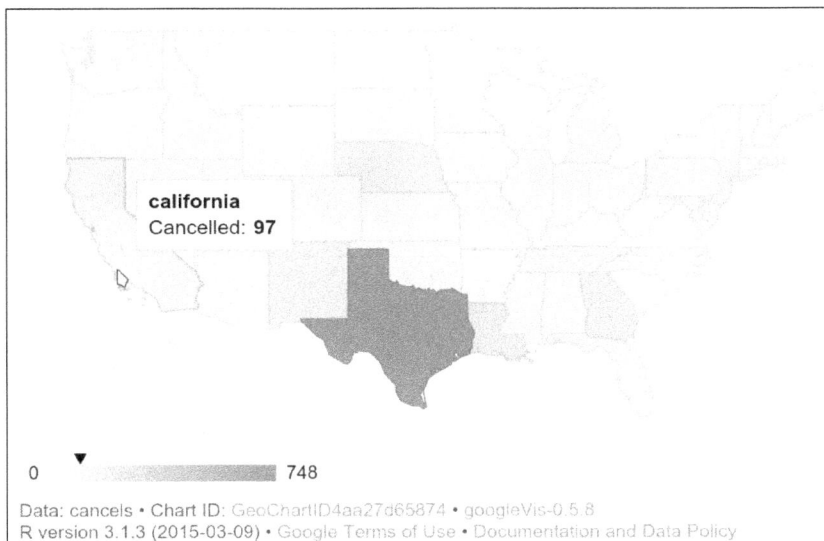

The package also offers opportunities to query the Google Map API via the `gvisMap` function. We will use this feature to render the airports from the `dt` dataset as points on a Google Map with an auto-generated tooltip of the variables.

But first, as usual, we have to do some data transformations again. The location argument of the `gvisMap` function takes the latitude and longitude values separated by a colon:

```
> dt$LatLong <- paste(dt$lat, dt$lon, sep = ':')
```

We also have to generate the tooltips as a new variable, which can be done easily with an `apply` call. We will concatenate the variable names and actual values separated by a HTML line break:

```
> dt$tip <- apply(dt, 1, function(x)
+                    paste(names(dt), x, collapse = '<br/ >'))
```

And now we just pass these arguments to the function for an instant interactive map:

```
> plot(gvisMap(dt, 'LatLong', tipvar = 'tip'))
```

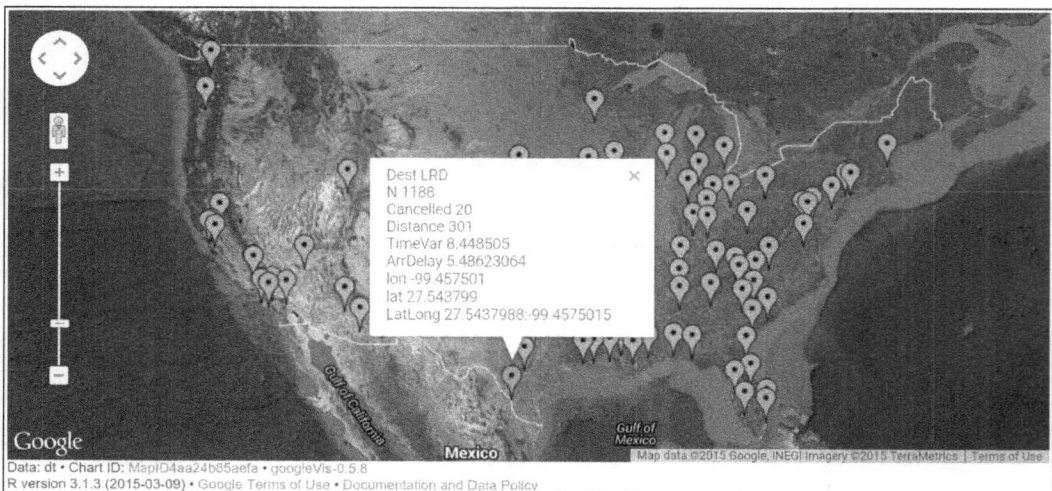

Another nifty feature of the `googleVis` package is that you can easily merge the different visualizations into one by using the `gvisMerge` function. The use of this function is quite simple: specify any two `gvis` objects you want to merge, and also whether they are to be placed horizontally or vertically.

JavaScript mapping libraries

The great success of the trending JavaScript data visualization libraries is only partly due to their great design. I suspect other factors also contribute to the general spread of such tools: it's very easy to create and deploy full-blown data models, especially since the release and on-going development of Mike Bostock's D3.js.

Although there are also many really useful and smart R packages to interact directly with D3 and topojson (see for example my R user activity compilation at `http://bit.ly/countRies`). Now we will only focus on how to use Leaflet— probably the most used JavaScript library for interactive maps.

What I truly love in R is that there are many packages wrapping other tools, so that R users can rely on only one programming language, and we can easily use C++ programs and Hadoop MapReduce jobs or build JavaScript-powered dashboards without actually knowing anything about the underlying technology. This is especially true when it comes to Leaflet!

There are at least two very nice packages that can generate a Leaflet plot from the R console, without a single line of JavaScript. The `Leaflet` reference class of the `rCharts` package was developed by Ramnath Vaidyanathan, and includes some methods to create a new object, set the viewport and zoom level, add some points or polygons to the map, and then render or print the generated HTML and JavaScript code to the console or to a file.

Unfortunately, this package is not on CRAN yet, so you will have to install it from GitHub:

```
> devtools::install_github('ramnathv/rCharts')
```

As a quick example, let's generate a Leaflet map of the airports with some tooltips, like we did with the Google Maps API in the previous section. As the `setView` method expects numeric geo-coordinates as the center of the map, we will use Kansas City's airport as a reference:

```
> library(rCharts)
> map <- Leaflet$new()
> map$setView(as.numeric(dt[which(dt$Dest == 'MCI'),
```

```
+    c('lat', 'lon')]), zoom = 4)
> for (i in 1:nrow(dt))
+      map$marker(c(dt$lat[i], dt$lon[i]), bindPopup = dt$tip[i])
> map$show()
```

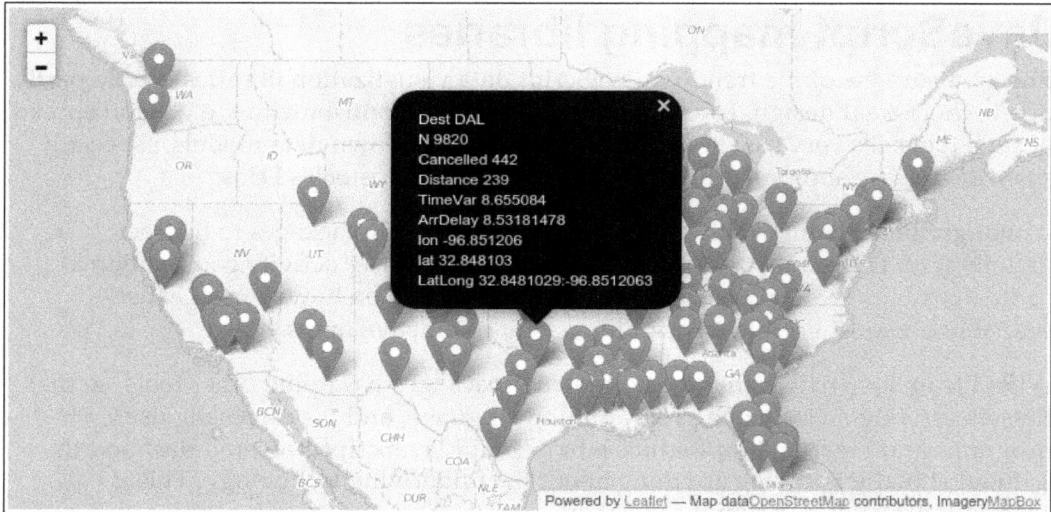

Similarly, RStudio's `leaflet` package and the more general `htmlwidgets` package also provide some easy ways to generate JavaScript-powered data visualizations. Let's load the library and define the steps one by one using the pipe operator from the `magrittr` package, which is pretty standard for all packages created or inspired by RStudio or Hadley Wickham:

```
> library(leaflet)
> leaflet(us) %>%
+    addProviderTiles("Acetate.terrain") %>%
+    addPolygons() %>%
+    addMarkers(lng = dt$lon, lat = dt$lat, popup = dt$tip)
```

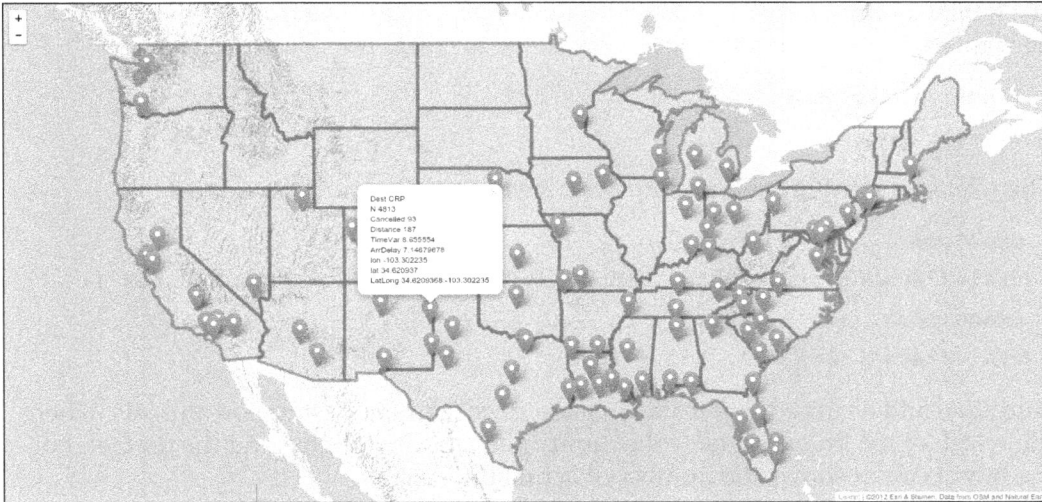

I especially like this preceding map, as we can load a third-party satellite map in the background, then render the states as polygons; we also added the original data points along with some useful tooltips on the very same map with literally a one-line R command. We could even color the state polygons based on the aggregated results we computed in the previous sections! Ever tried to do the same in Java?

Alternative map designs

Besides being able to use third-party tools, another main reason why I tend to use R for all my data analysis tasks is that R is extremely powerful in creating custom data exploration, visualization, and modeling designs.

As an example, let's create a flow-map based on our data, where we will highlight the flights from Houston based on the number of actual and cancelled flights. We will use lines and circles to render these two variables on a 2-dimensional map, and we will also add a contour plot in the background based on the average time delay.

But, as usual, let's do some data transformations first! To keep the number of flows at a minimal level, let's get rid of the airports outside the USA at last:

```
> dt <- dt[point.in.polygon(dt$lon, dt$lat,
+                           usa_data$x, usa_data$y) == 1, ]
```

We will need the `diagram` package (to render curved arrows from Houston to the destination airports) and the `scales` package to create transparent colors:

```
> library(diagram)
> library(scales)
```

Then, let's render the contour map described in the *Contour Lines* section:

```
> map("usa")
> title('Number of flights, cancellations and delays from Houston')
> image(look, add = TRUE)
> map("state", lwd = 3, add = TRUE)
```

And then add a curved line from Houston to each of the destination airports, where the width of the line represents the number of cancelled flights and the diameter of the target circles shows the number of actual flights:

```
> for (i in 1:nrow(dt)) {
+    curvedarrow(
+       from       = rev(as.numeric(h)),
+       to         = as.numeric(dt[i, c('lon', 'lat')]),
+       arr.pos    = 1,
+       arr.type   = 'circle',
+       curve      = 0.1,
+       arr.col    = alpha('black', dt$N[i] / max(dt$N)),
+       arr.length = dt$N[i] / max(dt$N),
+       lwd        = dt$Cancelled[i] / max(dt$Cancelled) * 25,
+       lcol       = alpha('black',
+                    dt$Cancelled[i] / max(dt$Cancelled)))
+ }
```

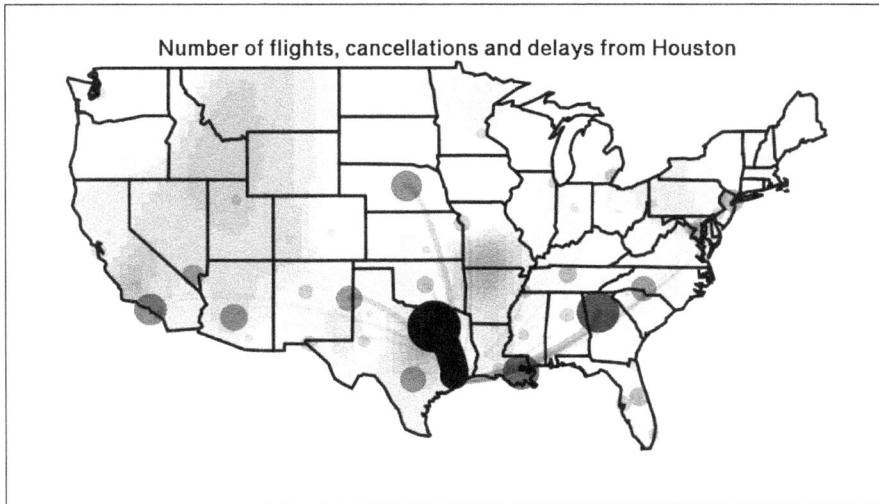

Well, this chapter ended up being about visualizing spatial data, and not really about analyzing spatial data by fitting models, filtering raw data, and looking for spatial effects. In the last section of the chapter, let's see how one can start using analytical approaches with spatial data.

Spatial statistics

Most exploratory data analysis projects dealing with spatial data start by looking for, and potentially filtering, spatial autocorrelation. In simple terms, this means that we are looking for spatial effects in the data—for instance, the similarities of some data points can be (partly) explained by the short distance between them; further points seem to differ a lot more. There is nothing surprising in this statement; probably all of you agree with this. But how can we test this on real data with analytical tools?

Moran's I index is a well-known and generally used measure to test whether spatial autocorrelation is present or not in the variable of interest. This is a quite simple statistical test with the null hypothesis that there is no spatial autocorrelation in the dataset.

With the current data structure we have, probably the easiest way to compute Moran's I is to load the `ape` package, and pass the similarity matrix along with the variable of interest to the `Moran.I` function. First, let's compute this similarity matrix by the inverse of the Euclidian distance matrix:

```
> dm <- dist(dt[, c('lon', 'lat')])
> dm <- as.matrix(dm)
> idm <- 1 / dm
> diag(idm) <- 0
> str(idm)
 num [1:88, 1:88] 0 0.0343 0.1355 0.2733 0.0467 ...
 - attr(*, "dimnames")=List of 2
  ..$ : chr [1:88] "1" "3" "6" "7" ...
  ..$ : chr [1:88] "1" "3" "6" "7" ...
```

Then let's replace all possible missing values (because the number of flights can be one as well, resulting in zero variance) in the `TimeVar` column, and let's see if there is any spatial autocorrelation in the variance of the actual elapsed time of the flights:

```
> dt$TimeVar[is.na(dt$TimeVar)] <- 0
> library(ape)
> Moran.I(dt$TimeVar, idm)
$observed
[1] 0.1895178

$expected
[1] -0.01149425

$sd
[1] 0.02689139

$p.value
[1] 7.727152e-14
```

This was pretty easy, wasn't it? Based on the returned P value, we can reject the null hypothesis, and the 0.19 Moran's I suggests that the variation in the elapsed flight time is affected by the location of the destination airports, probably due to the very different distances.

A reverse dependency of the previously mentioned `sp` package, the `spdep` package can also compute this index, although we have to first transform the similarity matrix into a list object:

```
> library(spdep)
> idml <- mat2listw(idm)
> moran.test(dt$TimeVar, idml)


  Moran's I test under randomisation


data:  dt$TimeVar
weights: idml


Moran I statistic standard deviate = 1.7157, p-value = 0.04311
alternative hypothesis: greater
sample estimates:
Moran I statistic      Expectation          Variance
     0.108750656      -0.011494253       0.004911818
```

Although the test results are similar to the previous run, and we can reject the null hypothesis of zero spatial autocorrelation in the data, the Moran's I index and the P values are not identical. This is mainly due to the fact that the `ape` package used weight matrix for the computation, while the `moran.test` function was intended to be used with polygon data, as it requires the neighbor lists of the data. Well, as our example included point data, this is not a clean-cut solution. Another main difference between the approaches is that the `ape` package uses normal approximation, while `spdep` implements randomization. But this difference is still way too high, isn't it?

Reading the function documentation reveals that we can improve the `spdep` approach: when converting the `matrix` into a `listw` object, we can specify the actual type of the originating matrix. In our case, as we are using the inverse distance matrix, a row-standardized style seems more appropriate:

```
> idml <- mat2listw(idm, style = "W")
> moran.test(dt$TimeVar, idml)


  Moran's I test under randomisation


data:  dt$TimeVar
weights: idml
```

```
Moran I statistic standard deviate = 7.475, p-value = 3.861e-14
alternative hypothesis: greater
sample estimates:
Moran I statistic        Expectation          Variance
     0.1895177587        -0.0114942529      0.0007231471
```

Now the differences between this and the `ape` results are in an acceptable range, right?

Unfortunately, this section cannot cover related questions or other statistical methods dealing with spatial data, but there are many really useful books out there dedicated to the topic. Please be sure to check the *Appendix* at the end of the book for some suggested titles.

Summary

Congratulations, you have just finished the last systematic chapter of the book! Here, we focused on how to analyze spatial data mainly with data visualization tools.

Now let's see how we can combine the methods learned in the previous chapters. In the final part of the book, we will analyze the R community with various data science tools. If you liked this chapter, I am sure you will enjoy the final one as well.

14
Analyzing the R Community

In this final chapter, I will try to summarize what you have learned in the past 13 chapters. To this end, we will create an actual case study, independent from the previously used `hflights` and `mtcars` datasets, and will now try to estimate the size of the R community. This is a rather difficult task as there is no list of R users around the world; thus, we will have to build some predicting models on a number of partial datasets.

To this end, we will do the following in this chapter:

- Collect live data from different data sources on the Internet
- Cleanse the data and transform it to a standard format
- Run some quick descriptive, exploratory analysis methods
- Visualize the extracted data
- Build some log-linear models on the number of R users based on an independent list of names

R Foundation members

One of the easiest things we can do is count the members of the R Foundation—the organization coordinating the development of the core R program. As the ordinary members of the Foundation include only the *R Development Core Team*, we had better check the supporting members. Anyone can become a supporting member of the Foundation by paying a nominal yearly fee— I highly suggest you do this, by the way. The list is available on the `http://r-project.org` site, and we will use the XML package (for more detail, see *Chapter 2, Getting Data from the Web*) to parse the HTML page:

```
> library(XML)
> page <- htmlParse('http://r-project.org/foundation/donors.html')
```

Now that we have the HTML page loaded into R, we can use the XML Path Language to extract the list of the supporting members of the Foundation, by reading the list after the `Supporting members` header:

```
> list <- unlist(xpathApply(page,
+    "//h3[@id='supporting-members']/following-sibling::ul[1]/li",
+    xmlValue))
> str(list)
 chr [1:279] "Klaus Abberger (Germany)" "Claudio Agostinelli (Italy)"
```

Form this character vector of 279 names and countries, let's extract the list of supporting members and the countries separately:

```
> supporterlist <- sub(' \\([a-zA-Z ]*\\)$', '', list)
> countrylist   <- substr(list, nchar(supporterlist) + 3,
+                              nchar(list) - 1)
```

So we first extracted the names by removing everything starting from the opening parenthesis in the strings, and then we matched the countries by the character positions computed from the number of characters in the names and the original strings.

Besides the name list of 279 supporting members of the R Foundation, we also know the proportion of the citizenship or residence of the members:

```
> tail(sort(prop.table(table(countrylist)) * 100), 5)
     Canada Switzerland          UK     Germany         USA
   4.659498    5.017921    7.168459   15.770609   37.992832
```

Visualizing supporting members around the world

Probably it's not that surprising that most supporting members are from the USA, and some European countries are also at the top of this list. Let's save this table so that we can generate a map on this count data after some quick data transformations:

```
> countries <- as.data.frame(table(countrylist))
```

As mentioned in *Chapter 13, Data Around Us,* the `rworldmap` package can render country-level maps in a very easy way; we just have to map the values with some polygons. Here, we will use the `joinCountryData2Map` function, first enabling the `verbose` option to see what country names have been missed:

```
> library(rworldmap)
> joinCountryData2Map(countries, joinCode = 'NAME',
+     nameJoinColumn = 'countrylist', verbose = TRUE)
32 codes from your data successfully matched countries in the map
4 codes from your data failed to match with a country code in the map
     failedCodes failedCountries
[1,] NA          "Brasil"
[2,] NA          "CZ"
[3,] NA          "Danmark"
[4,] NA          "NL"
213 codes from the map weren't represented in your data
```

So we tried to match the country names stored in the countries data frame, but failed for the previously listed four strings. Although we could manually fix this, in most cases it's better to automate what we can, so let's pass all the failed strings to the Google Maps geocoding API and see what it returns:

```
> library(ggmap)
> for (fix in c('Brasil', 'CZ', 'Danmark', 'NL')) {
+     countrylist[which(countrylist == fix)] <-
+         geocode(fix, output = 'more')$country
+ }
```

Now that we have fixed the country names with the help of the Google geocoding service, let's regenerate the frequency table and map those values to the polygon names with the `rworldmap` package:

```
> countries <- as.data.frame(table(countrylist))
> countries <- joinCountryData2Map(countries, joinCode = 'NAME',
+     nameJoinColumn = 'countrylist')
36 codes from your data successfully matched countries in the map
0 codes from your data failed to match with a country code in the map
211 codes from the map weren't represented in your data
```

These results are much more satisfying! Now we have the number of supporting members of the R Foundation mapped to the countries, so we can easily plot this data:

```
> mapCountryData(countries, 'Freq', catMethod = 'logFixedWidth',
+     mapTitle = 'Number of R Foundation supporting members')
```

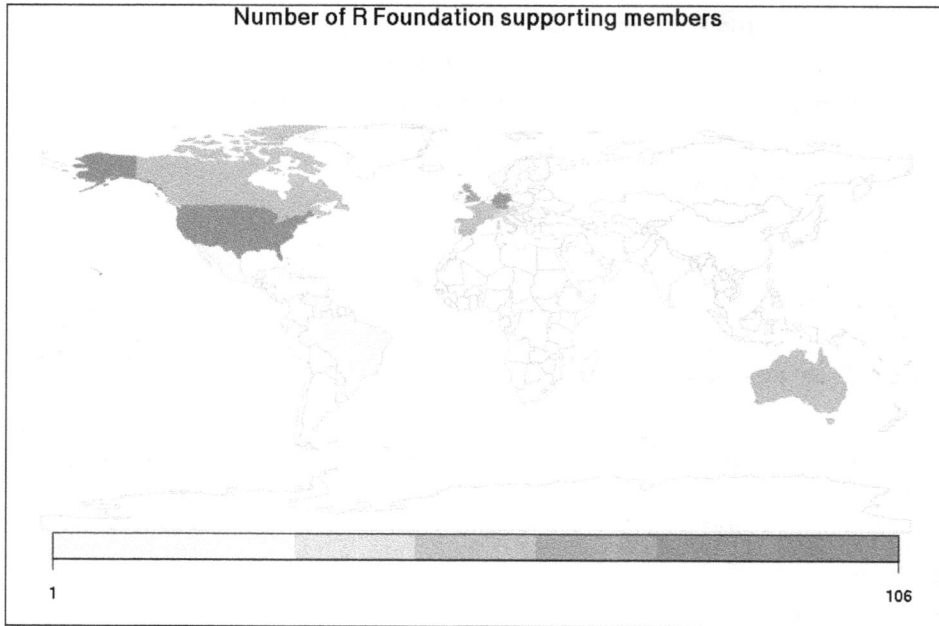

Well, it's clear that most supporting members of the R Foundation are based in the USA, Europe, Australia, and New Zealand (where R was born more than 20 years ago).

But the number of supporters is unfortunately really low, so let's see what other data sources we can find and utilize in order to estimate the number of R users around the world.

R package maintainers

Another similarly straightforward data source might be the list of R package maintainers. We can download the names and e-mail addresses of the package maintainers from a public page of CRAN, where this data is stored in a nicely structured HTML table that is extremely easy to parse:

```
> packages <- readHTMLTable(paste0('http://cran.r-project.org',
+    '/web/checks/check_summary.html'), which = 2)
```

Extracting the names from the `Maintainer` column can be done via some quick data cleansing and transformations, mainly using regular expressions. Please note that the column name starts with a space—that's why we quoted the column name:

```
> maintainers <- sub('(.*) <(.*)>', '\\1', packages$' Maintainer')
> maintainers <- gsub(' ', ' ', maintainers)
> str(maintainers)
 chr [1:6994] "Scott Fortmann-Roe" "Gaurav Sood" "Blum Michael" ...
```

This list of almost 7,000 package maintainers includes some duplicated names (they maintain multiple packages). Let's see the list of the top, most prolific R package developers:

```
> tail(sort(table(maintainers)), 8)
    Paul Gilbert       Simon Urbanek Scott Chamberlain    Martin Maechler
              22                  22                24                 25
        ORPHANED         Kurt Hornik   Hadley Wickham Dirk Eddelbuettel
              26                  29                31                 36
```

Although there's an odd name in the preceding list (orphaned packages do not have a maintainer—it's worth mentioning that having only 26 packages out of the 6,994 no longer actively maintained is a pretty good ratio), but the other names are indeed well known in the R community and work on a number of useful packages.

The number of packages per maintainer

On the other hand, there are a lot more names in the list associated with only one or a few R packages. Instead of visualizing the number of packages per maintainer on a simple bar chart or histogram, let's load the `fitdistrplus` package, which we will use on the forthcoming pages to fit various theoretical distributions on this analyzed dataset:

```
> N <- as.numeric(table(maintainers))
> library(fitdistrplus)
> plotdist(N)
```

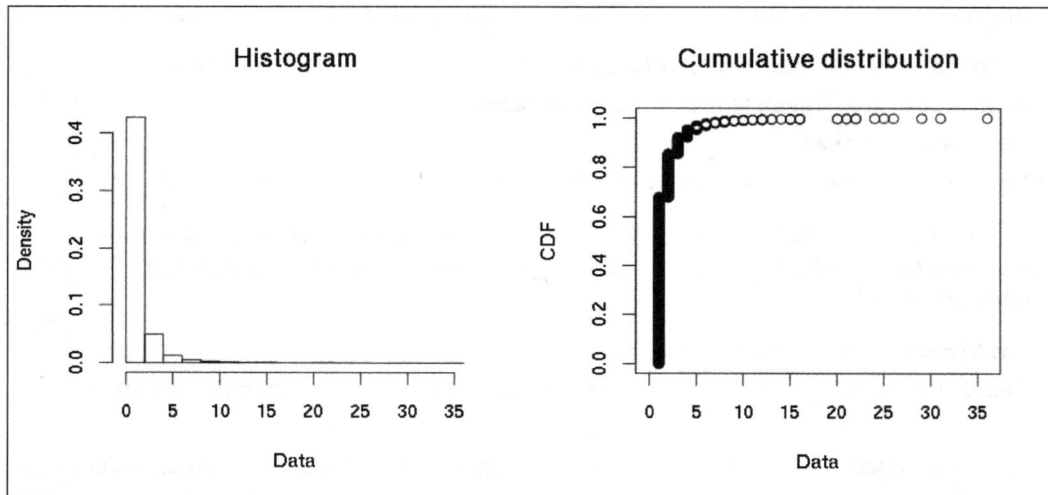

The preceding plots also show that most people in the list maintain only one, but no more than two or three, packages. If we are interested in how long/heavy tailed this distribution is, we might want to call the `descdist` function, which returns some important descriptive statistics on the empirical distribution and also plots how different theoretical distributions fit our data on a skewness-kurtosis plot:

```
> descdist(N, boot = 1e3)
summary statistics
------
min:  1    max:   36
median:  1
mean:  1.74327
estimated sd:  1.963108
estimated skewness:  7.191722
estimated kurtosis:  82.0168
```

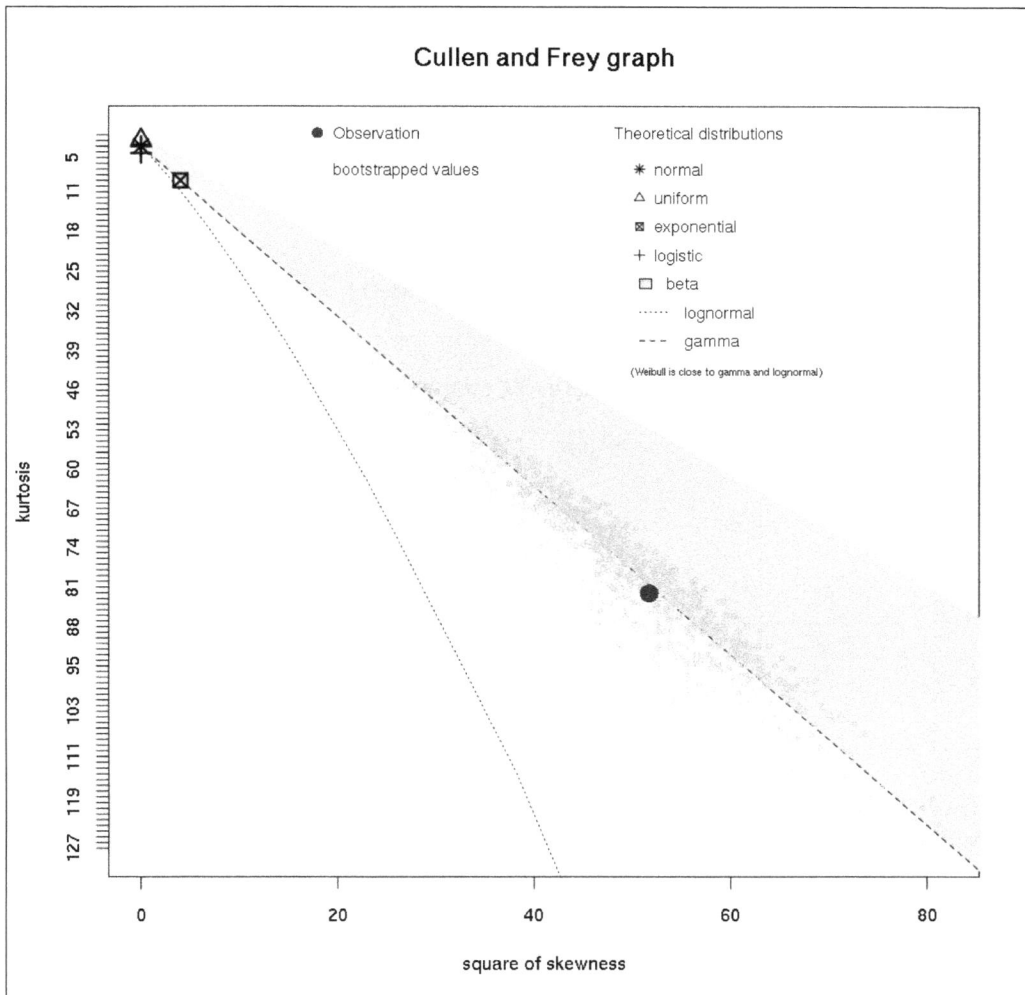

Cullen and Frey graph

Our empirical distribution seems to be rather long-tailed with a very high kurtosis, and it seems that the gamma distribution is the best fit for this dataset. Let's see the estimate parameters of this gamma distribution:

```
> (gparams <- fitdist(N, 'gamma'))
Fitting of the distribution ' gamma ' by maximum likelihood
Parameters:
      estimate Std. Error
shape 2.394869 0.05019383
rate  1.373693 0.03202067
```

We can use these parameters to simulate a lot more R package maintainers with the `rgamma` function. Let's see how many R packages would be available on CRAN with, for example, 100,000 package maintainers:

```
> gshape <- gparams$estimate[['shape']]
> grate  <- gparams$estimate[['rate']]
> sum(rgamma(1e5, shape = gshape, rate = grate))
[1] 173655.3
> hist(rgamma(1e5, shape = gshape, rate = grate))
```

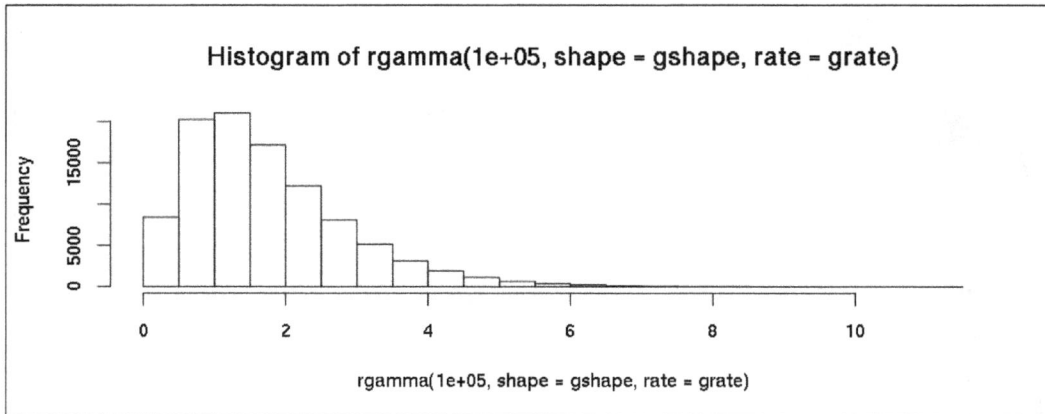

Histogram of rgamma(1e+05, shape = gshape, rate = grate)

It's rather clear that this distribution is not as long-tailed as our real dataset: even with 100,000 simulations, the largest number was below 10, as we can see in the preceding plot; in reality, though, the R package maintainers are a lot more productive with up to 20 or 30 packages.

Let's verify this by estimating the proportion of R package maintainers with no more than two packages based on the preceding gamma distribution:

```
> pgamma(2, shape = gshape, rate = grate)
[1] 0.6672011
```

But this percentage is a lot higher in the real dataset:

```
> prop.table(table(N <= 2))

    FALSE      TRUE
0.1458126 0.8541874
```

This may suggest trying to fit a longer-tailed distribution. Let's see for example how Pareto distribution would fit our data. To this end, let's follow the analytical approach by using the lowest value as the location of the distribution, and the number of values divided by the sum of the logarithmic difference of all these values from the location as the shape parameter:

```
> ploc <- min(N)
> pshp <- length(N) / sum(log(N) - log(ploc))
```

Unfortunately, there is no `ppareto` function in the base `stats` package, so we have to first load the `actuar` or `VGAM` package to compute the distribution function:

```
> library(actuar)
> ppareto(2, pshp, ploc)
[1] 0.9631973
```

Well, now this is even higher than the real proportion! It seems that none of the preceding theoretical distributions fit our data perfectly—which is pretty normal by the way. But let's see how these distributions fit our original data set on a joint plot:

```
> fg <- fitdist(N, 'gamma')
> fw <- fitdist(N, 'weibull')
> fl <- fitdist(N, 'lnorm')
> fp <- fitdist(N, 'pareto', start = list(shape = 1, scale = 1))
> par(mfrow = c(1, 2))
> denscomp(list(fg, fw, fl, fp), addlegend = FALSE)
> qqcomp(list(fg, fw, fl, fp),
+     legendtext = c('gamma', 'Weibull', 'Lognormal', 'Pareto'))
```

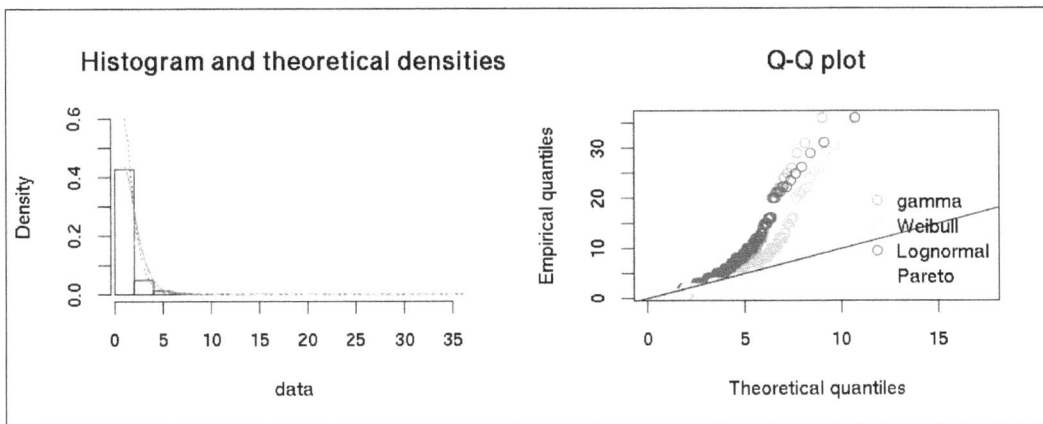

After all, it seems that the Pareto distribution is the closest fit to our long-tailed data. But more importantly, we know about more than 4,000 R users besides the previously identified 279 R Foundation supporting members:

```
> length(unique(maintainers))
[1] 4012
```

What other data sources can we use to find information on the (number of) R users?

The R-help mailing list

R-help is the official, main mailing list providing general discussion about problems and solutions using R, with many active users and several dozen e-mails every day. Fortunately, this public mailing list is archived on several sites, and we can easily download the compressed monthly files from, for example, ETH Zurich's R-help archives:

```
> library(RCurl)
> url <- getURL('https://stat.ethz.ch/pipermail/r-help/')
```

Now let's extract the URL of the monthly compressed archives from this page via an XPath query:

```
> R.help.toc <- htmlParse(url)
> R.help.archives <- unlist(xpathApply(R.help.toc,
+       "//table//td[3]/a", xmlAttrs), use.names = FALSE)
```

And now let's download these files to our computer for future parsing:

```
> dir.create('r-help')
> for (f in R.help.archives)
+       download.file(url = paste0(url, f),
+           file.path('help-r', f), method = 'curl'))
```

> Depending on your operating system and R version, the `curl` option that we used to download files via the HTTPS protocol might not be available. In such cases, you can try other another method or update the query to use the RCurl, `curl`, or `httr` packages.

Downloading these ~200 files takes some time and you might also want to add a `Sys.sleep` call in the loop so as not to overload the server. Anyway, after some time, you will have a local copy of the `R-help` mailing list in the `r-help` folder, ready to be parsed for some interesting data:

```
> lines <- system(paste0(
+       "zgrep -E '^From: .* at .*' ./help-r/*.txt.gz"),
+                   intern = TRUE)
> length(lines)
[1] 387218
> length(unique(lines))
[1] 110028
```

> Instead of loading all the text files into R and using `grep` there, I pre-filtered the files via the Linux command line `zgrep` utility, which can search in `gzipped` (compressed) text files efficiently. If you do not have `zgrep` installed (it is available on both Windows and the Mac), you can extract the files first and use the standard `grep` approach with the very same regular expression.

So we filtered for all lines of the e-mails and headers, starting with the `From` string, that hold information on the senders in the e-mail address and name. Out of the ~387,000 e-mails, we have found around ~110,000 unique e-mail sources. To understand the following regular expressions, let's see how one of these lines looks:

```
> lines[26]
[1] "./1997-April.txt.gz:From: pcm at ptd.net (Paul C. Murray)"
```

Now let's process these lines by removing the static prefix and extracting the names found between parentheses after the e-mail address:

```
> lines    <- sub('.*From: ', '', lines)
> Rhelpers <- sub('.*\\((.*)\\)', '\\1', lines)
```

And we can see the list of the most active `R-help` posters:

```
> tail(sort(table(Rhelpers)), 6)
       jim holtman      Duncan Murdoch          Uwe Ligges
            4284                6421                6455
Gabor Grothendieck  Prof Brian Ripley     David Winsemius
            8461                9287               10135
```

This list seems to be legitimate, right? Although my first guess was that Professor Brian Ripley with his brief messages will be the first one in this list. As a result of some earlier experiences, I know that matching names can be tricky and cumbersome, so let's verify that our data is clean enough and there's only one version of the Professor's name:

```
> grep('Brian( D)? Ripley', names(table(Rhelpers)), value = TRUE)
 [1] "Brian D Ripley"
 [2] "Brian D Ripley [mailto:ripley at stats.ox.ac.uk]"
 [3] "Brian Ripley"
 [4] "Brian Ripley <ripley at stats.ox.ac.uk>"
 [5] "Prof Brian D Ripley"
 [6] "Prof Brian D Ripley [mailto:ripley at stats.ox.ac.uk]"
 [7] "         Prof Brian D Ripley <ripley at stats.ox.ac.uk>"
 [8] "\"Prof Brian D Ripley\" <ripley at stats.ox.ac.uk>"
 [9] "Prof Brian D Ripley <ripley at stats.ox.ac.uk>"
[10] "Prof Brian Ripley"
[11] "Prof. Brian Ripley"
[12] "Prof Brian Ripley [mailto:ripley at stats.ox.ac.uk]"
[13] "Prof Brian Ripley [mailto:ripley at stats.ox.ac.uk] "
[14] "            \tProf Brian Ripley <ripley at stats.ox.ac.uk>"
[15] "  Prof Brian Ripley <ripley at stats.ox.ac.uk>"
[16] "\"Prof Brian Ripley\" <ripley at stats.ox.ac.uk>"
[17] "Prof Brian Ripley<ripley at stats.ox.ac.uk>"
[18] "Prof Brian Ripley <ripley at stats.ox.ac.uk>"
[19] "Prof Brian Ripley [ripley at stats.ox.ac.uk]"
[20] "Prof Brian Ripley <ripley at toucan.stats>"
[21] "Professor Brian Ripley"
[22] "r-help-bounces at r-project.org [mailto:r-help-bounces at
r-project.org] On Behalf Of Prof Brian Ripley"
[23] "r-help-bounces at stat.math.ethz.ch [mailto:r-help-bounces at stat.
math.ethz.ch] On Behalf Of Prof Brian Ripley"
```

Well, it seems that the Professor used some alternative `From` addresses as well, so a more valid estimate of the number of his messages should be something like:

```
> sum(grepl('Brian( D)? Ripley', Rhelpers))
[1] 10816
```

So using quick, regular expressions to extract the names from the e-mails returned most of the information we were interested in, but it seems that we have to spend a lot more time to get the whole information set. As usual, the Pareto rule applies: we can spend around 80 percent of our time on preparing data, and we can get 80 percent of the data in around 20 percent of the whole project timeline.

Due to page limitations, we will not cover data cleansing on this dataset in greater detail at this point, but I highly suggest checking Mark van der Loo's `stringdist` package, which can compute string distances and similarities to, for example, merge similar names in cases like this.

Volume of the R-help mailing list

But besides the sender, these e-mails also include some other really interesting data as well. For example, we can extract the date and time when the e-mail was sent—to model the frequency and temporal pattern of the mailing list.

To this end, let's filter for some other lines in the compressed text files:

```
> lines <- system(paste0(
+     "zgrep -E '^Date: [A-Za-z]{3}, [0-9]{1,2} [A-Za-z]{3} ",
+     "[0-9]{4} [0-9]{2}:[0-9]{2}:[0-9]{2} [-+]{1}[0-9]{4}' ",
+     "./help-r/*.txt.gz"),
+                     intern = TRUE)
```

This returns fewer lines when compared to the previously extracted `From` lines:

```
> length(lines)
[1] 360817
```

This is due to the various date and time formats used in the e-mail headers, as sometimes the day of the week was not included in the string or the order of year, month, and day was off compared to the vast majority of other mails. Anyway, we will only concentrate on this significant portion of mails with the standard date and time format but, if you are interested in transforming these other time formats, you might want to check Hadley Wickham's `lubridate` package to help your workflow. But please note that there's no general algorithm to guess the order of decimal year, month, and day—so you will end up with some manual data cleansing for sure!

Let's see how these (subset of) lines look:

```
> head(sub('.*Date: ', '', lines[1]))
[1] "Tue, 1 Apr 1997 20:35:48 +1200 (NZST)"
```

Then we can simply get rid of the `Date` prefix and parse the time stamps via `strptime`:

```
> times <- strptime(sub('.*Date: ', '', lines),
+               format = '%a, %d %b %Y %H:%M:%S %z')
```

Now that the data is in a parsed format (even the local time-zones were converted to UTC), it's relatively easy to see, for example, the number of e-mails on the mailing list per year:

```
> plot(table(format(times, '%Y')), type = 'l')
```

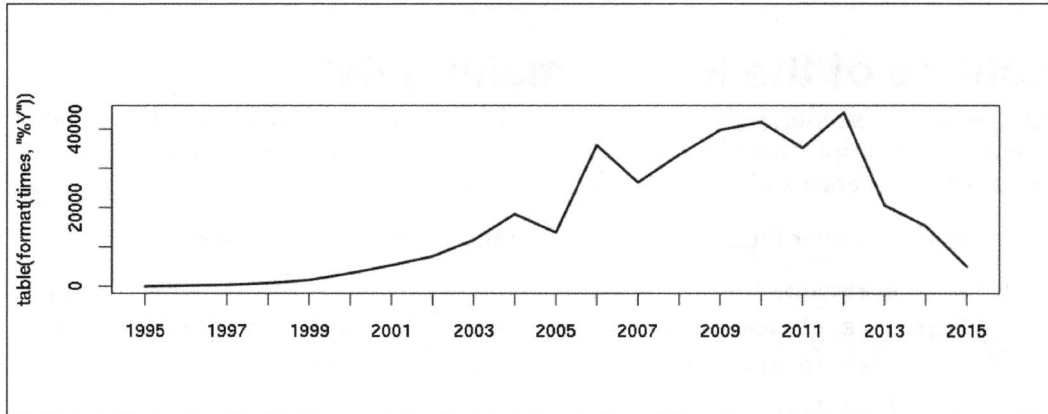

Although the volume on the `R-help` mailing list seems to have decreased in the past few years, it's not due to the lower R activity: R users, okay as is or no/. others on the Internet, nowadays tend to use other information channels more often than e-mail—for example: StackOverflow and GitHub (or even Facebook and LinkedIn). For a related research, please see the paper of Bogdan Vasilescu at al at http://web.cs.ucdavis.edu/~filkov/papers/r_so.pdf.

Well, we can do a lot better than this, right? Let's massage our data a bit and visualize the frequency of mails based on the day of week and hour of the day via a more elegant graph—inspired by GitHub's punch card plot:

```
> library(data.table)
> Rhelp <- data.table(time = times)
> Rhelp[, H := hour(time)]
> Rhelp[, D := wday(time)]
```

Visualizing this dataset is relatively straightforward with `ggplot`:

```
> library(ggplot2)
> ggplot(na.omit(Rhelp[, .N, by = .(H, D)]),
+        aes(x = factor(H), y = factor(D), size = N)) + geom_point() +
+        ylab('Day of the week') + xlab('Hour of the day') +
+        ggtitle('Number of mails posted on [R-help]') +
+        theme_bw() + theme('legend.position' = 'top')
```

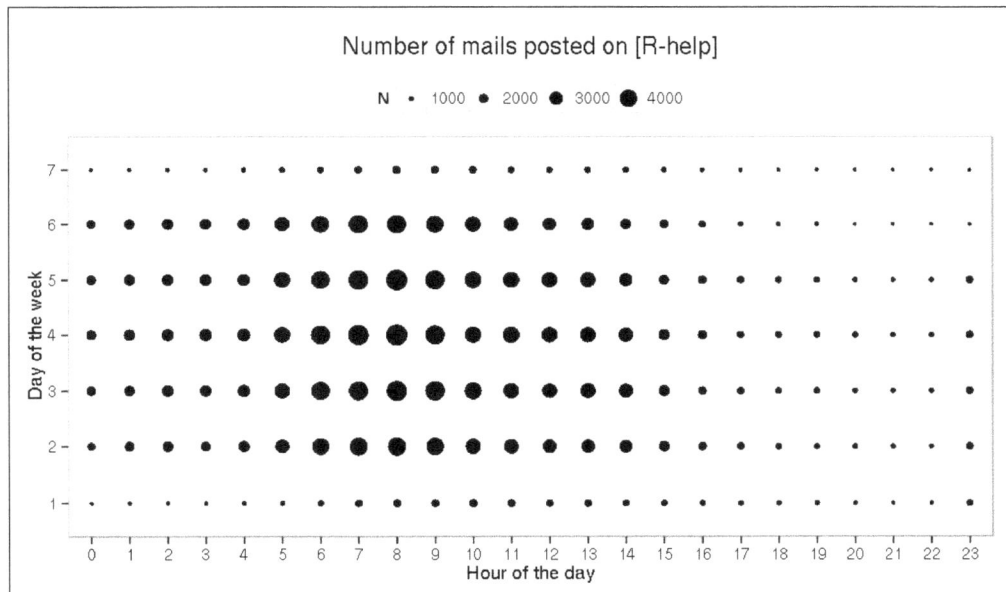

As the times are by UTC, the early morning mails might suggest that where most `R-help` posters live has a positive GMT offset—if we suppose that most e-mails were written in business hours. Well, at least the lower number of e-mails on the weekends seems to suggest this statement.

And it seems that the UTC, UTC+1, and UTC+2 time zones are indeed rather frequent, but the US time zones are also pretty common for the `R-help` posters:

```
> tail(sort(table(sub('.*([+-][0-9]{4}).*', '\\1', lines))), 22)

-1000 +0700 +0400 -0200 +0900 -0000 +0300 +1300 +1200 +1100 +0530
  164   352   449  1713  1769  2585  2612  2917  2990  3156  3938
-0300 +1000 +0800 -0600 +0000 -0800 +0200 -0500 -0400 +0100 -0700
 4712  5081  5493 14351 28418 31661 42397 47552 50377 51390 55696
```

Forecasting the e-mail volume in the future

And we can also use this relatively clean dataset to forecast the future volume of the `R-help` mailing list. To this end, let's aggregate the original dataset to count data daily, as we saw in *Chapter 3, Filtering and Summarizing Data*:

```
> Rhelp[, date := as.Date(time)]
> Rdaily <- na.omit(Rhelp[, .N, by = date])
```

Now let's transform this `data.table` object into a time-series object by referencing the actual mail counts as values and the dates as the index:

```
> Rdaily <- zoo(Rdaily$N, Rdaily$date)
```

Well, this daily dataset is a lot spikier than the previously rendered yearly graph:

```
> plot(Rdaily)
```

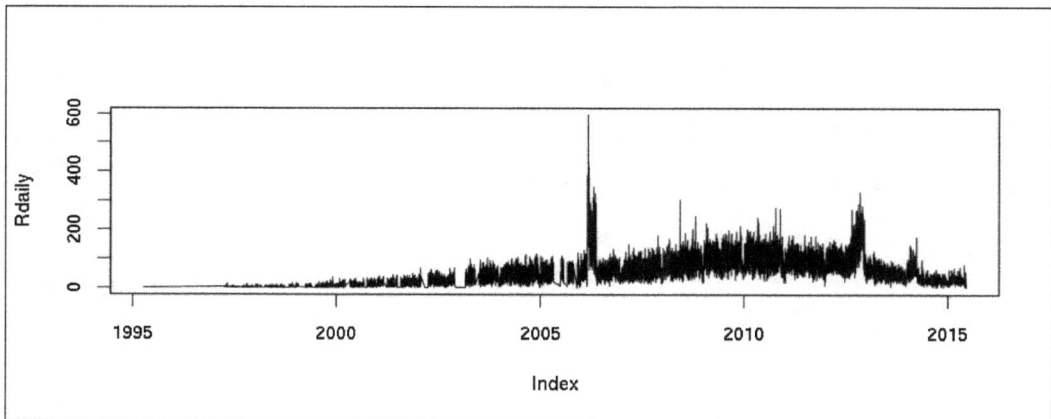

But instead of smoothing or trying to decompose this time-series, like we did in *Chapter 12, Analyzing Time-series*, let's rather see how we can provide some quick estimates (based on historical data) on the forthcoming number of mails on this mailing list with some automatic models. To this end, we will use the `forecast` package:

```
> library(forecast)
> fit <- ets(Rdaily)
```

The `ets` function implements a fully automatic method that can select the optimal trend, season, and error type for the given time-series. Then we can simply call the `predict` or `forecast` function to see the specified number of estimates, only for the next day in this case:

```
> predict(fit, 1)
     Point Forecast    Lo 80    Hi 80        Lo 95     Hi 95
5823        28.48337  9.85733 47.10942 -0.002702251  56.96945
```

So it seems that, for the next day, our model estimated around 28 e-mails with a confidence interval of 80 percent being somewhere between 10 and 47. Visualizing predictions for a slightly longer period of time with some historical data can be done via the standard `plot` function with some useful new parameters:

```
> plot(forecast(fit, 30), include = 365)
```

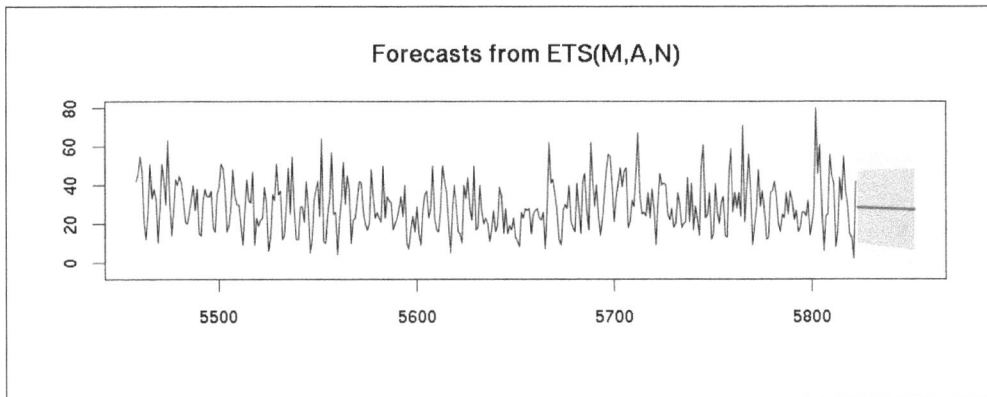

Analyzing overlaps between our lists of R users

But our original idea was to predict the number of R users around the world and not to focus on some minor segments, right? Now that we have multiple data sources, we can start building some models combining those to provide estimates on the global number of R users.

The basic idea behind this approach is the capture-recapture method, which is well known in ecology, where we first try to identify the probability of capturing a unit from the population, and then we use this probability to estimate the number of not captured units.

In our current study, units will be R users and the samples are the previously captured name lists on the:

- Supporters of the *R Foundation*
- R package maintainers who submitted at least one package to *CRAN*
- *R-help* mailing list e-mail senders

Let's merge these lists with a tag referencing the data source:

```
> lists <- rbindlist(list(
+      data.frame(name = unique(supporterlist), list = 'supporter'),
+      data.frame(name = unique(maintainers),   list = 'maintainer'),
+      data.frame(name = unique(Rhelpers),      list = 'R-help')))
```

Next let's see the number of names we can find in one, two or all three groups:

```
> t <- table(lists$name, lists$list)
> table(rowSums(t))

    1     2     3
44312   860    40
```

So there are (at least) 40 persons who support the R Foundation, maintain at least one R package on CRAN, and have posted at least one mail to `R-help` since 1997! I am happy and proud to be one of these guys -- especially with an accent in my name, which often makes matching of strings more complex.

Now, if we suppose these lists refer to the same population, namely R users around the world, then we can use these common occurrences to predict the number of R users who somehow missed supporting the R Foundation, maintaining a package on CRAN, and writing a mail to the R-help mailing list. Although this assumption is obviously off, let's run this quick experiment and get back to these outstanding questions later.

One of the best things in R is that we have a package for almost any problem. Let's load the `Rcapture` package, which provides some sophisticated, yet easily accessible, methods for capture-recapture models:

```
> library(Rcapture)
> descriptive(t)

Number of captured units: 45212

Frequency statistics:
            fi      ui      vi      ni
```

```
i = 1   44312     279     157     279
i = 2     860    3958    3194    4012
i = 3      40   40975   41861   41861
fi: number of units captured i times
ui: number of units captured for the first time on occasion i
vi: number of units captured for the last time on occasion i
ni: number of units captured on occasion i
```

These numbers from the first `fi` column are familiar from the previous table, and represent the number of R users identified on one, two, or all three lists. It's a lot more interesting to fit some models on this data with a simple call such as:

```
> closedp(t)
```

```
Number of captured units: 45212
```

```
Abundance estimations and model fits:
                   abundance        stderr   deviance df         AIC         BIC
M0                  750158.4       23800.7  73777.800  5   73835.630   73853.069
Mt                  192022.2        5480.0    240.278  3     302.109     336.986
Mh Chao (LB)        806279.2       26954.8  73694.125  4   73753.956   73780.113
Mh Poisson2        2085896.4      214443.8  73694.125  4   73753.956   73780.113
Mh Darroch         5516992.8     1033404.9  73694.125  4   73753.956   73780.113
Mh Gamma3.5       14906552.8     4090049.0  73694.125  4   73753.956   73780.113
Mth Chao (LB)       205343.8        6190.1     30.598  2      94.429     138.025
Mth Poisson2       1086549.0      114592.9     30.598  2      94.429     138.025
Mth Darroch        6817027.3     1342273.7     30.598  2      94.429     138.025
Mth Gamma3.5      45168873.4    13055279.1     30.598  2      94.429     138.025
Mb                     -36.2           6.2    107.728  4     167.559     193.716
Mbh                   -144.2          25.9     84.927  3     146.758     181.635
```

Once again, I have to emphasize that these estimates are not actually on the abundance of all R users around the world, because:

- Our non-independent lists refer to far more specific groups
- The model assumptions do not stand
- The R community is definitely not a closed population and some open-population models would be more reliable
- We missed some very important data-cleansing steps, as noted

Further ideas on extending the capture-recapture models

Although this playful example did not really help us to find out the number of R users around the world, with some extensions the basic idea is definitely viable. First of all, we might consider analyzing the source data in smaller chunks—for example, looking for the same e-mail addresses or names in different years of the R-help archives. This might help with estimating the number of persons who were thinking about submitting a question to R-help, but did not actually send the e-mail after all (for example, because another poster's question had already been answered or she/he resolved the problem without external help).

On the other hand, we could also add a number of other data sources to the models, so that we can do more reliable estimates on some other R users who do not contribute to the R Foundation, CRAN, or R-help.

I have been working on a similar study over the past 2 years, collecting data on the number of:

- R Foundation ordinary and supporting members, donators and benefactors
- Attendees at the annual R conference between 2004 and 2015
- CRAN downloads per package and country in 2013 and 2014
- R User Groups and meet-ups with the number of members
- The http://www.r-bloggers.com visitors in 2013
- GitHub users with at least one repository with R source code
- Google search trends on R-related terms

You can find the results on an interactive map and the country-level aggregated data in a CSV file at http://rapporter.net/custom/R-activity and an offline data visualization presented in the past two *useR!* conferences at http://bit.ly/useRs2015.

The number of R users in social media

An alternative way to try to estimate the number of R users could be to analyze the occurrence of the related terms on social media. This is relatively easy on Facebook, where the marketing API allows us to query the size of the so-called target audiences, which we can use to define targets for some paid ads.

Well, we are not actually interested in creating a paid advertisement on Facebook right now, although this can be easily done with the `fbRads` package, but we can use this feature to see the estimated size of the *target* group of persons interested in R:

```
> library(fbRads)
> fbad_init(FB account ID, FB API token)
> fbad_get_search(q = 'rstats', type = 'adinterest')
            id                     name audience_size path description
6003212345926 R (programming language)       1308280 NULL          NA
```

Of course, to run this quick example you will need to have a (free) Facebook developer account, a registered application, and a generated token (please see the package docs for more details), but it is definitely worth it: we have just found out that there are more than 1.3M users around the world interested in R! That's really impressive, although it seems to be rather high to me, especially when compared with some other statistical software, such as:

```
> fbad_get_search(fbacc = fbacc, q = 'SPSS', type = 'adinterest')
             id       name audience_size path description
1 6004181236095       SPSS        203840 NULL          NA
2 6003262140109 SPSS Inc.          2300 NULL          NA
```

Having said this, comparing R with other programming languages suggests that the audience size might actually be correct:

```
> res <- fbad_get_search(fbacc = fbacc, q = 'programming language',
+                        type = 'adinterest')
> res <- res[order(res$audience_size, decreasing = TRUE), ]
> res[1:10, 1:3]
              id                           name audience_size
1   6003030200185          Programming language     295308880
71  6004131486306                           C++      27812820
72  6003017204650                           PHP      23407040
73  6003572165103               Lazy evaluation      18251070
74  6003568029103   Object-oriented programming      14817330
2   6002979703120     Ruby (programming language)    10346930
75  6003486129469                      Compiler      10101110
76  6003127967124                    JavaScript       9629170
3   6003437022731     Java (programming language)     8774720
4   6003682002118   Python (programming language)     7932670
```

There are many programmers around the world, it seems! But what are they talking about and what are the trending topics? We will cover these questions in the next section.

R-related posts in social media

One option to collect posts from the past few days of social media is processing Twitter's global stream of Tweet data. This stream data and API provides access to around 1 percent of all tweets. If you are interested in all this data, then a commercial Twitter Firehouse account is needed. In the following examples, we will use the free Twitter search API, which provides access to no more than 3,200 tweets based on any search query—but this will be more than enough to do some quick analysis on the trending topics among R users.

So let's load the twitteR package and initialize the connection to the API by providing our application tokens and secrets, generated at https://apps.twitter.com:

```
> library(twitteR)
> setup_twitter_oauth(...)
```

Now we can start using the searchTwitter function to search tweets for any keywords, including hashtags and mentions. This query can be fine-tuned with a couple of arguments. Since, until, and *n* set the beginning and end date, also the number of tweets to return respectively. Language can be set with the lang attribute by the ISO 639-1 format—for example, use en for English.

Let's search for the most recent tweet with the official R hashtag:

```
> str(searchTwitter("#rstats", n = 1, resultType = 'recent'))
Reference class 'status' [package "twitteR"] with 17 fields
  $ text          : chr "7 #rstats talks in 2014"| __truncated__
  $ favorited     : logi FALSE
  $ favoriteCount : num 2
  $ replyToSN     : chr(0)
  $ created       : POSIXct[1:1], format: "2015-07-21 19:31:23"
  $ truncated     : logi FALSE
  $ replyToSID    : chr(0)
  $ id            : chr "623576019346280448"
  $ replyToUID    : chr(0)
  $ statusSource  : chr "Twitter Web Client"
  $ screenName    : chr "daroczig"
```

```
$ retweetCount : num 2
$ isRetweet    : logi FALSE
$ retweeted    : logi FALSE
$ longitude     : chr(0)
$ latitude      : chr(0)
$ urls          :'data.frame':   2 obs. of  5 variables:
 ..$ url        : chr [1:2]
    "http://t.co/pStTeyBr2r" "https://t.co/5L4wyxtooQ"
 ..$ expanded_url: chr [1:2] "http://budapestbiforum.hu/2015/en/cfp"
    "https://twitter.com/BudapestBI/status/623524708085067776"
 ..$ display_url : chr [1:2] "budapestbiforum.hu/2015/en/cfp"
    "twitter.com/BudapestBI/sta…"
 ..$ start_index : num [1:2] 97 120
 ..$ stop_index  : num [1:2] 119 143
```

This is quite an impressive amount of information for a character string with no more than 140 characters, isn't it? Besides the text including the actual tweet, we got some meta-information as well—for example, the author, post time, the number of times other users favorited or retweeted the post, the Twitter client name, and the URLs in the post along with the shortened, expanded, and displayed format. The location of the tweet is also available in some cases, if the user enabled that feature.

Based on this piece of information, we could focus on the Twitter R community in very different ways. Examples include:

* Counting the users mentioning R
* Analyzing social network or Twitter interactions
* Time-series analysis on the time of posts
* Spatial analysis on the location of tweets
* Text mining of the tweet contents

Probably a mixture of these (and other) methods would be the best approach, and I highly suggest you do that as an exercise to practice what you have learned in this book. However, in the following pages we will only concentrate on the last item.

So first, we need some recent tweets on the R programming language. To search for #rstats posts, instead of providing the related hashtag (like we did previously), we can use the Rtweets wrapper function as well:

```
> tweets <- Rtweets(n = 500)
```

This function returned 500 reference classes similar to those we saw previously. We can count the number of original tweets excluding retweets:

```
> length(strip_retweets(tweets))
[1] 149
```

But, as we are looking for the trending topics, we are interested in the original list of tweets, where the retweets are also important as they give a natural weight to the trending posts. So let's transform the list of reference classes to a data.frame:

```
> tweets <- twListToDF(tweets)
```

This dataset consists of 500 rows (tweets) and 16 variables on the content, author, and location of the posts, as described previously. Now, as we are only interested in the actual text of the tweets, let's load the tm package and import our corpus as seen in *Chapter 7, Unstructured Data*:

```
> library(tm)
Loading required package: NLP
> corpus <- Corpus(VectorSource(tweets$text))
```

As the data is in the right format, we can start to clean the data from the common English words and transform everything into lowercase format; we might also want to remove any extra whitespace:

```
> corpus <- tm_map(corpus, removeWords, stopwords("english"))
> corpus <- tm_map(corpus, content_transformer(tolower))
> corpus <- tm_map(corpus, removePunctuation)
> corpus <- tm_map(corpus, stripWhitespace)
```

It's also wise to remove the R hashtag, as this is part of all tweets:

```
> corpus <- tm_map(corpus, removeWords, 'rstats')
```

And then we can use the wordcloud package to plot the most important words:

```
> library(wordcloud)
Loading required package: RColorBrewer
> wordcloud(corpus)
```

Summary

In the past few pages, I have tried to cover a variety of data science and R programming topics, although many important methods and questions were not addressed due to page limitation. To this end, I've compiled a short reading list in the *References* chapter of the book. And don't forget: now it's your turn to practice everything you learned in the previous chapters. I wish you a lot of fun and success in this journey!

And once again, thanks for reading this book; I hope you found it useful. If you have any questions, comments, or any kind of feedback, please feel free to get in touch, I'm looking forward to hearing from you!

References

Although there are quite a number of good and free resources on the Internet on R and data science (such as StackOverflow, GitHub wikis, http://www.r-bloggers. com/, and some free e-books), sometimes it's better to buy a book with structured content—just like you did.

In this appendix, I've listed a few books and other references that I've found useful in the past while learning R. I suggest that you at least skim through these materials if you want to become a professional data scientist with a decent R background, and are not fond of the autodidact way.

For the sake of reproducibility, all the R packages used in this book are listed with the actual package versions and sources of installation as well.

General good readings on R

Although the forthcoming lists are related to the different chapters of the book, the following is a list of a few general references that are very good resources on introductory and advanced R topics:

- *Quick-R* by *Robert I. Kabacoff* at http://www.statmethods.net
- The official R manuals at https://cran.r-project.org/manuals.html
- *An R "meta" book* by *Joseph Ricker* at http://blog.revolutionanalytics. com/2014/03/an-r-meta-book.html
- *R For Dummies, Wiley, 2012* by *Andrie de Vries* and *Joris Meys*
- *R in Action, Manning, 2015* by *Robert I. Kabacoff*
- *R in a Nutshell, O'Reilly, 2010* by *Joseph Adler*
- *Art of R Programming, 2011,* by *Norman Matloff*

- *The R Inferno*, by *Partrick Burns* available at `http://www.burns-stat.com/documents/books/the-r-inferno/`
- *Advanced R*, by *Hadley Wickham, 2015* at `http://adv-r.had.co.nz`

Chapter 1 – Hello, Data!

The loaded R package versions (in the order mentioned in the chapter):

- hflights 0.1 (CRAN)
- microbenchmark 1.4-2 (CRAN)
- R.utils 2.0.2 (CRAN)
- sqldf 0.4-10 (CRAN)
- ff 2.2-13 (CRAN)
- bigmemory 4.4.6 (CRAN)
- data.table 1.9.4 (CRAN)
- RMySQL 0.10.3 (CRAN)
- RPostgreSQL 0.4 (CRAN)
- ROracle 1.1-12 (CRAN)
- dbConnect 1.0 (CRAN)
- XLConnect 0.2-11 (CRAN)
- xlsx 0.5.7 (CRAN)

The related R packages:

- mongolite 0.4 (CRAN)
- MonetDB.R 0.9.7 (CRAN)
- RcppRedis 0.1.5 (CRAN)
- RCassandra 0.1-3 (CRAN)
- RSQLite 1.0.0 (CRAN)

Related reading:

- R Data Import/Export manual available at `https://cran.r-project.org/doc/manuals/r-release/R-data.html`
- High-Performance and Parallel Computing with R, CRAN Task View available at `http://cran.r-project.org/web/views/HighPerformanceComputing.html`

- Hadley Wickham's dplyr vignette on databases available at `https://cran.r-project.org/web/packages/dplyr/vignettes/databases.html`
- RODBC vignette available at `https://cran.r-project.org/web/packages/RODBC/vignettes/RODBC.pdf`
- Docker Docs available at `http://docs.docker.com`
- VirtualBox manual available at `http://www.virtualbox.org/manual`
- MySQL downloads available at `https://dev.mysql.com/downloads/mysql`

Chapter 2 – Getting Data from the Web

The loaded R package versions (in the order mentioned in the chapter):

- RCurl 1.95-4.1 (CRAN)
- rjson 0.2.13 (CRAN)
- plyr 1.8.1 (CRAN)
- XML 3.98-1.1 (CRAN)
- wordcloud 2.4 (CRAN)
- RSocrata 1.4 (CRAN)
- quantmod 0.4 (CRAN)
- Quandl 2.3.2 (CRAN)
- devtools 1.5 (CRAN)
- GTrendsR (BitBucket @ d507023f81b17621144a2bf2002b845ffb00ed6d)
- weatherData 0.4 (CRAN)

The related R packages:

- jsonlite 0.9.16 (CRAN)
- curl 0.6 (CRAN)
- bitops 1.0-6 (CRAN)
- xts 0.9-7 (CRAN)
- RJSONIO 1.2-0.2 (CRAN)
- RGoogleDocs 0.7 (OmegaHat.org)

Related reading:

- Chrome Devtools manual at `https://developer.chrome.com/devtools`
- Chrome DevTools course on CodeSchool from `http://discover-devtools.codeschool.com/`
- XPath on Mozilla Developer Network from `https://developer.mozilla.org/en-US/docs/Web/XPath`
- Firefox Developer Tools from `https://developer.mozilla.org/en-US/docs/Tools`
- Firebug for Firefox available at `http://getfirebug.com/`
- *XML and Web Technologies for Data Sciences with R* by *Deborah Nolan, Duncan Temple Lang (2014), Springer*
- *The jsonlite Package: A Practical and Consistent Mapping Between JSON Data and R Objects* by *Jeroen Ooms* (2014) available at `http://arxiv.org/abs/1403.2805`
- *Web Technologies and Services CRAN Task View* by *Scott Chamberlain, Karthik Ram, Christopher Gandrud*, and *Patrick Mair* (2014) available at `http://cran.r-project.org/web/views/WebTechnologies.html`

Chapter 3 – Filtering and Summarizing Data

The loaded R package versions (in the order mentioned in the chapter):

- sqldf 0.4-10 (CRAN)
- hflights 0.1 (CRAN)
- dplyr 0.4.1 (CRAN)
- data.table 1.9.4. (CRAN)
- plyr 1.8.2 (CRAN)
- microbenchmark 1.4-2 (CRAN)

Further reading:

- The data.table manuals, vignettes and other documentation at `https://github.com/Rdatatable/data.table/wiki/Getting-started`
- *Introduction to dplyr, vignette* at `https://cran.rstudio.com/web/packages/dplyr/vignettes/introduction.html`

Chapter 4 – Restructuring Data

The loaded R package versions (in the order mentioned in the chapter):

- hflights 0.1 (CRAN)
- dplyr 0.4.1 (CRAN)
- data.table 1.9.4. (CRAN)
- pryr 0.1 (CRAN)
- reshape 1.4.2 (CRAN)
- ggplot2 1.0.1 (CRAN)
- tidyr 0.2.0 (CRAN)

Further R packages:

- jsonlite 0.9.16 (CRAN)

Further reading:

- *Introduction to data.table, Tutorial slides at the useR!* 2014 conference by *Matt Dowle* at `http://user2014.stat.ucla.edu/files/tutorial_Matt.pdf`
- *Reshaping data with the reshape package, Hadley Wickham, 2006* at `http://had.co.nz/reshape/introduction.pdf`
- *Practical tools for exploring data and models* by *Hadley Wickham, 2008* at `http://had.co.nz/thesis/`
- *Two-table verbs, Package vignette,* at `https://cran.r-project.org/web/packages/dplyr/vignettes/two-table.html`
- *Introduction to dplyr, Package vignette,* `http://cran.r-project.org/web/packages/dplyr/vignettes/introduction.html`
- *Data Wrangling cheat sheet, RStudio, 2015* `https://www.rstudio.com/wp-content/uploads/2015/02/data-wrangling-cheatsheet.pdf`
- *Data manipulation with dplyr, Tutorial slides and materials at the useR!* 2014 conference by *Hadley Wickham* at `http://bit.ly/dplyr-tutorial`
- *Tidy data, The Journal of Statistical Software. 59(10): 1:23* by *Hadley Wickham, 2014* at `http://vita.had.co.nz/papers/tidy-data.html`

Chapter 5 – Building Models
(authored by Renata Nemeth and Gergely Toth)

The loaded R package versions (in the order mentioned in the chapter):

- gamlss.data 4.2-7 (CRAN)
- scatterplot3d 0.3-35 (CRAN)
- Hmisc 3.16-0 (CRAN)
- ggplot2 1.0.1 (CRAN)
- gridExtra 0.9.1 (CRAN)
- gvlma 1.0.0.2 (CRAN)
- partykit 1.0-1 (CRAN)
- rpart 4.1-9 (CRAN)

Further reading:

- *Applied regression analysis and other multivariable methods. Duxbury Press* in 2008 by *David G. Kleinbaum, Lawrence L. Kupper, Azhar Nizam, Keith E. Muller*
- *An R Companion to Applied Regression, Sage, Web companion* by *John Fox* in *2011* at `http://socserv.socsci.mcmaster.ca/jfox/Books/Companion/appendix.html`
- *Practical Regression and Anova using R Julian J, Faraway* in 2002 at `https://cran.r-project.org/doc/contrib/Faraway-PRA.pdf`
- *Linear Models with R* by *Julian J, CRC, Faraway, 2014* at `http://www.maths.bath.ac.uk/~jjf23/LMR/`

Chapter 6 – Beyond the Linear Trend Line
(authored by Renata Nemeth and Gergely Toth)

The loaded R package versions (in the order mentioned in the chapter):

- catdata 1.2.1 (CRAN)
- vcdExtra 0.6.8 (CRAN)
- lmtest 0.9-33 (CRAN)
- BaylorEdPsych 0.5 (CRAN)
- ggplot2 1.0.1 (CRAN)
- MASS 7.3-40 (CRAN)

- broom 0.3.7 (CRAN)
- data.table 1.9.4. (CRAN)
- plyr 1.8.2 (CRAN)

Further R packages:

- LogisticDx 0.2 (CRAN)

Further reading:

- *Applied regression analysis and other multivariable methods. Duxbury Press, David G. Kleinbaum, Lawrence L. Kupper, Azhar Nizam, and Keith E. Muller in 2008*
- *An R Companion to Applied Regression, Sage, Web companion* by *John Fox* in *2011* at `http://socserv.socsci.mcmaster.ca/jfox/Books/Companion/appendix.html`

Chapter 7 – Unstructured Data

The loaded R package versions (in the order mentioned in the chapter):

- tm 0.6-1 (CRAN)
- wordcloud 2.5 (CRAN)
- SnowballC 0.5.1 (CRAN)

Further R packages:

- coreNLP 0.4-1 (CRAN)
- topicmodels 0.2-2 (CRAN)
- textcat 1.0-3 (CRAN)

Further reading:

- *Christopher D. Manning, Hinrich Schütze* (1999): *Foundations of Statistical Natural Language Processing. MIT.*
- *Daniel Jurafsky, James H. Martin* (2009): *Speech and Language Processing. Prentice Hall.*
- *Christopher D. Manning, Prabhakar Raghavan, Hinrich Schütze* (2008): *Introduction to Information Retrieval. Cambridge University Press.* `http://nlp.stanford.edu/IR-book/html/htmledition/irbook.html`
- *Ingo Feinerer: Introduction to the tm Package Text Mining in R.* `https://cran.r-project.org/web/packages/tm/vignettes/tm.pdf`

- Ingo Feinerer (2008): A Text Mining Framework in R and Its Applications. http://epub.wu.ac.at/1923/1/document.pdf
- Yanchang Zhao: Text Mining with R: Twitter Data Analysis. http://www.rdatamining.com/docs/text-mining-with-r-of-twitter-data-analysis
- Stefan Thomas Gries (2009): Quantitative Corpus Linguistics with R: A Practical Introduction. Routledge.

Chapter 8 – Polishing Data

The loaded R package versions (in the order mentioned in the chapter):

- hflights 0.1 (CRAN)
- rapportools 1.0 (CRAN)
- Defaults 1.1-1 (CRAN)
- microbenchmark 1.4-2 (CRAN)
- Hmisc 3.16-0 (CRAN)
- missForest 1.4 (CRAN)
- outliers 0.14 (CRAN)
- lattice 0.20-31 (CRAN)
- MASS 7.3-40 (CRAN)

Further R packages:

- imputeR 1.0.0 (CRAN)
- VIM 4.1.0 (CRAN)
- mvoutlier 2.0.6 (CRAN)
- randomForest 4.6-10 (CRAN)
- AnomalyDetection 1.0 (GitHub @ c78f0df02a8e34e37701243faf79a6c00120e797)

Further reading:

- *Inference and Missing Data, Biometrika 63(3), 581-592, Donald B. Rubin in 1976*
- *Statistical Analysis with Missing Data, Wiley Roderick, J. A. Little in 2002*
- *Flexible Imputation of Missing Data, CRC, Stef van Buuren in 2012*
- *Robust Statistical Methods CRAN Task View, Martin Maechler* at https://cran.r-project.org/web/views/Robust.html

Chapter 9 – From Big to Smaller Data

The loaded R package versions (in the order mentioned in the chapter):

- hflights 0.1 (CRAN)
- MVN 3.9 (CRAN)
- ellipse 0.3-8 (CRAN)
- psych 1.5.4 (CRAN)
- GPArotation 2014.11-1 (CRAN)
- jpeg 0.1-8 (CRAN)

Further R packages:

- mvnormtest 0.1-9 (CRAN)
- corrgram 1.8 (CRAN)
- MASS 7.3-40 (CRAN)
- sem 3.1-6 (CRAN)
- ca 0.58 (CRAN)

Further reading:

- *FactoMineR: An R Package for Multivariate Analysis, JSS, Sebastien Le, Julie Josse, Francois Husson* in *2008* at `http://factominer.free.fr/docs/article_FactoMineR.pdf`
- *Exploratory Multivariate Analysis by Example using R, CRC, Francois Husson, Sebastien Le, Jerome Pages* in *2010*
- *An index of factor simplicity, Psychometrika 39, 31–36 Kaiser, H. F.* in *1974*
- *Principal Component Analysis in R, Gregory B. Anderson,* at `http://www.ime.usp.br/~pavan/pdf/MAE0330-PCA-R-2013`
- *Structural Equation Modeling With the sem Package in R, John Fox* in *2006* at `http://socserv.mcmaster.ca/jfox/Misc/sem/SEM-paper.pdf`
- *Correspondence Analysis in R, with Two- and Three-dimensional Graphics: The ca Package, JSS* by *Oleg Nenadic, Michael Greenacre* in *2007* at `http://www.jstatsoft.org/v20/i03/paper`
- *PCA explained visually, Victor Powell* at `http://setosa.io/ev/principal-component-analysis/`

Chapter 10 – Classification and Clustering

The loaded R package versions (in the order mentioned in the chapter):

- NbClust 3.0 (CRAN)
- cluster 2.0.1 (CRAN)
- poLCA 1.4.1 (CRAN)
- MASS 7.3-40 (CRAN)
- nnet 7.3-9 (CRAN)
- dplyr 0.4.1 (CRAN)
- class 7.3-12 (CRAN)
- rpart 4.1-9 (CRAN)
- rpart.plot 1.5.2 (CRAN)
- partykit 1.0-1 (CRAN)
- party 1.0-2- (CRAN)
- randomForest 4.6-10 (CRAN)
- caret 6.0-47 (CRAN)
- C50 0.1.0-24 (CRAN)

Further R packages:

- glmnet 2.0-2 (CRAN)
- gbm 2.1.1 (CRAN)
- xgboost 0.4-2 (CRAN)
- h2o 3.0.0.30 (CRAN)

Further reading:

- *The Elements of Statistical Learning. Data Mining, Inference, and Prediction, Springer* by *Trevor Hastie, Robert Tibshirani, Jerome Friedman* in *2009* at `http://statweb.stanford.edu/~tibs/ElemStatLearn/`
- *An Introduction to Statistical Learning, Springer* by *Gareth James, Daniela Witten, Trevor Hastie, Robert Tibshirani* in *2013* at `http://www-bcf.usc.edu/~gareth/ISL/`

- *R and Data Mining: Examples and Case Studies* by *Yanchang Zhao* at `http://www.rdatamining.com/docs/r-and-data-mining-examples-and-case-studies`

- *Machine learning benchmarks* by *Szilard Pafka* in *2015* at `https://github.com/szilard/benchm-ml`

Chapter 11 – Social Network Analysis of the R Ecosystem

The loaded R package versions (in the order mentioned in the chapter):

- tools 3.2
- plyr 1.8.2 (CRAN)
- igraph 0.7.1 (CRAN)
- visNetwork 0.3 (CRAN)
- miniCRAN 0.2.4 (CRAN)

Further reading:

- *Statistical Analysis of Network Data with R, Springer* by *Eric D. Kolaczyk, Gábor Csárdi* in *2014*

- *Linked, Plume Publishing, Albert-László Barabási* in *2003*

- *Social Network Analysis Labs in R and SoNIA* by *Sean J. Westwood* in *2010* at `http://sna.stanford.edu/rlabs.php`

Chapter 12 – Analyzing Time-series

The loaded R package versions (in the order mentioned in the chapter):

- hflights 0.1 (CRAN)
- data.table 1.9.4 (CRAN)
- forecast 6.1 (CRAN)
- tsoutliers 0.6 (CRAN)
- AnomalyDetection 1.0 (GitHub)
- zoo 1.7-12 (CRAN)

Further R packages:

- xts 0.9-7 (CRAN)

Further reading:

- *Forecasting: principles and practice, OTexts, Rob J Hyndman, George Athanasopoulos in 2013* at `https://www.otexts.org/fpp`
- *Time Series Analysis and Its Applications, Springer, Robert H. Shumway, David S. Stoffer in 2011* at `http://www.stat.pitt.edu/stoffer/tsa3/`
- *Little Book of R for Time Series, Avril Coghlan in 2015* at `http://a-little-book-of-r-for-time-series.readthedocs.org/en/latest/`
- *Time Series Analysis CRAN Task View* by *Rob J Hyndman* at `https://cran.r-project.org/web/views/TimeSeries.html`

Chapter 13 – Data Around Us

The loaded R package versions (in the order mentioned in the chapter):

- hflights 0.1 (CRAN)
- data.table 1.9.4 (CRAN)
- ggmap 2.4 (CRAN)
- maps 2.3-9 (CRAN)
- maptools 0.8-36 (CRAN)
- sp 1.1-0 (CRAN)
- fields 8.2-1 (CRAN)
- deldir 0.1-9 (CRAN)
- OpenStreetMap 0.3.1 (CRAN)
- rCharts 0.4.5 (GitHub @ 389e214c9e006fea0e93d73621b83daa8d3d0ba2)
- leaflet 0.0.16 (CRAN)
- diagram 1.6.3 (CRAN)
- scales 0.2.4 (CRAN)
- ape 3.2 (CRAN)
- spdep 0.5-88 (CRAN)

Further R packages:

- raster 2.3-40 (CRAN)
- rgeos 0.3-8 (CRAN)
- rworldmap 1.3-1 (CRAN)
- countrycode 0.18 (CRAN)

Further reading:

- *Applied Spatial Data Analysis with R, Springer* by *Roger Bivand, Edzer Pebesma, Virgilio Gómez-Rubio* in *2013*
- *Spatial Data Analysis in Ecology and Agriculture Using R, CRC, Richard E. Plant* in *2012*
- *Numerical Ecology with R, Springer* by *Daniel Borcard, Francois Gillet,* and *Pierre Legendre* in *2012*
- *An Introduction to R for Spatial Analysis and Mapping, Sage, Chris Brunsdon, Lex Comber* in *2015*
- *Geocomputation, A Practical Primer, Sage* by *Chris Brunsdon, Alex David Singleton* in *2015*
- *Analysis of Spatial Data CRAN Task View* by *Roger Bivand* at `https://cran.r-project.org/web/views/Spatial.html`

Chapter 14 – Analysing the R Community

The loaded R package versions (in the order mentioned in the chapter):

- XML 3.98-1.1 (CRAN)
- rworldmap 1.3-1 (CRAN)
- ggmap 2.4 (CRAN)
- fitdistrplus 1.0-4 (CRAN)
- actuar 1.1-9 (CRAN)
- RCurl 1.95-4.6 (CRAN)
- data.table 1.9.4 (CRAN)
- ggplot2 1.0.1 (CRAN)
- forecast 6.1 (CRAN)
- Rcapture 1.4-2 (CRAN)
- fbRads 0.1 (GitHub @ 4adbfb8bef2dc49b80c87de604c420d4e0dd34a6)

- twitteR 1.1.8 (CRAN)
- tm 0.6-1 (CRAN)
- wordcloud 2.5 (CRAN)

Further R packages:

- jsonlite 0.9.16 (CRAN)
- curl 0.6 (CRAN)
- countrycode 0.18 (CRAN)
- VGAM 0.9-8 (CRAN)
- stringdist 0.9.0 (CRAN)
- lubridate 1.3.3 (CRAN)
- rgithub 0.9.6 (GitHub @ 0ce19e539fd61417718a664fc1517f9f9e52439c)
- Rfacebook 0.5 (CRAN)

Further reading:

- *How social Q&A sites are changing knowledge sharing in open source software communities, ACM* by *Bogdan Vasilescu, Alexander Serebrenik, Prem Devanbu,* and *Vladimir Filkov* in *2014* at `http://web.cs.ucdavis.edu/~filkov/papers/r_so.pdf`
- *Where is the R Activity?* by *James Cheshire* in *2013* at `http://spatial.ly/2013/06/r_activity/`
- *Seven quick facts about R,* by *David Smith* in *2014* at `http://blog.revolutionanalytics.com/2014/04/seven-quick-facts-about-r.html`
- *The attendants of useR! 2013 around the world* by *Gergely Daroczi* in *2013* at `http://blog.rapporter.net/2013/11/the-attendants-of-user-2013-around-world.html`
- *R users all around the world* by *Gergely Daroczi* in *2014* at `http://blog.rapporter.net/2014/07/user-at-los-angeles-california.html`
- *R activity around the world* by *Gergely Daroczi* in *2014* at `http://rapporter.net/custom/R-activity`
- *R users all around the world* (updated) by *Gergely Daroczi* in *2015* `https://www.scribd.com/doc/270254924/R-users-all-around-the-world-2015`

Index

dependent variable 127
deviance (Gsq) 250
devtools package 34, 60
diagram package 318
dimension reduction 193, 200
discrete predictors 121-128
discriminant analysis 250-254
Discriminant Function Analysis (DA) 250
dismo package 311
dist function 236
Docker 14
documents
  segmentation 166-168
dplyr
  versus data.table 91
dplyr package 68, 75
  anti_join 97
  inner_join 97
  left_join 97
  semi_join 97
dummy variables 235, 256

E

eigenvalue 210
Elbow-rule 212
ellipse package 201
e-mail volume
  forecasting 338
Equimax 219, 220
Excel spreadsheets
  loading 35, 36
explanatory variables 127
exploratory data analysis 207
Extensible Markup Language (XML) 46
extract, transform, and load (ETL) 11
extrapolation 113
extreme values
  about 118, 185, 186
  testing 187, 188

F

Factor Analysis (FA) 193
fbRads package 343
feature extraction 193
ff package 5

fields package 307
filter function 86
Finance APIs 57, 58
fitdistrplus package 328
forecast package 287, 289, 338
foreign package 35
formula notation 110
F-test 118, 119
FTP 39

G

gamlss.data package 109
gbm package 265
gdata package 35
Generalized Linear Models (GLM) 107
geocodes 298
geocoding 297-299
geospatial data 297
ggmap package 298, 311
ggplot2 package 101, 137, 272
GLM 127
goodness-of-fit 134
Google BigQuery 34
Google documents and analytics 60
Google Maps
  querying 313, 314
Google Maps API
  about 298
  accessing 313, 314
googlesheets package 60
googleVis package 313, 315
Google Visualization API 312
GPArotation package 220
Gradient Boosting 265
graphical user interface
  used, for connecting databases 32, 33
graphics package 300
GTrendsR package 60
gvlma package 116

H

Hard Drive Data Sets
  reference link, for downloading dataset 136
HBase 14, 34
hclust function 236

helper functions 73-76
heteroscedasticity 115
hflights package 3
hierarchical cluster algorithm 166
hierarchical clustering 236-240
Hmisc package 179, 180
Holt-Winters filtering 286-288
homoscedasticity 115
hot-deck method 179
HTML tables
  data, reading from 48, 49
htmlwidgets package 316
HTTP headers 49
httr package 41
Hypertable 14
Hypertext Transfer Protocol Secure
        (HTTPS) 39

I

ID3 265
igraph package 269, 274-276, 280
Impala 34
imputeR package 180
independent variables 127
interactive maps 312
interactive network plots 277
Internet
  datasets, loading into 38-41
isopleth 306

J

Java Database Connectivity (JDBC) 32
JavaScript mapping libraries 315-317
JSON 42
jsonlite package 45

K

Kaiser criterion 212
Kaiser-Meyer-Olkin (KMO) 205
K-means clustering 243-245
kmeans function 243
K-Nearest Neighbors (k-NN) 258-260
knn function 258

L

Latent Class Analysis (LCA) 247-250
latent class models 247-250
Latent Class Regression (LCR) model 247
latitude 299
lattice package 272
law of large numbers 170
LCR model 250
leaflet package 316
least-squares approach 111
legend 305
lemma 163
lemmatisation 163
level plot 306
likelihood ratio 134
line
  fitting, in data 118-120
linear discriminant 250
linear regression 107
linear regression models 127
linear regression, with continuous
        predictors
  about 109
  model interpretation 109-112
  multiple predictors 112-115
lists, of R users
  overlaps, analyzing between 339-341
lmtest library 135
lmtest package 133
LogisticDx package 133
logistic regression
  about 107, 129-257
  data considerations 133
  model comparison 135
  model fit 133, 134
logistic regression model 130
logit 129
longitude 299
lubridate package 335

M

machine learning algorithms
  classification trees 260-263
  K-Nearest Neighbors (k-NN) 258-260
  random forest 264

[PACKT] PUBLISHING | open source*
community experience distilled

Thank you for buying
Mastering Data Analysis with R

About Packt Publishing

Packt, pronounced 'packed', published its first book, *Mastering phpMyAdmin for Effective MySQL Management*, in April 2004, and subsequently continued to specialize in publishing highly focused books on specific technologies and solutions.

Our books and publications share the experiences of your fellow IT professionals in adapting and customizing today's systems, applications, and frameworks. Our solution-based books give you the knowledge and power to customize the software and technologies you're using to get the job done. Packt books are more specific and less general than the IT books you have seen in the past. Our unique business model allows us to bring you more focused information, giving you more of what you need to know, and less of what you don't.

Packt is a modern yet unique publishing company that focuses on producing quality, cutting-edge books for communities of developers, administrators, and newbies alike. For more information, please visit our website at www.packtpub.com.

About Packt Open Source

In 2010, Packt launched two new brands, Packt Open Source and Packt Enterprise, in order to continue its focus on specialization. This book is part of the Packt Open Source brand, home to books published on software built around open source licenses, and offering information to anybody from advanced developers to budding web designers. The Open Source brand also runs Packt's Open Source Royalty Scheme, by which Packt gives a royalty to each open source project about whose software a book is sold.

Writing for Packt

We welcome all inquiries from people who are interested in authoring. Book proposals should be sent to author@packtpub.com. If your book idea is still at an early stage and you would like to discuss it first before writing a formal book proposal, then please contact us; one of our commissioning editors will get in touch with you.

We're not just looking for published authors; if you have strong technical skills but no writing experience, our experienced editors can help you develop a writing career, or simply get some additional reward for your expertise.

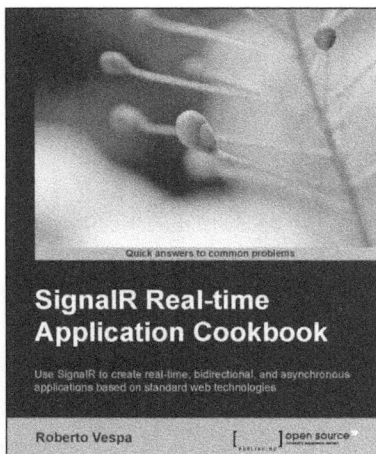

SignalR Real-time Application Cookbook

ISBN: 978-1-78328-595-2   Paperback: 292 pages

Use SignalR to create real-time, bidirectional, and asynchronous applications based on standard web technologies

1. Build high performance real-time web applications.

2. Broadcast messages from the server to many clients simultaneously.

3. Implement complex and reactive architectures.

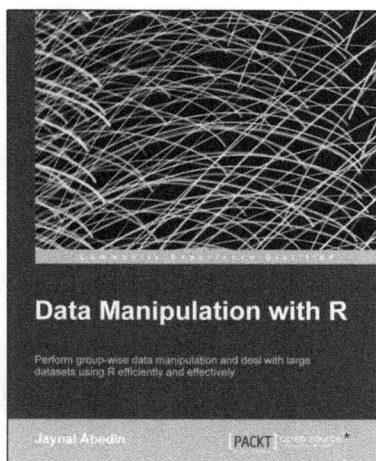

Data Manipulation with R

ISBN: 978-1-78328-109-1   Paperback: 102 pages

Perform group-wise data manipulation and deal with large datasets using R efficiently and effectively

1. Perform factor manipulation and string processing.

2. Learn group-wise data manipulation using plyr.

3. Handle large datasets, interact with database software, and manipulate data using sqldf.

Please check **www.PacktPub.com** for information on our titles

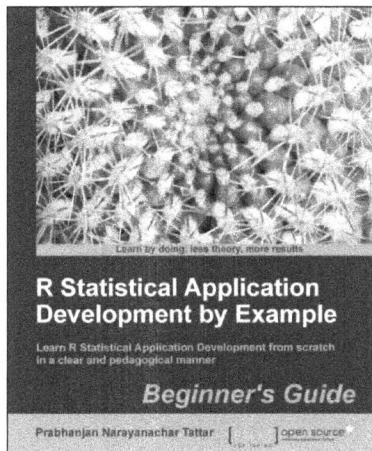

R Statistical Application Development by Example Beginner's Guide

ISBN: 978-1-84951-944-1   Paperback: 344 pages

Learn R Statistical Application Development from scratch in a clear and pedagogical manner

1. A self-learning guide for the user who needs statistical tools for understanding uncertainty in computer science data.

2. Essential descriptive statistics, effective data visualization, and efficient model building.

3. Every method explained through real data sets enables clarity and confidence for unforeseen scenarios.

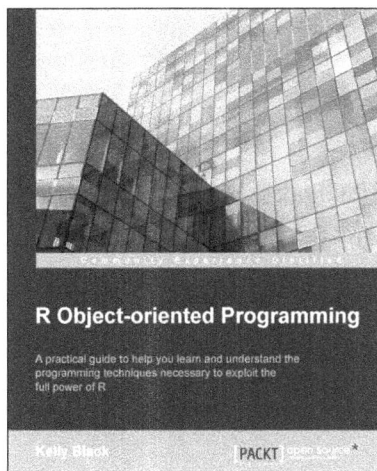

R Object-oriented Programming

ISBN: 978-1-78398-668-2   Paperback: 190 pages

A practical guide to help you learn and understand the programming techniques necessary to exploit the full power of R

1. Learn and understand the programming techniques necessary to solve specific problems and speed up development processes for statistical models and applications.

2. Explore the fundamentals of building objects and how they program individual aspects of larger data designs.

3. Step-by-step guide to understand how OOP can be applied to application and data models within R.

Please check **www.PacktPub.com** for information on our titles

www.ingramcontent.com/pod-product-compliance
Lightning Source LLC
Chambersburg PA
CBHW080703220326
41598CB00033B/5298